Tribology of 2D Nanomaterials

Tribology of 2D Nanomaterials

Yanfei Liu
Xiangyu Ge

Basel • Beijing • Wuhan • Barcelona • Belgrade • Novi Sad • Cluj • Manchester

Yanfei Liu
School of Mechanical Engineering
Beijing Institute of Technology
Beijing
China

Xiangyu Ge
School of Mechanical Engineering
Beijing Institute of Technology
Beijing
China

Editorial Office
MDPI AG
Grosspeteranlage 5
4052 Basel, Switzerland

This is a reprint of articles from the Special Issue published online in the open access journal *Lubricants* (ISSN 2075-4442) (available at: www.mdpi.com/journal/lubricants/special_issues/tribology_2D_nanomaterials).

For citation purposes, cite each article independently as indicated on the article page online and using the guide below:

Lastname, A.A.; Lastname, B.B. Article Title. *Journal Name* **Year**, *Volume Number*, Page Range.

ISBN 978-3-7258-1850-1 (Hbk)
ISBN 978-3-7258-1849-5 (PDF)
https://doi.org/10.3390/books978-3-7258-1849-5

© 2024 by the authors. Articles in this book are Open Access and distributed under the Creative Commons Attribution (CC BY) license. The book as a whole is distributed by MDPI under the terms and conditions of the Creative Commons Attribution-NonCommercial-NoDerivs (CC BY-NC-ND) license (https://creativecommons.org/licenses/by-nc-nd/4.0/).

Contents

Yanfei Liu and Xiangyu Ge
Tribology of 2D Nanomaterials
Reprinted from: *Lubricants* 2024, 12, 199, doi:10.3390/lubricants12060199 1

Yunqi Fang, Yang Sun, Fengqin Shang, Jing Zhang, Jiayu Yao and Zihan Yan et al.
Preparation and Tribological Performance of Multi-Layer van der Waals Heterostructure WS_2/h-BN
Reprinted from: *Lubricants* 2024, 12, 163, doi:10.3390/lubricants12050163 2

Zhongnan Wang, Hui Guo, Ji Zhang, Yi Qian and Yanfei Liu
Two-Dimensional Nanomaterials in Hydrogels and Their Potential Bio-Applications
Reprinted from: *Lubricants* 2024, 12, 149, doi:10.3390/lubricants12050149 17

Shuo Xiang, Xinghao Zhi, Hebin Bao, Yan He, Qinhui Zhang and Shigang Lin et al.
Tribological Behavior of GTL Base Oil Improved by Ni-Fe Layered Double Hydroxide Nanosheets
Reprinted from: *Lubricants* 2024, 12, 146, doi:10.3390/lubricants12050146 35

Yi Dong, Biao Ma, Cenbo Xiong, Yong Liu and Qin Zhao
Study on the Lubricating Characteristics of Graphene Lubricants
Reprinted from: *Lubricants* 2023, 11, 506, doi:10.3390/lubricants11120506 52

Changling Tian, Haichao Cai, Yujun Xue, Lulu Pei and Yongjian Yu
Effect of Argon Flow Rate on Tribological Properties of Rare Earth Ce Doped MoS_2 Based Composite Coatings by Magnetron Sputtering
Reprinted from: *Lubricants* 2023, 11, 432, doi:10.3390/lubricants11100432 67

Feng Nan and Dong Wang
Tribological Properties of Attapulgite Nanofiber as Lubricant Additive for Electric-Brush Plated Ni Coating
Reprinted from: *Lubricants* 2023, 11, 204, doi:10.3390/lubricants11050204 79

Kishan Nath Sidh, Dharmender Jangra and Harish Hirani
An Experimental Investigation of the Tribological Performance and Dispersibility of 2D Nanoparticles as Oil Additives
Reprinted from: *Lubricants* 2023, 11, 179, doi:10.3390/lubricants11040179 91

Umair Khan, Aurang Zaib, Sakhinah Abu Bakar, Siti Khuzaimah Soid, Anuar Ishak and Samia Elattar et al.
Non-Similar Solutions of Dissipative Buoyancy Flow and Heat Transfer Induced by Water-Based Graphene Oxide Nanofluid through a Yawed Cylinder
Reprinted from: *Lubricants* 2023, 11, 60, doi:10.3390/lubricants11020060 116

Qian Wu, Honglin Li, Liangbin Wu, Zihan Bo, Changge Wang and Lei Cheng et al.
Synergistic Lubrication and Antioxidation Efficacies of Graphene Oxide and Fullerenol as Biological Lubricant Additives for Artificial Joints
Reprinted from: *Lubricants* 2022, 11, 11, doi:10.3390/lubricants11010011 139

Kean Pin Ng, Kia Wai Liew and Elaine Lim
Comparative Study of Tribological Properties of Modified and Non-modified Graphene-Oil Nanofluids under Heated and Non-heated Conditions
Reprinted from: *Lubricants* 2022, 10, 288, doi:10.3390/lubricants10110288 153

Yanfei Liu, Shengtao Yu, Qiuyu Shi, Xiangyu Ge and Wenzhong Wang
Graphene-Family Lubricant Additives: Recent Developments and Future Perspectives
Reprinted from: *Lubricants* **2022**, *10*, 215, doi:10.3390/lubricants10090215 . . . . **169**

Editorial

Tribology of 2D Nanomaterials

Yanfei Liu and Xiangyu Ge *

School of Mechanical Engineering, Beijing Institute of Technology, Beijing 100081, China; liuyanfei@bit.edu.cn
* Correspondence: gexy@bit.edu.cn

Tribology is the science and engineering of interacting surfaces in relative motion. It includes the study and application of the principles of friction, lubrication, and wear. Tribology is a branch of mechanical engineering and materials science that plays a critical role in the design and maintenance of mechanical systems, as it helps to reduce friction, wear, and energy consumption, as well as prevent damage and failure of mechanical components. Recently, 2D nanomaterials have garnered considerable attention as solid lubricants, fillers of composites, and nanoadditives for lubricants to enhance lubrication and wear protection performances. The use of 2D nanomaterials in tribology is still an active area of research, and researchers are exploring ways to optimize these materials for various applications, including improving their durability and scalability for industrial use.

The nine research articles and two review articles covered in this Special Issue embrace a wide range of topics, from the application of 2D nanomaterials as lubrication nanoadditives for mechanical and biomedical applications to the use of nanofluids with 2D nanomaterials for reduced shear stress and enhanced heat transfer capacity and the tribological performance of thin film and nanocomposites based on 2D nanomaterials. The dispersion stability of 2D nanomaterials as nanoadditives is crucial for lubrication and anti-wear performance, and the synergetic effect between different 2D nanomaterials or between 2D nanomaterials and other types of additives can also lead to enhanced tribological performance. When applied in nanocomposites, the interaction between 2D nanomaterials and other components plays a significant role in the enhancement of mechanical and tribological properties. Overall, 2D nanomaterials provide opportunities for further improvement in lubrication and reduced wear and present an ideal platform for the fundamental understanding of frictional behaviors.

The Guest Editors extend their gratitude to all authors and reviewers for their contributions, as well as to the editorial staff of MDPI journal *Lubricants* for their support and guidance.

Citation: Liu, Y.; Ge, X. Tribology of 2D Nanomaterials. *Lubricants* **2024**, *12*, 199. https://doi.org/10.3390/lubricants12060199

Received: 30 May 2024
Accepted: 31 May 2024
Published: 1 June 2024

Copyright: © 2024 by the authors. Licensee MDPI, Basel, Switzerland. This article is an open access article distributed under the terms and conditions of the Creative Commons Attribution (CC BY) license (https://creativecommons.org/licenses/by/4.0/).

Conflicts of Interest: The authors declare no conflicts of interest.

Disclaimer/Publisher's Note: The statements, opinions and data contained in all publications are solely those of the individual author(s) and contributor(s) and not of MDPI and/or the editor(s). MDPI and/or the editor(s) disclaim responsibility for any injury to people or property resulting from any ideas, methods, instructions or products referred to in the content.

Article

Preparation and Tribological Performance of Multi-Layer van der Waals Heterostructure WS$_2$/h-BN

Yunqi Fang [1], Yang Sun [1,2,*], Fengqin Shang [1], Jing Zhang [1], Jiayu Yao [1], Zihan Yan [1] and Hangyan Shen [1,*]

[1] College of Materials and Chemistry, China Jiliang University, Hangzhou 310018, China
[2] School of Materials Science and Engineering, Tianjin University, Tianjin 300354, China
* Correspondence: yangsun11@cjlu.edu.cn (Y.S.); shenhangyan@cjlu.edu.cn (H.S.)

Abstract: Van der Waals heterostructures with incommensurate contact interfaces show excellent tribological performance, which provides solutions for the development of new solid lubricants. In this paper, a facile electrostatic layer-by-layer self-assembly (LBL) technique was proposed to prepare multi-layer van der Waals heterostructures tungsten disulfide/hexagonal boron nitride (vdWH WS$_2$/h-BN). The h-BN and WS$_2$ were modified with poly (diallyldimethylammonium chloride) (PDDA) and sodium dodecyl benzene sulfonate (SDBS) to obtain the positively charged PDDA@h-BN and the negatively charged SDBS@WS$_2$, respectively. When the mass ratio of PDDA to h-BN and SDBS to WS$_2$ were both 1:1 and the pH was 3, the zeta potential of PDDA@h-BN and SDBS@WS$_2$ were 60.0 mV and −50.1 mV, respectively. Under the electrostatic interaction, the PDDA@h-BN and SDBS@WS$_2$ attracted each other and stacked alternately along the (002) crystal plane forming the multi-layer (four-layer) vdWH WS$_2$/h-BN. The addition of the multi-layer vdWH WS$_2$/h-BN (1.0 wt%) to the base oil resulted in a significant reduction of 33.8% in the friction coefficient (0.104) and 16.8% in the wear rate (4.43 × 10^{-5} mm^3/(N·m)). The excellent tribological property of the multi-layer vdWH WS$_2$/h-BN arose from the lattice mismatch (26.0%), a 15-fold higher interlayer slip possibility, and the formation of transfer film at the contact interface. This study provided an easily accessible method for the multi-layer vdWH with excellent tribological properties.

Keywords: van der Waals heterostructure; WS$_2$/h-BN; electrostatic interaction; layer-by-layer self-assembly; tribological property

Citation: Fang, Y.; Sun, Y.; Shang, F.; Zhang, J.; Yao, J.; Yan, Z.; Shen, H. Preparation and Tribological Performance of Multi-Layer van der Waals Heterostructure WS$_2$/h-BN. Lubricants 2024, 12, 163. https://doi.org/10.3390/lubricants12050163

Received: 3 March 2024
Revised: 29 April 2024
Accepted: 2 May 2024
Published: 7 May 2024

Copyright: © 2024 by the authors. Licensee MDPI, Basel, Switzerland. This article is an open access article distributed under the terms and conditions of the Creative Commons Attribution (CC BY) license (https://creativecommons.org/licenses/by/4.0/).

1. Introduction

Two-dimensional layered materials (2DLMs), such as tungsten disulfide (WS$_2$) and hexagonal boron nitride (h-BN), are connected via weak van der Waals forces between adjacent layers, making them prone to interlayer slip [1,2]. Based on this property, WS$_2$ and h-BN are widely used as lubricant additives in fields such as aerospace appliances [1], automobiles [3], and mechanical equipment [4]. By stacking different 2DLMs along the (002) crystal plane, a van der Waals heterostructure (vdWH) can be constructed [5,6]. When sliding along the heterogeneous interface, the vdWH exhibits excellent tribological properties, with a friction coefficient of as low as 0.001 and near-zero wear [7–10]. Given the ultra-low friction coefficient and high wear resistance, the vdWH is expected to become a new lubricant additive.

Presently, the dry transfer method [11], wet transfer method [12], physical vapor deposition (PVD) [13], chemical vapor deposition (CVD) [14], spray-coating process [15], and hydrothermal method [16] are usually used to construct vdWHs. The dry transfer and wet transfer methods were employed for the layer-by-layer transfer of 2DLMs in air and liquid to achieve vdWH stacking. Castellanos et al. [11] proposed a fully dry transfer method that utilizes a viscoelastic stamp to transfer and adhere a single layer of MoS$_2$ to h-BN, constructing the vdWH MoS$_2$/h-BN with a thickness 1.5 nm. Dean et al. [12] used poly (methyl methacrylate) as the support layer to transfer graphene to the h-BN sheet in aqueous solution to form a single-layer vdWH graphene/h-BN. Both the dry transfer and wet transfer methods require the 2DLMs to be transferred and stacked layer by layer, resulting in extremely low preparation efficiency [17]. Dai et al. [13] successfully constructed single-layer vdWH $MoSe_2$/graphene using PVD to deposit $MoSe_2$ directly evaporated on the graphene surface at 650 °C. For the PVD method, $MoSe_2$ layers tend to raise the wrinkles and cracks at high temperatures. Han et al. [18] used CVD to deposit MoS_2 on an h-BN substrate at 1050 °C to construct a single-layer vdWH MoS_2/h-BN. The PVD and CVD methods are primarily suitable for constructing single-layer vdWHs [19], whereas the preparation of multi-layer vdWHs typically necessitates multiple adjustments in reaction sources and pressure. Macknojia et al. [15] sprayed $Ti_3C_2T_X$-MoS_2 onto a steel surface to form a 3.1 μm thick vdWH $Ti_3C_2T_X$-MoS_2 coating. A friction coefficient of 0.003 was demonstrated under test conditions of 20 N and 0.2 m/s. However, the spray method was used to prepare coatings, limiting vdWH to a wide range of applications. Zhao et al. [20] prepared an irregular flower-shaped vdWH WS_2/MoS_2 in a high-pressure reactor at 200 °C using the hydrothermal synthesis method. The hydrothermal synthesis method is challenging in achieving the precise stacking of layered 2DLMs along the direction of the (002) crystal plane due to the difficulty in controlling the growth direction. Furthermore, these methods face challenges in efficiently producing large quantities of multi-layer vdWHs, limiting their applicability in the field of tribology. Therefore, it is necessary to develop simpler and more efficient methods to construct a multi-layer vdWH with multiple heterostructure interfaces.

In this study, electrostatic LBL self-assembly technology was proposed to simply and efficiently prepare a multi-layer vdWH WS_2/h-BN with excellent lubrication properties. PDDA and SDBS as modifiers were used to modify h-BN and WS_2 particles to obtain positively charged PDDA@h-BN and negatively charged SDBS@WS_2, respectively. To enhance the electrostatic interaction between PDDA@h-BN and SDBS@WS_2, the effects of modifier dosage and pH on zeta potential were systematically investigated. Then, the PDDA@h-BN dispersion was mixed with the SDBS@WS_2 dispersion and formed the multi-layer vdWH WS_2/h-BN in a short time. Finally, the tribological behavior of the multi-layer vdWH WS_2/h-BN was investigated, and the lubrication mechanism was revealed.

2. Materials and Methods

2.1. Materials

Hexagonal boron nitride (h-BN, 99.9% pure, 3–5 μm) and tungsten disulfide (WS_2, 99.9% pure, 3–4 μm), provided by Aladdin Reagent Co., Ltd. (Shanghai, China), were used as raw materials for the preparation of a multi-layer vdWH WS_2/h-BN. Figure 1 shows the microstructure and elemental composition of h-BN and WS_2 powders. It can be seen from Figure 1a,b, that the h-BN and WS_2 both had smooth surfaces. In addition, both h-BN and WS_2 had typical layered structures (as indicated with the white arrow) with a thickness of about 0.1–0.2 μm as shown in the inset of Figure 1a,b. Figure 1c,d show the EDS results of the A and B regions, respectively. The h-BN particles were mainly composed of the elements B and N, while the primary elements of WS_2 were W and S. In addition, a certain amount of C was detected on the surfaces of h-BN (10.75 wt%) and WS_2 (12.81 wt%). In addition, poly (diallyldimethylammonium chloride) (PDDA, Ron reagent) and sodium dodecyl benzene sulfonate (SDBS, Aladdin reagent) were used as modifiers to adjust the surface charge of h-BN and WS_2 for a suitable zeta potential.

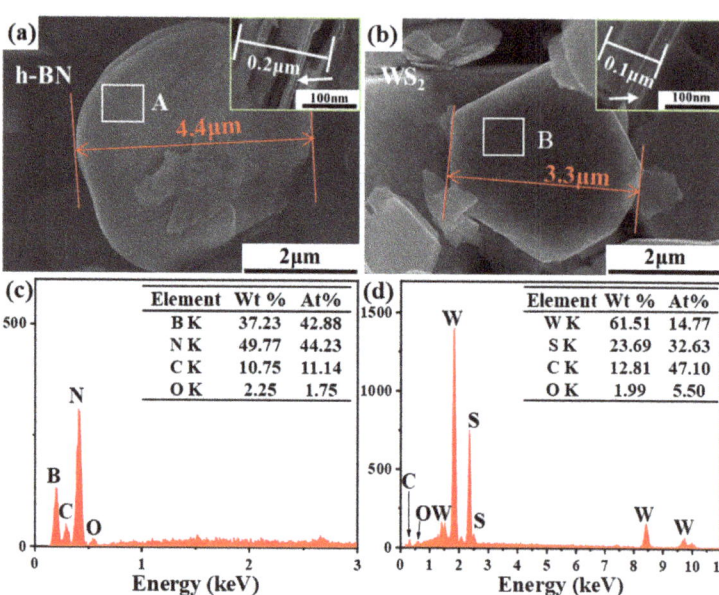

Figure 1. SEM images of (**a**) h-BN and (**b**) WS$_2$, and EDS results of (**c**) h-BN and (**d**) WS$_2$. Insets in (**a**,**b**) are the typical layered structure.

2.2. Preparation of Multi-Layer vdWH WS$_2$/h-BN

The preparation process of a multi-layer vdWH WS$_2$/h-BN is shown in Figure 2. The preparation included the following two parts:

(1) Synthesis of PDDA@h-BN and SDBS@WS.

The h-BN powder was dispersed in deionized water and a uniform dispersion of 0.2 g/L was formed via sonication for 30 min. Then, PDDA was slowly added to the h-BN dispersion, and the mass ratio of h-BN to PDDA was 1:1, 1:5, 1:10, and 1:15. At the same time, the pH value of the h-BN dispersion was accurately adjusted from 3 to 11 with HCl or NaOH solution, and the reaction system was stirred at 25 °C for 2 h. The obtained product was collected, and the h-BN modified via PDDA was named PDDA@h-BN. WS$_2$ was modified using SDBS with the same method, and the product was SDBS@WS$_2$.

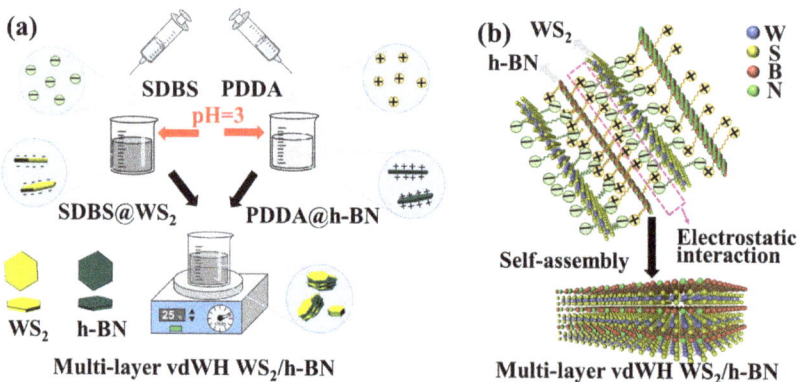

Figure 2. (**a**) The preparation process of a multi-layer vdWH WS$_2$/h-BN; (**b**) schematic diagram of the process of electrostatic LBL self-assembly of WS$_2$ and h-BN under electrostatic interaction.

(2) Synthesis of multi-layer vdWH WS_2/h-BN.

As shown in Figure 2b, the positively charged PDDA@h-BN and the negatively charged SDBS@WS_2 realized electrostatic LBL self-assembly through electrostatic interaction. First, PDDA@h-BN and SDBS@WS_2 were dispersed into deionized water to achieve a homogeneous dispersion (0.2 g/L) under sonication. Then, the PDDA@h-BN dispersion was slowly added dropwise to the SDBS@WS_2 dispersion and continuously stirred at 300 rpm for 15 min. Subsequently, it was observed that the negatively charged SDBS@WS_2 and the positively charged PDDA@h-BN flocculated under electrostatic interaction, and quickly settled to the bottom after the stirring was stopped. The precipitate was collected and dried at 60 °C for 6 h to obtain a multi-layer vdWH WS_2/h-BN.

2.3. Characterization

Fourier-transform infrared spectroscopy (FTIR, TENSOR27, Bruker Corporation, Ettlingen, Germany) was used to characterize the surface functional groups before and after the modification of WS_2 and h-BN in the wavenumber range of 4000 to 500 cm^{-1}. The zeta potential of PDDA@h-BN, SDBS@WS_2, and a multi-layer vdWH WS_2/h-BN were measured using the nanoparticle size and a zeta potential analyzer (DLS, Zetasizer Nano ZS90, Malvern Instruments Ltd., Malvern, UK). Before this measurement, the samples were uniformly dispersed in water. The crystal structure of PDDA@h-BN, SDBS@WS_2, and a multi-layer vdWH WS_2/h-BN was examined on an X-ray diffractometer (XRD, X'Pert Pro, PANalytical, Almelo, The Netherlands) with Cu Kα radiation (λ = 0.154 nm) at a step size of 0.06° in the 2θ range of 10–80°. A scanning electron microscope (SEM, SU8010, Hitachi, Ltd., Tokyo, Japan) and an energy-dispersive X-ray spectrometer (EDS, Oxford Instruments, Abingdon, UK) were used to study the microstructure and elemental distribution of a multi-layer vdWH WS_2/h-BN.

2.4. Tribological Tests

In the friction test, commercial 5W40 engine oil (Tianjin Nisseki Lubricating Grease Co., Tianjin, China) was used as the base oil. The multi-layer vdWH WS_2/h-BN (1 wt%) was added to the base oil as an additive, and a uniform dispersion system was obtained via ultrasonication for 10 min and stirring for 30 min. The tribological behavior was examined using a high-speed reciprocating friction and wear tester (HSR-2M, Lanzhou Zhongke Kaihua Technology Development Co., Lanzhou, China) with an applied load of 20 N and a rotating speed of 500 rpm (the corresponding frequency was 8.3 Hz), corresponding with a 2000 m sliding distance. Throughout the test, the upper silicon nitride (Si_3N_4) ball with a diameter of 6 mm slid against the lower fixed 304 stainless steels, and the length of the wear track was kept at 5 mm (*l*). Then, a probe-type surface profiler (P-6, KLA-Tencor) was utilized to acquire a cross-sectional profile (*s*) of the wear track. The value of the wear volume (V) was obtained with V = s × l. The wear rate was given via W = V/(F × L), where W was the wear rate, V was the value of the wear volume, F was the normal load (20 N), and L was the sliding distance [21]. The microstructure and elemental distribution of the wear track were characterized using SEM and EDS to analyze the lubrication mechanism.

3. Results and Discussion

3.1. Subsection Characterization of PDDA@h-BN and SDBS@WS_2

Figure 3 illustrates the impact of modifier dosage and pH on the zeta potential of PDDA@h-BN and SDBS@WS_2, respectively. The relationship between the zeta potential curve and the modifier dosage is shown in Figure 3a. The original zeta potential of h-BN was −26.9 mV. When the mass ratio of h-BN to PDDA was 1:1, h-BN was modified with PDDA to form PDDA@h-BN particles and the surface changed from negatively charged to positively charged with a zeta potential of 53.6 mV. The transformation from a negatively charged surface to a positively charged surface indicated that PDDA had been successfully adsorbed to the h-BN surface. When the mass ratio increased from 1:1 to 1:15, the zeta potential rose from 53.6 mV to 57.3 mV. The original zeta potential of WS_2 was −18.6 mV and dropped after modification with SDBS. The zeta potential of SDBS@WS$_2$ was −31.6 mV when the mass ratio reached 1:1. Then, the zeta potential stabilized at about −36.6 mV as the amount of SDBS increased (the mass ratio of WS$_2$ to SDBS was 1:10 and 1:15). The stable zeta potential of SDBS@WS$_2$ could be attributed to the saturation of SDBS adsorption on the surface of WS$_2$ [22].

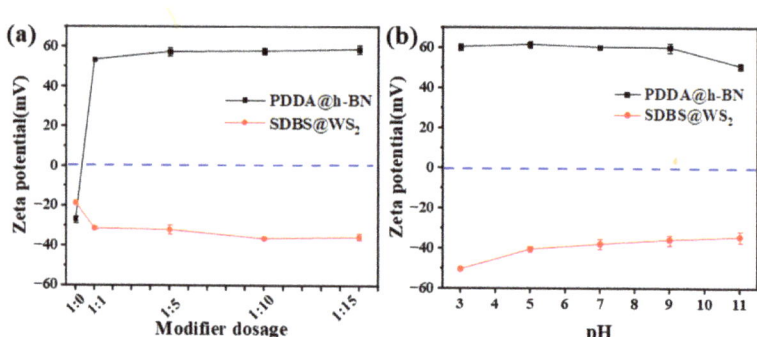

Figure 3. The effect of (**a**) modifier dosage and (**b**) pH on the zeta potential of PDDA@h-BN and SDBS@WS$_2$.

The above results showed that when the mass ratio reached 1:1, the zeta potential tended to stabilize. Based on this result, the effect of pH on the zeta potential of PDDA@h-BN and SDBS@WS$_2$, both at a mass ratio of 1:1, is shown in Figure 3b. Interestingly, the zeta potential of PDDA@h-BN and SDBS@WS$_2$ showed opposite trends with the pH range from 3 to 11. When pH increased from 3 to 9, the zeta potential of PDDA@h-BN remained around 60.0 mV, while the zeta potential rapidly decreased to 51.5 mV at pH = 11. However, the zeta potential of SDBS@WS$_2$ showed an increasing trend in the pH range from 3 to 11, with the highest electronegativity reaching −50.1 mV at pH = 3. And the zeta potential of SDBS@WS$_2$ increased to −34.2 mV at pH = 11. Since the pH value of the solution controlled the degree of the protonation of the modifier group, the charge density of PDDA@h-BN and SDBS@WS$_2$ was greatly increased when pH = 3. In addition, the zeta potential difference between PDDA@h-BN and SDBS@WS$_2$ reached a maximum value of 110.1 mV at pH = 3, which was beneficial for promoting the attraction and assembly of PDDA@h-BN and SDBS@WS$_2$ under electrostatic interaction.

The microstructure, EDS, and FTIR spectroscopic results of PDDA@h-BN and SDBS@WS$_2$ are shown in Figure 4. It can be clearly seen from Figure 4a that a number of irregular particles appeared on the surface of PDDA@h-BN, which was due to the deposition of PDDA on the surface of h-BN (Figure 1a) [23,24]. The EDS results for region A in Figure 4c indicate that PDDA@h-BN was still composed of four elements, B, N, C, and O. Compared with the EDS results of h-BN (Figure 1c), the elemental C content increased significantly from 10.75 wt% (Figure 1c) to 21.66 wt% (Figure 4c), suggesting that PDDA successfully adsorbed on the surface of h-BN. In addition, the elemental C content on the surface of the SDBS@WS$_2$ increased significantly from 12.81 wt% (Figure 1d) to 21.52 wt% (Figure 4d), which confirmed the adsorption of SDBS particles on the WS$_2$ surface (Figure 4b).

To confirm that the modifier had been successfully adsorbed on the particle surface, PDDA@h-BN and SDBS@WS$_2$ were characterized by FTIR. As shown in Figure 4e, h-BN exhibited strong absorption peaks at 1377 cm^{-1} and 817 cm^{-1}, which were attributed to the tensile and bending vibrations of the B-N bond, respectively [25]. PDDA had an unsaturated C=C stretching vibration peak at 1635 cm^{-1}, where the peak at 1473 cm^{-1} belonged to the asymmetric vibration of C-H from N-CH$_3$, and the absorption peak at 1122 cm^{-1} was caused by C-N bending vibration [26]. The peaks belonging to PDDA appeared at PDDA@h-BN, which indicated that PDDA had been adsorbed onto the surface of h-BN. As shown in Figure 4f, the peak of WS$_2$ at 1062 cm^{-1} was attributed to the S-S bond and the peak at 659 cm^{-1} was ascribed to the W-S bond [27]. The peaks at 1010 cm^{-1} and 1190 cm^{-1} corresponded to the S=O and S-O stretching vibrations of SDBS [28]. In addition, the peaks that observed at 1450 cm^{-1} and 835 cm^{-1} were assigned to C=C bonding and C-H out-of-plane bending vibrations in the benzene ring, respectively [28]. Compared to WS$_2$, peaks belonging to SDBS also appeared in the FTIR spectra of SDBS@WS$_2$, confirming the adsorption of SDBS on the surface of WS$_2$. The EDS and FTIR results indicated that PDDA and SDBS had been firmly attached to the surfaces of h-BN and WS$_2$, making PDDA@h-BN particles positively charged and SDBS@WS$_2$ particles negatively charged, respectively (Figure 3).

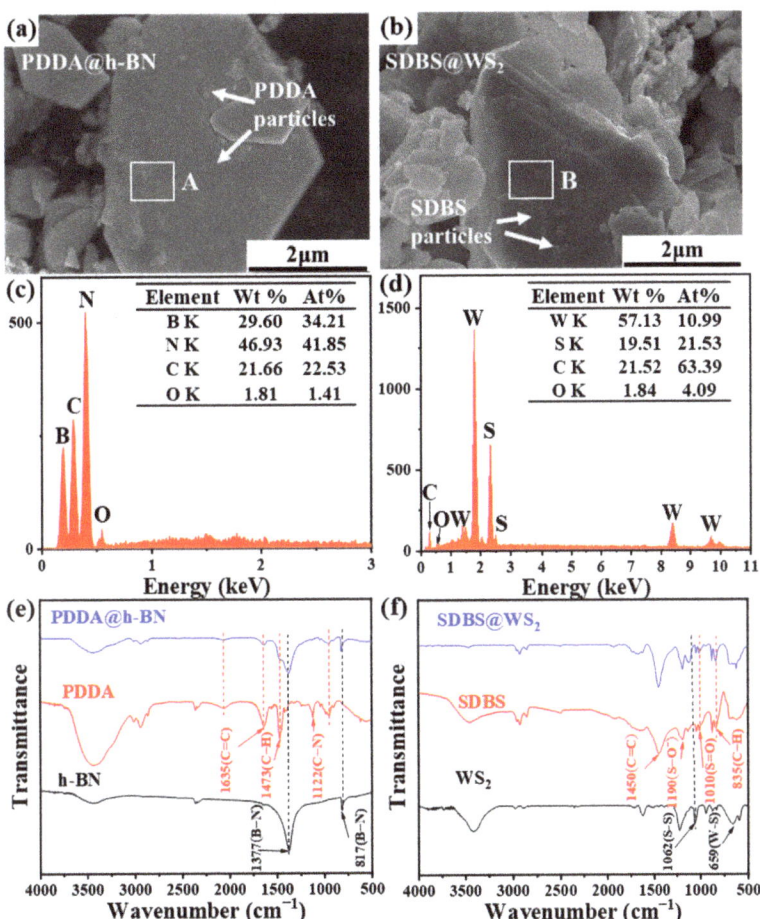

Figure 4. SEM images of (**a**) PDDA@h-BN and (**b**) SDBS@WS$_2$; EDS results of (**c**) PDDA@h-BN and (**d**) SDBS@WS$_2$; and FTIR spectra of (**e**) PDDA@h-BN and (**f**) SDBS@WS$_2$.

3.2. Characterization of Multi-Layer vdWH WS$_2$/h-BN

Figure 5 presents the dispersion state; zeta potential; and XRD patterns of PDDA@h-BN, SDBS@WS$_2$, and the multi-layer vdWH WS$_2$/h-BN. The dispersion states of PDDA@h-BN, SDBS@WS$_2$, and the multi-layer vdWH WS$_2$/h-BN in water are shown in Figure 5a. PDDA@h-BN and SDBS@WS$_2$ were homogeneously dispersed in water for 24 h, showing good dispersion properties. When PDDA@h-BN was added dropwise to SDBS@WS$_2$, the positively charged PDDA@h-BN and negatively charged SDBS@WS$_2$ were attracted to each other through electrostatic interaction, leading to agglomeration. And then, the multi-layer vdWH WS$_2$/h-BN tended to precipitate to the bottom of the container within 15 min.

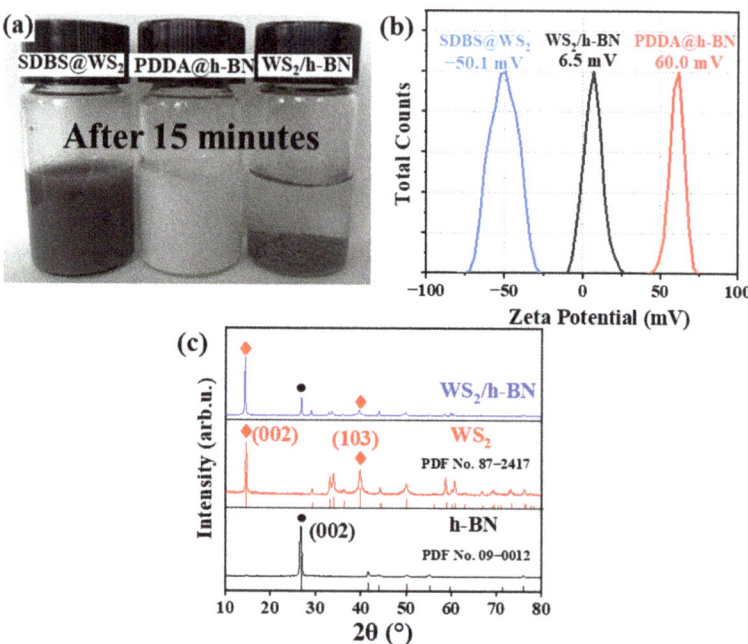

Figure 5. SDBS@WS$_2$, PDDA@h-BN, and multi-layer vdWH WS$_2$/h-BN (**a**) dispersion in water, (**b**) zeta potential, and (**c**) XRD pattern.

Zeta potential was an essential parameter for the stability of reactive dispersed systems [29,30]. To identify the reasons for the different dispersion states of PDDA@h-BN, SDBS@WS$_2$, and the multi-layer vdWH WS$_2$/h-BN, zeta potential tests were conducted. In Figure 5b, both PDDA@h-BN and SDBS@WS$_2$ possessed high zeta potential, which were −50.1 mV and 60.0 mV, respectively. Due to the electrostatic repulsion, PDDA@h-BN and SDBS@WS$_2$ were well dispersed in water to form a stable dispersion. When PDDA@h-BN was mixed with SDBS@WS$_2$, these two particles self-assembled under the action of electrostatic interaction and formed flocculent precipitates. This phenomenon arose from the electrical neutralization between PDDA@h-BN and SDBS@WS$_2$, resulting in a decrease in zeta potential to 6.5 mV (Figure 5b). Figure 5c shows the XRD patterns of SDBS@WS$_2$, PDDA@h-BN, and the multi-layer vdWH WS$_2$/h-BN. The diffraction peaks at 14.6° and 26.7° corresponded to the (002) crystal plane of WS$_2$ (PDF No. 87-2417) and the (002) crystal plane of h-BN (PDF No. 09-0012) [20,24], respectively. In addition, all characteristic peaks belonging to WS$_2$ and h-BN appeared in the multi-layer vdWH WS$_2$/h-BN.

SEM and EDS were used to investigate the microstructure and elemental composition of the multi-layer vdWH WS$_2$/h-BN, and the results are shown in Figure 6. It can be seen from Figure 6a that the multi-layer vdWH WS$_2$/h-BN had a well-stacked layered structure, in which WS$_2$ was stacked with h-BN along the vertical direction (the (002) crystal plane), forming a sandwich-like structure with four layers of heterogeneous interface. To study the distribution of h-BN and WS$_2$ in detail, an EDS line scan of the AB line in Figure 6a was performed for the multi-layer vdWH WS$_2$/h-BN. Figure 6b showed that the distribution of WS$_2$ and h-BN in the multi-layer vdWH WS$_2$/h-BN was derived from the variation of the S-element curve and the N-element curve. The line scan result showed that in the structure of the multi-layer vdWH WS$_2$/h-BN, WS$_2$ and h-BN were stacked alternately along the

(002) crystal plane with a thickness of 2 μm. According to the above analysis results, the multi-layer vdWHs WS_2/h-BN stacked alternately with WS_2 and h-BN were successfully synthesized under electrostatic interaction between SDBS@WS_2 and PDDA@h-BN.

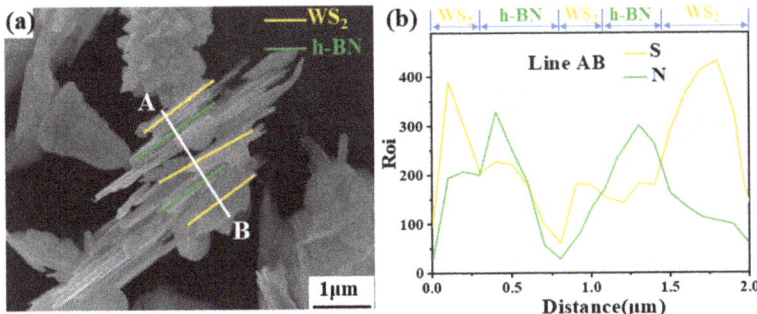

Figure 6. (a) SEM image and (b) line scan analysis of multi-layer vdWH WS_2/h-BN.

3.3. Tribological Property

The dispersion of the additive in the base oil was essential to ensure its uninterrupted supply to the contact interface. Figure 7 shows the dispersion state of the h-BN, WS_2, the WS_2+h-BN mixture, and the multi-layer vdWH WS_2/h-BN in the base oil after sonication for 10 min and stirring for 30 min. All the additives were evenly dispersed in the base oil for 24 h without precipitation or stratification, indicating the excellent dispersibility of the additives in the base oil. The excellent dispersion stability of the multi-layer vdWH WS_2/h-BN in the base oil ensured its continuous access to the contact interfaces.

Figure 7. The dispersion states of h-BN, WS_2, WS_2+ h-BN mixture, and vdWH WS_2 h-BN in the base oil for 24 h.

To evaluate the tribological properties of different additives, the multi-layer vdWH WS_2/h-BN, the WS_2, h-BN, and the WS_2+h-BN mixture were added to the base oil, respectively. The tribological property analysis was carried out using a high-speed reciprocating friction and wear tester under a load of 20 N. Figure 8 shows the friction coefficient and wear rate with the additive content of 1.0 wt%, respectively. As shown in Figure 8a, the average friction coefficient of the base oil, WS_2, h-BN, the WS_2+h-BN mixture, and the multi-layer vdWH WS_2/h-BN were 0.157, 0.121, 0.122, 0.147, and 0.104, respectively. After adding WS_2 and h-BN to the base oil, a significant decrease in the friction coefficient was observed. This phenomenon could be attributed to the formation of a continuous transfer film composed of WS_2 and h-BN at the friction interface [31]. What was more important was that the multi-layer vdWH WS_2/h-BN exhibited the lowest friction coefficient, which was reduced by 33% compared to the base oil.

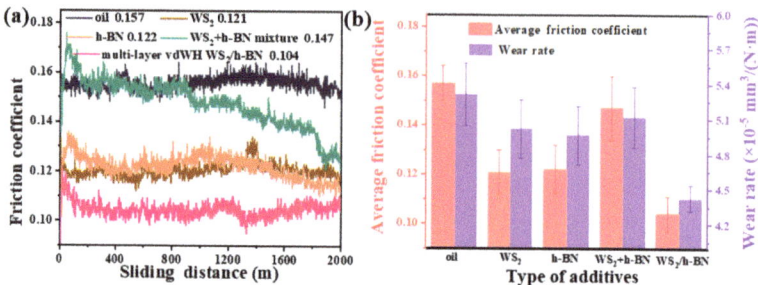

Figure 8. (**a**) The friction coefficient curve and the (**b**) average friction coefficient and wear rate of adding WS$_2$, h-BN, the WS$_2$+h-BN mixture, and the multi-layer vdWH WS$_2$/h-BN to the base oil.

Figure 8b shows the average friction coefficient and wear rate of the base oil and the base oil with four kinds of additives. After adding additives to the base oil, the friction coefficient and wear rate showed a decreasing trend. The wear rates with the additives of WS$_2$, h-BN, the WS$_2$+h-BN mixture, and the multi-layer vdWH WS$_2$/h-BN were 5.02×10^{-5}, 4.97×10^{-5}, 5.12×10^{-5}, and 4.43×10^{-5} mm^3/(N·m), respectively. Compared with the base oil, the wear rates were decreased by 5.6%, 6.5%, 3.7%, and 16.8%, respectively. The results suggested that the multi-layer vdWH WS$_2$/h-BN additive exhibited the lowest friction coefficient and wear rate, significantly improving the tribological properties of the oil lubrication system.

The surface profile and microstructure of the stainless steel (the counter disc), which was lubricated with the base oil, WS$_2$, h-BN, the WS$_2$+h-BN mixture, and the multi-layer vdWH WS$_2$/h-BN, are shown in Figure 9. The wear track lubricated with base oil was the deepest and widest, reaching 28.1 µm and 830.1 µm (Figure 9a$_1$). There were many grooves and spallings on the wear surface (Figure 9a$_3$), showing serious abrasive wear. When WS$_2$, h-BN, and the WS$_2$+h-BN mixture were added to the base oil, the width of the wear track was reduced to 812.1, 809.8, and 828.7 µm, respectively. At the same time, the depth of the wear track decreased to 27.3, 27.4, and 27.9 µm, respectively. In addition, the deep grooves on the surface of the wear track evolved into shallow grooves, showing slight abrasive wear. Compared with other additives, the wear track of the multi-layer vdWH WS$_2$/h-BN had the smoothest wear surface. The specific performance was that there were a small number of grooves on the surface of the wear track (Figure 9e$_3$), and it showed the smallest wear track width (770.9 µm) and wear track depth (25.1 µm), which were 7.1% and 10.7% lower than those of base oil, respectively. The result of the wear track confirmed that multi-layer vdWH WS$_2$/h-BN had an excellent anti-friction and wear-resistant performance.

In order to understand the evolution of the transfer film, Figure 10 displays the SEM and EDS mappings of the wear track lubricated with the multi-layer vdWH WS$_2$/h-BN (1.0 wt%) at sliding distances of 200 m and 2000 m. Figure 10a shows a uniform wear track with a smooth surface and the formation of a small amount of transfer film. As the sliding distance increased to 2000 m, the coverage area of the dark area increased significantly and connected together, which indicated the formation of a continuous transfer film, as shown in Figure 10c. The B and N in the transfer film derived from the h-BN in the multi-layer vdWH WS$_2$/h-BN where the W and S were attributed to WS$_2$ [32]. As the sliding distance increased from 200 m to 2000 m, the content of WS$_2$ and h-BN increased from 24.57 wt% to 57.99 wt%, indicating that a continuous transfer film composed of the multi-layer vdWH WS$_2$/h-BN formed [4,32–34]. During the friction process, the multi-layer vdWH WS$_2$/h-BN deposits on the surface of the stainless steel and forms a continuous transfer film, thereby reducing friction and the wear of the contact interface [35].

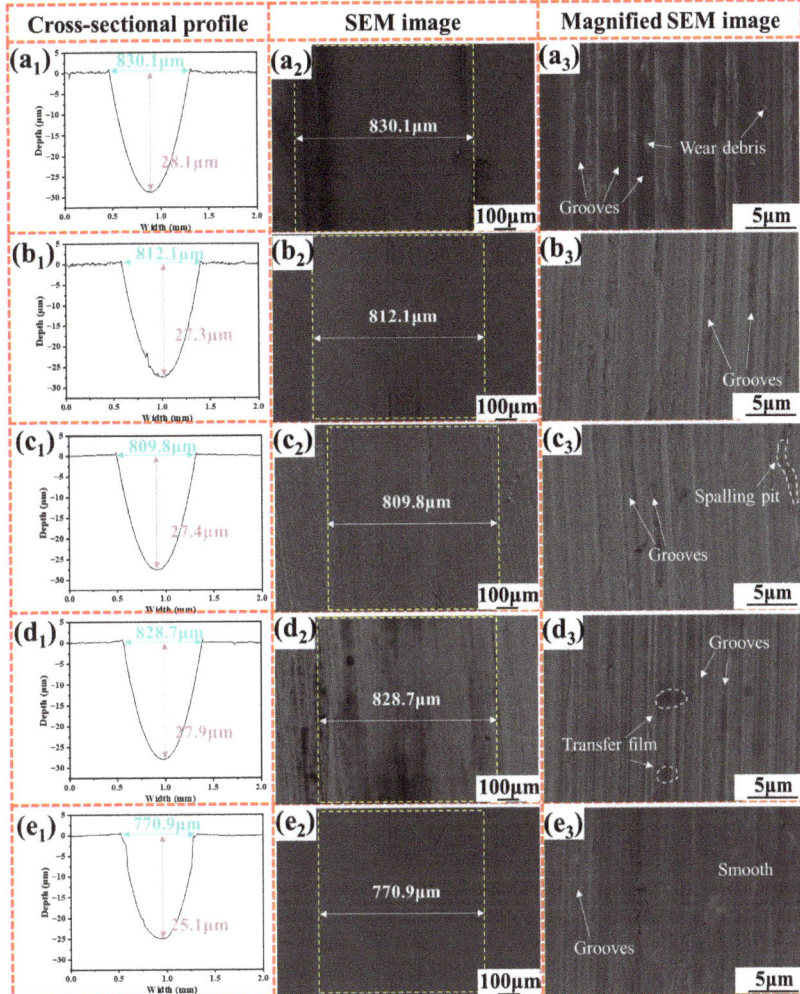

Figure 9. Cross-sectional profile and SEM images of wear tracks of (a_1–a_3) oil, (b_1–b_3) WS_2, (c_1–c_3) h-BN, (d_1–d_3) WS_2+h-BN mixture, and (e_1–e_3) multi-layer vdWH WS_2/h-BN under a load of 20 N.

3.4. Analysis of Lubrication Mechanism

The lubrication mechanism of the multi-layer vdWH WS_2/h-BN was analyzed as shown in Figure 11a. Firstly, the multi-layer vdWH WS_2/h-BN within the oil is evenly dispersed across the contact area. Under shear forces, these multi-layer vdWHs WS_2/h-BN adhered to the surface of stainless steel and gradually formed a transfer film within the contact area [33]. The EDS analysis of the wear track in Figure 10 confirmed the formation of a transfer film based on the multi-layer vdWH WS_2/h-BN during friction. The transfer film of the multi-layer vdWH WS_2/h-BN with a graphite-like lamellar structure greatly reduced frictional force and improved lubrication performance through interlayer slip [36].

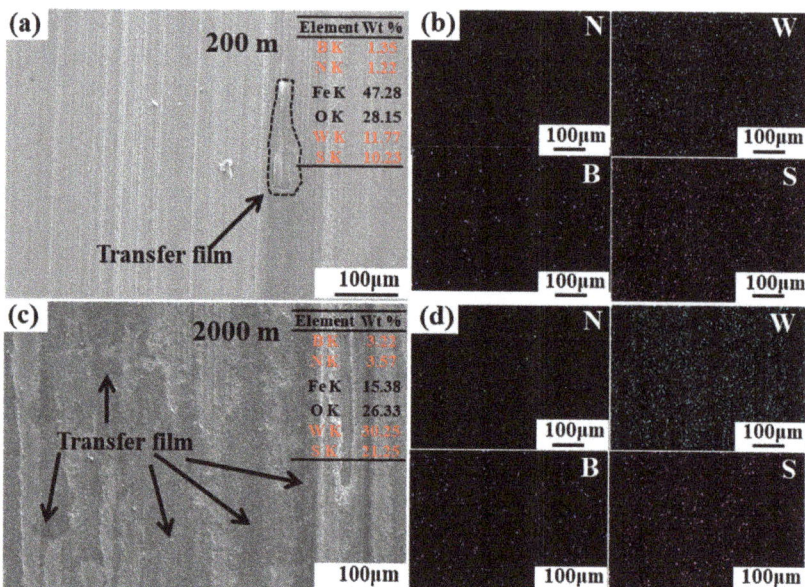

Figure 10. SEM and EDS mappings show the transfer film of multi-layer vdWH WS$_2$/h-BN after sliding from (**a**,**b**) 200 m to (**c**,**d**) 2000 m.

Furthermore, the lubrication performance of the multi-layer vdWH WS$_2$/h-BN was much better than that of h-BN, WS$_2$, and the WS$_2$+h-BN mixture. It was attributed to the ultra-low interfacial interactions of the multi-layer vdWH WS$_2$/h-BN. When sliding took place between the layers, WS$_2$ and h-BN were often firmly locked in a commensurate contact state with strong interfacial interactions due to the no-lattice mismatch characteristics between layers, which results in greater sliding resistance [37,38]. Figure 11b showed that there was a natural lattice mismatch (26.0%) in the multi-layer vdWH WS$_2$/h-BN due to the different lattice constants between WS$_2$ (a_1 = 0.315 nm) and h-BN (a_2 = 0.250 nm), which brought the typical incommensurate contact state between layers. This incommensurate contact state induced weak interfacial interactions, making the multi-layer vdWH WS$_2$/h-BN more prone to interlayer sliding, greatly improving the lubrication performance.

Since interfacial slip was the source of the lubricating properties of layered materials [4,39], the number of slip interfaces would inevitably affect the lubricating properties of layered materials. In order to explore the influence of the number of slip interfaces on the tribological properties of vdWH WS$_2$/h-BN, the friction and wear properties of single-layer and multi-layer vdWH WS$_2$/h-BN were shown in Figure 12. Compared with the single-layer vdWH WS$_2$/h-BN (0.114), the friction coefficient of the multi-layer vdWH WS$_2$/h-BN (0.104) decreased by 8.7%. Besides, the wear rate of multi-layer vdWH WS$_2$/h-BN (4.43 × 10^{-5} mm^3/(N·m)) was reduced by 8.1% compared to that of single-layer (4.82 × 10^{-5} mm^3/(N·m)). The multi-layer vdWH WS$_2$/h-BN with alternating stacks of WS$_2$ and h-BN had more heterogeneous interfaces than single-layer vdWH WS$_2$/h-BN which allowed to undergo multi-slip in the slip process. Figure 13 shows the slip probability of the single-layer and multi-layer vdWH WS$_2$/h-BN slip system. It could be seen that single-layer vdWH WS$_2$/h-BN had just one heterogeneous interface, and a single slip could be performed during the slip process. Multi-layer (4 layers) vdWH WS$_2$/h-BN with 4 heterogeneous interfaces could slide along one or more heterogeneous interfaces at the same time during the slip process, which owned a total of 15 slip states. When interlayer slip occurred, the slip probability of the multi-layer vdWH WS$_2$/h-BN was 15 times higher than that of the single-layer vdWH WS$_2$/h-BN. Therefore, the multi-layer vdWH WS$_2$/h-

BN with multiple heterogeneous interfaces exhibited a greater possibility of interlayer sliding [36,40], and had superior tribological property.

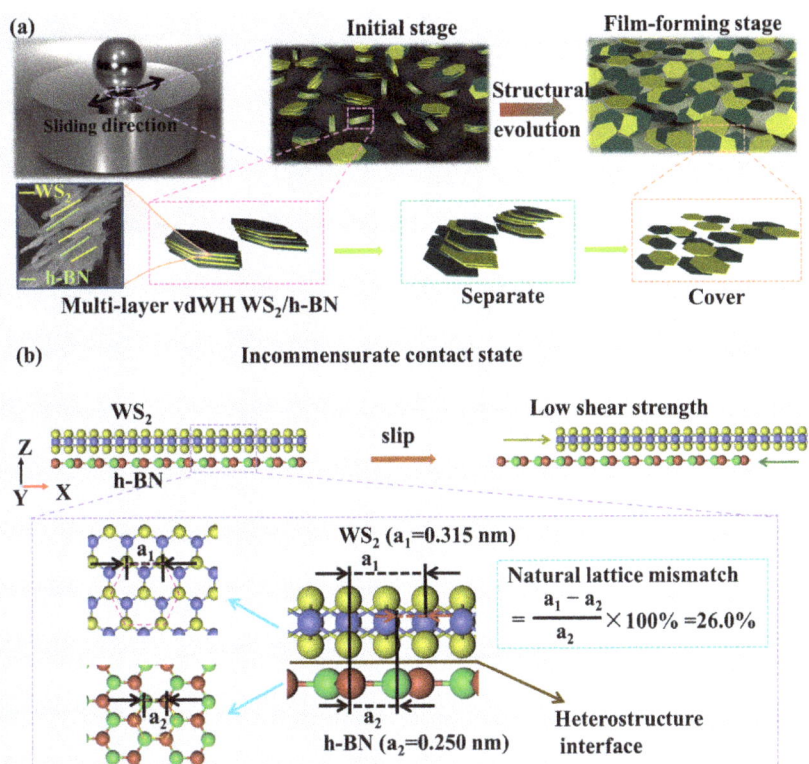

Figure 11. (a) Schematic diagram of the lubrication mechanisms of the multi-layer vdWH WS_2/h-BN and (b) the state of interlayer slips of the multi-layer vdWH WS_2/h-BN.

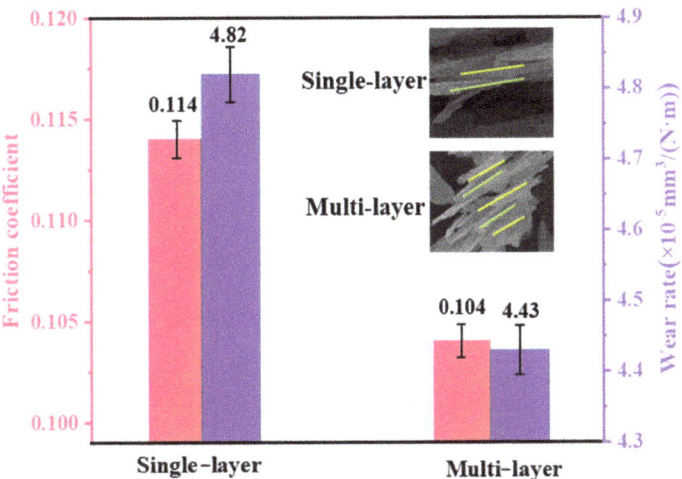

Figure 12. The friction and wear properties of single-layer and multi-layer vdWH WS_2/h-BN.

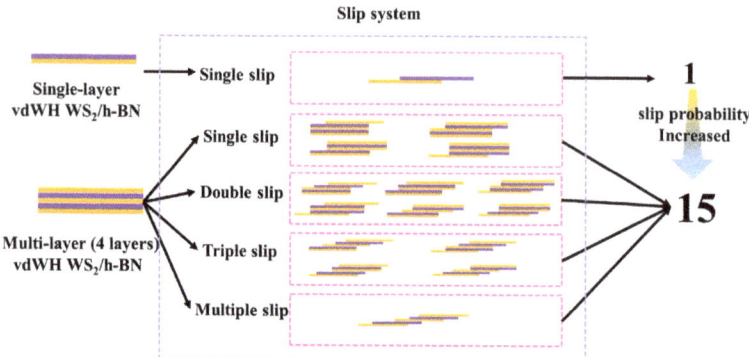

Figure 13. Schematic diagram of single-layer and multi-layer vdWH WS$_2$/h-BN slip system.

4. Conclusions

The multi-layer vdWH WS$_2$/h-BN with alternating stacking of WS$_2$ and h-BN was successfully prepared by electrostatic LBL self-assembly technology. The effects of modifier dosage and pH on the zeta potential of h-BN and WS$_2$ were systematically investigated. In addition, the tribological behavior and lubricating mechanism of multi-layer vdWH WS$_2$/h-BN were studied, respectively.

(1) SDBS and PDDA were used to modify WS$_2$ and h-BN, which the negatively charged SDBS@WS$_2$ and the positively charged PDDA@h-BN were successfully prepared. When the mass ratio of h-BN to PDDA was 1:1 and pH = 3, the zeta potential of PDDA@h-BN was 60.0 mV. Under the same conditions, the zeta potential of SDBS@WS$_2$ was −50.1 mV.

(2) The multi-layer (4 layers) vdWH WS$_2$/h-BN synthesized by electrostatic interaction method had a sandwich-like structure, in which WS$_2$ and h-BN were stacked alternately along the (002) crystal plane.

(3) Compared with the base oil, the addition of multi-layer vdWH WS$_2$/h-BN (1.0 wt%) reduced the friction coefficient (0.104) by 33.8% and the wear rate (4.43 × 10^{-5} mm^3/(N·m)) by 16.8%. This arose from a natural lattice mismatch of 26.0% between the heterogeneous interfaces in multi-layer vdWH WS$_2$/h-BN.

(4) The multi-layer (4 layers) vdWH WS$_2$/h-BN had better tribological property than single-layer vdWH WS$_2$/h-BN. The friction coefficient (0.104) and wear rate (4.43 × 10^{-5} mm^3/(N·m)) were 8.7% and 8.1% lower than those of single-layer vdWH WS$_2$/h-BN, respectively. This was due to the fact that multi-layer vdWH WS$_2$/h-BN provided more heterogeneous interfaces, which made its slip probability 15 times higher than that of single-layer vdWH WS$_2$/h-BN.

Author Contributions: Y.F.: Methodology, Investigation, Data Curation, Writing—Original Draft; Y.S.: Funding Acquisition, and Writing—Review and Editing; F.S.: Methodology and Supervision; J.Z.: Investigation and Methodology; J.Y.: Validation. Z.Y.: Conceptualization; H.S.: Conceptualization, Resources, Supervision, and Writing—Review and Editing. All authors have read and agreed to the published version of the manuscript.

Funding: The research was supported by the National College Students Innovation and Entrepreneurship Training Program (202310356034), the Zhejiang Provincial Natural Science Foundation (LQ24E040001) and Science and Technology Innovation Activity Program for College Students in Zhejiang Province (New Seedling Talent Program) Project (2024R409043).

Data Availability Statement: The data presented in this study are available in the article.

Conflicts of Interest: The authors declare no conflicts of interest.

References

1. Tu, Y.; Zhang, L.; Zhang, X.; Kang, X. Improving the mechanical and tribological behavior of Cu-WS$_2$ self-lubricating composite with the addition of WS$_2$ nanosheet. *Wear* **2023**, *530–531*, 205013. [CrossRef]
2. Zhang, D.; Li, Z.; Klausen, L.H.; Li, Q.; Dong, M. Friction behaviors of two-dimensional materials at the nanoscale. *Mater. Today Phys.* **2022**, *27*, 100771. [CrossRef]
3. Aldana, P.U.; Dassenoy, F.; Vacher, B.; Mogne, T.L.; Thiebaut, B.; Bouffet, A. Antispalling Effect of WS$_2$ Nanoparticles on the Lubrication of Automotive Gearboxes. *Tribol. Trans.* **2016**, *59*, 178–188. [CrossRef]
4. Kumari, S.; Chouhan, A.; Siva Kumar Konathala, L.N.; Sharma, O.P.; Ray, S.S.; Ray, A.; Khatri, O.P. Chemically functionalized 2D/2D hexagonal boron Nitride/Molybdenum disulfide heterostructure for enhancement of lubrication properties. *Appl. Surf. Sci.* **2022**, *579*, 152157–152168. [CrossRef]
5. Novoselov, K.S.; Mishchenko, A.; Carvalho, A.; Neto, A.H.C. 2D materials and van der Waals heterostructures. *Science* **2016**, *353*, 9439. [CrossRef] [PubMed]
6. Geim, A.K.; Grigorieva, I.V. Van der Waals heterostructures. *Nature* **2013**, *499*, 419–425. [CrossRef] [PubMed]
7. Song, Y.; Mandelli, D.; Hod, O.; Urbakh, M.; Ma, M.; Zheng, Q. Robust microscale superlubricity in graphite/hexagonal boron nitride layered heterojunctions. *Nat. Mater.* **2018**, *17*, 894–899. [CrossRef] [PubMed]
8. Mandelli, D.; Leven, I.; Hod, O.; Urbakh, M. Sliding friction of graphene/hexagonal-boron nitride heterojunctions: A route to robust superlubricity. *Sci. Rep.* **2017**, *7*, 10851. [CrossRef] [PubMed]
9. Liao, M.; Nicolini, P.; Du, L.; Yuan, J.; Wang, S.; Yu, H.; Tang, J.; Cheng, P.; Watanabe, K.; Taniguchi, T.; et al. Ultra-low friction and edge-pinning effect in large-lattice-mismatch van der Waals heterostructures. *Nat. Mater.* **2022**, *21*, 47–53. [CrossRef]
10. Tian, J.; Yin, X.; Li, J.; Qi, W.; Huang, P.; Chen, X.; Luo, J. Tribo-Induced Interfacial Material Transfer of an Atomic Force Microscopy Probe Assisting Superlubricity in a WS$_2$/Graphene Heterojunction. *ACS Appl. Mater. Interfaces* **2019**, *12*, 4031–4040. [CrossRef]
11. Wei, Z.; Li, B.; Xia, C.; Cui, Y.; He, J.; Xia, J.-B.; Li, J. Various Structures of 2D Transition-Metal Dichalcogenides and Their Applications. *Small Methods* **2018**, *2*, 1800094. [CrossRef]
12. Dean, C.R.; Young, A.F.; Meric, I.; Lee, C.; Wang, L.; Sorgenfrei, S.; Watanabe, K.; Taniguchi, T.; Kim, P.; Shepard, K.L.; et al. Boron nitride substrates for high-quality graphene electronics. *Nat. Nanotechnol.* **2010**, *5*, 722–726. [CrossRef] [PubMed]
13. Dai, T.-J.; Chen, Y.-Q.; Zhou, Z.-Y.; Sun, J.; Peng, X.-S.; Liu, X.-Z. Two-dimensional MoSe$_2$/graphene heterostructure thin film with wafer-scale continuity via van der Waals epitaxy. *Chem. Phys. Lett.* **2020**, *755*, 137762. [CrossRef]
14. Han, T.; Liu, H.; Wang, S.; Chen, S.; Yang, K. The Large-Scale Preparation and Optical Properties of MoS$_2$/WS$_2$ Vertical Hetero-Junction. *Molecules* **2020**, *25*, 1857. [CrossRef] [PubMed]
15. Macknojia, A.; Ayyagari, A.; Zambrano, D.; Rosenkranz, A.; Shevchenko, E.V.; Berman, D. Macroscale Superlubricity Induced by MXene/MoS$_2$ Nanocomposites on Rough Steel Surfaces under High Contact Stresses. *ACS Nano* **2023**, *17*, 2421–2430. [CrossRef] [PubMed]
16. Man, X.; Liang, P.; Shu, H.; Zhang, L.; Wang, D.; Chao, D.; Liu, Z.; Du, X.; Wan, H.; Wang, H. Interface Synergistic Effect from Layered Metal Sulfides of MoS$_2$/SnS$_2$ van der Waals Heterojunction with Enhanced Li-Ion Storage Performance. *J. Phys. Chem. C* **2018**, *122*, 24600–24608. [CrossRef]
17. Yu, X.; Wang, X.; Zhou, F.; Qu, J.; Song, J. 2D van der Waals Heterojunction Nanophotonic Devices: From Fabrication to Performance. *Adv. Funct. Mater.* **2021**, *31*, 2104260. [CrossRef]
18. Han, T.; Liu, H.; Chen, S.; Chen, Y.; Wang, S.; Li, Z. Fabrication and Characterization of MoS$_2$/h-BN and WS$_2$/h-BN Heterostructures. *Micromachines* **2020**, *11*, 1114–1130. [CrossRef]
19. Purbayanto, M.A.K.; Chandel, M.; Birowska, M.; Rosenkranz, A.; Jastrzębska, A.M. Optically Active MXenes in Van der Waals Heterostructures. *Adv. Mater.* **2023**, *35*, e2301850. [CrossRef]
20. Zhao, Y.; Liu, J.; Zhang, X.; Wang, C.; Zhao, X.; Li, J.; Jin, H. Convenient Synthesis of WS$_2$/MoS$_2$ Heterostructures with Enhanced Photocatalytic Performance. *J. Phys. Chem. C* **2019**, *123*, 27363–27368. [CrossRef]
21. Ren, K.; Yu, G.; Zhang, Z.; Wu, W.; Tian, P.; Chhattal, M.; Gong, Z.; Li, Y.; Zhang, J. Self-organized transfer film-induced ultralow friction of Graphene/MoWS$_4$ heterostructure nanocomposite. *Appl. Surf. Sci.* **2022**, *572*, 151443. [CrossRef]
22. Ni, X.; Li, Z.; Wang, Y. Adsorption Characteristics of Anionic Surfactant Sodium Dodecylbenzene Sulfonate on the Surface of Montmorillonite Minerals. *Front. Chem.* **2018**, *6*, 390–400. [CrossRef]
23. Wang, X.; Wang, X.; Zhao, J.; Song, J.; Su, C.; Wang, Z. Surface modified TiO$_2$ floating photocatalyst with PDDA for efficient adsorption and photocatalytic inactivation of Microcystis aeruginosa. *Water Res.* **2018**, *131*, 320–333. [CrossRef] [PubMed]
24. Shen, H.; Guo, J.; Wang, H.; Zhao, N.; Xu, J. Bioinspired modification of h-BN for high thermal conductive composite films with aligned structure. *ACS Appl. Mater. Interfaces* **2015**, *7*, 5701–5709. [CrossRef]
25. Sui, Y.; Li, P.; Dai, X.; Zhang, C. Green self-assembly of h-BN@PDA@MoS$_2$ nanosheets by polydopamine as fire hazard suppression materials. *React. Funct. Polym.* **2021**, *165*, 105965–105977. [CrossRef]
26. Lv, F.; Lu, X.; Song, J.; Zhu, M.; Wang, S.; Xu, Y.; Chang, X. Enhanced Aramid/Al$_2$O$_3$ interfacial properties by PDDA modification for the preparation of composite insulating paper. *Res. Chem. Intermed.* **2022**, *48*, 4815–4835. [CrossRef]
27. Vaziri, H.S.; Shokuhfar, A.; Afghahi, S.S.S. Synthesis of WS$_2$/CNT hybrid nanoparticles for fabrication of hybrid aluminum matrix nanocomposite. *Mater. Res. Express* **2020**, *7*, 025034–025048. [CrossRef]
28. Allahbakhsh, A.; Haghighi, A.H.; Sheydaei, M. Poly(ethylene trisulfide)/graphene oxide nanocomposites. *J. Therm. Anal. Calorim.* **2016**, *128*, 427–442. [CrossRef]

29. Cho, E.-C.; Chang-Jian, C.-W.; Zheng, J.-H.; Huang, J.-H.; Lee, K.-C.; Ho, B.-C.; Hsiao, Y.-S. Microwave-assisted synthesis of TiO_2/WS_2 heterojunctions with enhanced photocatalytic activity. *J. Taiwan Inst. Chem. Eng.* **2018**, *91*, 489–498. [CrossRef]
30. Shu, D.; Li, Y.; Liu, A.; Zhang, H.; Zhou, Y. Facile route to preparation of positively charged GO using poly (diallyldimethylammoniumchloride). *Fuller. Nanotub. Carbon Nanostruct.* **2019**, *28*, 394–401. [CrossRef]
31. Zhang, X.; Xu, H.; Wang, J.; Ye, X.; Lei, W.; Xue, M.; Tang, H.; Li, C. Synthesis of ultrathin WS_2 nanosheets and their tribological properties as lubricant additives. *Nanoscale Res. Lett.* **2016**, *11*, 442–451. [CrossRef] [PubMed]
32. Fan, X.; Li, X.; Zhao, Z.; Yue, Z.; Feng, P.; Ma, X.; Li, H.; Ye, X.; Zhu, M. Heterostructured rGO/MoS_2 nanocomposites toward enhancing lubrication function of industrial gear oils. *Carbon* **2022**, *191*, 84–97. [CrossRef]
33. Feng, P.; Ren, Y.; Li, Y.; He, J.; Zhao, Z.; Ma, X.; Fan, X.; Zhu, M. Synergistic lubrication of few-layer $Ti_3C_2T_x/MoS_2$ heterojunction as a lubricant additive. *Friction* **2022**, *10*, 2018–2032. [CrossRef]
34. Neves, G.O.; Araya, N.; Salvaro, D.B.; Lamim, T.d.S.; Giacomelli, R.O.; Binder, C.; Klein, A.N.; de Mello, J.D.B. Tribologically induced nanostructural evolution of carbon materials: A new perspective. *Friction* **2023**, *12*, 144–163. [CrossRef]
35. Shao, X.; Wang, L.; Yang, Y.; Yang, T.; Deng, G.; He, Y.; Dong, L.; Wang, H.; Yang, J. Influence of preload on the tribological performance of MoS_2/GO composite lubricating coating. *Tribol. Int.* **2023**, *181*, 108306. [CrossRef]
36. Kong, S.; Wang, J.; Hu, W.; Li, J. Effects of Thickness and Particle Size on Tribological Properties of Graphene as Lubricant Additive. *Tribol. Lett.* **2020**, *68*, 112–122. [CrossRef]
37. Hod, O.; Meyer, E.; Zheng, Q.; Urbakh, M. Structural superlubricity and ultralow friction across the length scales. *Nature* **2018**, *563*, 485–492. [CrossRef] [PubMed]
38. Liu, Y.; Grey, F.; Zheng, Q. The high-speed sliding friction of graphene and novel routes to persistent superlubricity. *Sci. Rep.* **2014**, *4*, 4875–4882. [CrossRef]
39. Li, C.; Yu, Y.; Ding, Q.; Yang, L.; Liu, B.; Bai, L. Enhancement on lubrication performances of water lubricants by multilayer graphene. *Tribol. Lett.* **2023**, *72*, 5–25. [CrossRef]
40. Liang, H.; Chen, X.; Bu, Y.; Xu, M.; Zheng, G.; Gao, K.; Hua, X.; Fu, Y.; Zhang, J. Macroscopic superlubricity of potassium hydroxide solution achieved by incorporating in-situ released graphene from friction pairs. *Friction* **2022**, *11*, 567–579. [CrossRef]

Disclaimer/Publisher's Note: The statements, opinions and data contained in all publications are solely those of the individual author(s) and contributor(s) and not of MDPI and/or the editor(s). MDPI and/or the editor(s) disclaim responsibility for any injury to people or property resulting from any ideas, methods, instructions or products referred to in the content.

Review

Two-Dimensional Nanomaterials in Hydrogels and Their Potential Bio-Applications

Zhongnan Wang [1], Hui Guo [1], Ji Zhang [2], Yi Qian [1,3] and Yanfei Liu [4,*]

[1] School of Mechanical, Electronic and Control Engineering, Beijing Jiaotong University, Beijing 100044, China; zhn.wang@bjtu.edu.cn (Z.W.); 21121292@bjtu.edu.cn (H.G.); laird.qian@zimmerbiomet.com (Y.Q.)
[2] Beijing Jishuitan Hospital, Capital Medical University, Beijing 100035, China; drzhangji@126.com
[3] Beijing Montagne Medical Device Co., Ltd., Beijing 100176, China
[4] School of Mechanical Engineering, Beijing Institute of Technology, Beijing 100081, China
* Correspondence: liuyanfei@bit.edu.cn

Abstract: Hydrogels with high hydrophilicity and excellent biocompatibility have been considered as potential candidates for various applications, including biomimetics, sensors and wearable devices. However, their high water content will lead to poor load-bearing and high friction. Currently, two-dimensional (2D) materials have been widely investigated as promising nanofillers to improve the mechanical and lubrication performances of hydrogels because of their unique physical–chemical properties. On one hand, 2D materials can participate in the cross-linking of hydrogels, leading to enhanced load-bearing capacity and fatigue resistance, etc.; on the other hand, using 2D materials as nanofillers also brings unique biomedical properties. The combination of hydrogels and 2D materials shows bright prospects for bioapplications. This review focusses on the recent development of high-strength and low-friction hydrogels with the addition of 2D nanomaterials. Functional properties and the underlying mechanisms of 2D nanomaterials are firstly overviewed. Subsequently, the mechanical and friction properties of hydrogels with 2D nanomaterials including graphene oxide, black phosphorus, MXenes, boron nitride, and others are summarized in detail. Finally, the current challenges and potential applications of using 2D nanomaterials in hydrogel, as well as future research, are also discussed.

Keywords: hydrogel; 2D nanomaterial; mechanical strength; lubrication behavior; nanofiller

Citation: Wang, Z.; Guo, H.; Zhang, J.; Qian, Y.; Liu, Y. Two-Dimensional Nanomaterials in Hydrogels and Their Potential Bio-Applications. *Lubricants* **2024**, *12*, 149. https://doi.org/10.3390/lubricants12050149

Received: 15 February 2024
Revised: 19 April 2024
Accepted: 25 April 2024
Published: 27 April 2024

Copyright: © 2024 by the authors. Licensee MDPI, Basel, Switzerland. This article is an open access article distributed under the terms and conditions of the Creative Commons Attribution (CC BY) license (https://creativecommons.org/licenses/by/4.0/).

1. Introduction

Hydrogels have attracted lots of attention for various applications, such as medical devices and micro-electrochemical systems (MEMS), because of their solid–liquid biphasic structures [1–4]. However, common polymer hydrogels such as polyvinyl alcohol (PVA), polyethylene glycol (PEG) and polyacrylic acid (PAA) have low mechanical strength (approximately 0.2 MPa~3 MPa), and poor antifouling and antibacterial ability [5–7]. Zwitterionic materials contain both cationic and anionic groups, with the overall charge being neutral. Zwitterionic hydrogels, like polymethylacrylamide sulfonate betaine (PSBMA), have a three-dimensional polymer network rich in cation and anion groups, and are electrically neutral at the macro level. They have a high water content and good resistance to bacterial adhesion, but their compressive strength (0.08 MPa) and lubricating ability ($\mu \approx 0.03$) are still insufficient [8]. Human cartilage with a complex lubrication mechanism exhibits a super-low sliding friction coefficient (below 0.01) and compressive strength (around 18 MPa). Therefore, there is still a significant gap in the mechanical and lubricating properties between traditional hydrogels and natural cartilage. Introducing nanofillers into hydrogels is an effective strategy to improve the mechanical and lubrication properties.

Two-dimensional (2D) nanomaterials are connected by strong covalent or ionic bonds within the layers and weak van der Waals forces between the layers [9,10]. Recently, 2D nanomaterials have been systematically investigated in the fields of bionics and wearable medical devices because of their unique layered structures, physicochemical characteristics and good biocompatibility [11,12]. Moreover, graphene and MXene materials also exhibit extremely high mechanical strength [13,14]. Also, MXene and graphene oxide (GO) can offer reduced friction and wear of counterparts [15]. Subsequently, some studies on new classes of 2D nanomaterials with unique properties have been reported. Black phosphorus (BP) and hexagonal boron nitride (h-BN), used as typical 2D nanomaterials, show also excellent performance in terms of strength improvement and friction reduction [16–19]. When BP nanosheets are deposited on the silicon nitride (Si_3N_4) surface, the shear action of the contact interface promotes the formation of phosphorus oxide with low shear on the BP surface, leading to a stable sliding friction coefficient of 0.001 in a pure water environment [20]. As a lubricating additive, 2D nanomaterials including GO, BP, MXene and h-BN will form a tribochemical or physical adsorption layer at the solid–liquid interface, which enhances the anti-friction and anti-wear properties.

Various 2D nanomaterials (Figure 1) exhibit different functionalized properties for medical applications, such as use in biosensors, cartilage repair, skin tissue engineering and bone reconstruction [21]. Marian et al. [22] reviewed the unique structural properties of 2D materials, including their large surface to volume ratio, adjustable surface chemistry, inherent biocompatibility, antibacterial/antiviral activity, and non-cytotoxicity, which show great potential for biomedical applications, such as load-bearing implants, dental implants, bone fracture fixation, invasive surgical devices, cardiovascular devices, contact lenses, and bio-MEMS/NEMS. Two-dimensional materials can be used as protective coatings, fillers for composite materials, or additives for fluid mixtures, allowing for the effective adjustment and control of biotribological behavior. Specifically, GO might be used as a reinforcement of artificial cartilage, while BP has the potential to promote bone growth. The unique properties of 2D materials make them promising nanofillers for use in hydrogels to improve the mechanical and tribological performances in bioengineering applications.

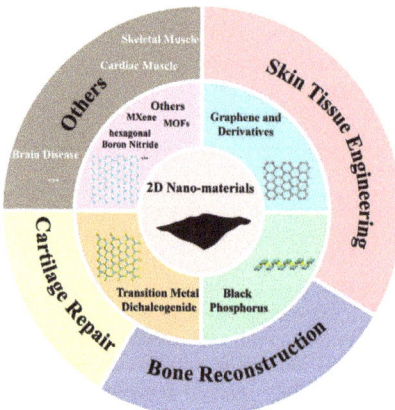

Figure 1. Various types and potential applications of 2D nanomaterials used in tissue engineering [21], copyright (2021), with permission from Wiley.

The introduction of GO at the nanoscale can significantly improve the mechanical and tribological properties of zwitterionic hydrogels, indicating their great potential for numerous applications [23]. Also, the combination of hydrogels and other 2D materials has yielded high-performance bioelectronic devices, which has promoted the development of biosensors and bioelectronics. However, 2D nanomaterials still have some inherent shortages, such as low synthesis efficiency and high economic cost, restricting their actual bioapplications. Moreover, the excellent mechanical and tribological behaviors of 2D nanomaterials with a feasible functional design mean they are often used as potential nano-additives for different load-bearing and lubrication systems. As a functional additive, the dispersion of 2D nanomaterials in hydrogels affects their interaction with polymer molecular chains, thus reducing the homogeneity of the overall network structure.

Within this work, the authors review the most recent research achievements of typical 2D materials used as nanofillers of hydrogel. Moreover, the fundamental mechanisms and functional properties of these 2D materials used as nanofillers of hydrogel are reviewed in detail. In addition, the unsolved issues and optimistic outlooks are also discussed. This review aims to provide useful insights and guidelines related to the design and development of hydrogels for various biomedical applications, and facilitate a continuously updated understanding of the influence of the typical 2D nanomaterials on mechanical and tribological performances.

2. Mechanical and Friction Properties of Hydrogels with 2D Nanomaterials

In 2004, Geim and Novoselov [24] obtained graphene via tape stripping; this material has excellent electron transport properties and electrical conductivity at room temperature. Graphene has excellent electronic properties, an extremely high specific surface area, and great potential applicability in the field of optoelectronics and catalytic reactions. h-BN, BP, MXene, and transition metal sulfides (TMDs) have graphene-like structures, which are stacked into massive crystals of different structures by van der Waals force. They show excellent optical, electrical, and chemical stability on a macro level, enriching the properties of 2D materials and promoting their applications as functional nano-additives. Table 1 gives the key properties of typical 2D materials.

Table 1. Mechanical properties of hydrogels reinforced by two-dimensional nanomaterials.

Nanoparticles	Elastic Modulus (GPa)	Tensile Strength (GPa)	Elastic Strain
GO	207.6 ± 23.4 [25]		0.4% [26]
BN	865 ± 73 [25]	70.5 ± 5.5 [27]	2.5 ± 3.0% [27]
BP	19.5~41.3 [28]	4.09~8.42 [29]	0.48% [28]
$Ti_3C_2T_x$ [13]	483.5 ± 13.2	15.4 ± 1.92	3.2%

2.1. Graphene Oxide

GO has more oxygen-containing functional groups, and can more easily physically react with other substances to yield the various characteristics of polymers, colloids or amphoteric molecules. GO has a special physically layered structure and stable chemical properties, and it can fill in the scratches and wear pores on friction surfaces, repair the damaged surfaces of friction pairs, enhance the anti-wear ability of concave and convex contacting interfaces, and lower the friction coefficient of material surfaces. Therefore, GO can be used as a nanofiller to enhance the lubrication and anti-wear properties of polymer hydrogels for a wide range of applications [23,30,31]. The interlayer structure and the number of functional groups in GO affect the shear resistance of the interface [32]. GO nanosheets have been annealed at different temperatures to obtain h-GO with different numbers of hydroxyl functional groups. It is proven that the number of oxygen-containing functional groups and the interlamellar distance will limit the hydrogen bonding of the molecules, and thus affect the lubrication ability of GO [33]. Moreover, the excellent lubrication performance of GO in metallic contact contexts means that the slip of the GO layer with low shear strength reduces the friction resistance at the interface, compared to a single physical deposition film [34].

The excellent biocompatibility of GO means it can also be used as a functional filler to enhance the properties of biological materials [35,36]. Hydrogel is a type of polymer material containing a hydrophilic three-dimensional network, which has a wide range of applications in biomedical research fields such as bionic skin, and bionic tissues and organs [37,38]. However, the traditional hydrogels lack sufficient mechanical and lubrication properties to completely meet the practical needs of bioapplications [39–41]. As a functional nanofiller, GO can be combined into a polyvinyl alcohol (PVA)/poly (N-isopropylacrylamide) (PNIPA) hydrogel to yield a good thermal response ability through free radical cross-linking. The mechanical properties of GO-PVA/PNIPA hydrogels can be enhanced with an increase in GO mass fraction, which has a potential application in temperature sensor research [39]. The diversity of pore sizes within the hydrogel's architecture dramatically influences its mechanical attributes. As the content of graphene oxide (GO) increases to a threshold level, the pores transition from a microscale to a nanoscale uniform distribution, resulting in a more compact structure. It is this intricate porous morphology, characterized by interconnected pores at the nanoscale, that markedly enhances the hydrogel's mechanical robustness [39]. The composite hydrogel synthesized by adding GO to chitosan shows good shape memory and self-healing properties. The change of solution pH will affect the hydrogen bond and hydrophobic action of polymer chains, resulting in a change of the state of the cross-linked network and the pH response ability of the composite hydrogel [42,43]. In addition, the shape and structure of GO will also affect the dispersibility in solvent and the lubrication performance of the composite hydrogel [44]. Compared to other fillers (carbon fiber [45], metal nanoparticles [46], ceramics [47]), GO [48] shows excellent self-lubrication capabilities and mechanical strength. A polyacrylamide (PAAm) hydrogel was combined with GO to significantly reduce its friction coefficient and wear rate with supramolecular polyethylene [49] (Figure 2). By enhancing the layer spacing of GO with polyethylene glycol (PEG), hydroxyapatite (HA) particles can be uniformly distributed on the GO surface in order to obtain a GO-PEG-HA hybrid, which is added to the PVA polymer as a filler to prepare a PVA/GO-PEG-HA hydrogel with excellent biocompatibility. This nanocomposite hydrogel exhibited an enhanced compression strength (4.49 MPa) and a reduced sliding friction coefficient (0.06), compared with PVA hydrogel [50]. The introduction of GO into a PVA/polyacrylic acid (PAA)/polydopamine multifunctional coating could reduce the contact angle of the Ti6Al4V alloy surface and achieve better wettability, as well as improving the corrosion resistance and biocompatibility [51]. The improvement of the anti-corrosion efficiency of hydrogel coatings is attributed to the nucleophilic reaction between the epoxy and amino groups from GO and the active ions of SBF solution (Figure 3a). Also, the GO sheets at the interface of the PVA/PAA/GO/PDA hydrogel coating and a cortical bone sample might enable a significant reduction in friction coefficient (Figure 3b).

2.2. Boron Nitride

Recently, BN materials have attracted widespread attention from researchers as an outstanding filler to enhance the mechanical strength and lubrication of various functional hydrogels, due to their unique ultimate thermal stability, chemical inertness, and resistance to oxidation.

BN has a graphite-like lamellar structure and is thought to facilitate cartilage movement, just like in bearing systems [52]. Many studies suggest its promising potential use in biomedical applications such as artificial cartilages, drug delivery, tissue engineering, biosensors and actuators [53].

Jing et al. [54] prepared a hydroxylated boron nitride nanosheet (OH-BNNS)/PVA interpenetrating hydrogel that exhibited controllable reinforcements in mechanical responses. Impressive 45% and 43% increases in compressive and tensile strengths, respectively, could be achieved when the addition of OH-BNNS is only 0.12 wt. % (Figure 4). The superior intrinsic properties of distributed OH-BNNS and strong hydrogen bonding interactions between the OH-BNNS and PVA chains collectively contribute to the efficient load transfer. Yang et al. [55] found that hexagonal boron nitride nanoplatelets (BNNPs) grafted with amino acid could be used as functional fillers for simultaneously enhancing the mechanical and self-healing properties of the PVA hydrogel composites. The incorporated fillers provided hydrogen-bonding interactions between -OH groups on the PVA chain and the -COOH groups originating from AA moieties to offer excellent tensile strength and healing efficiency.

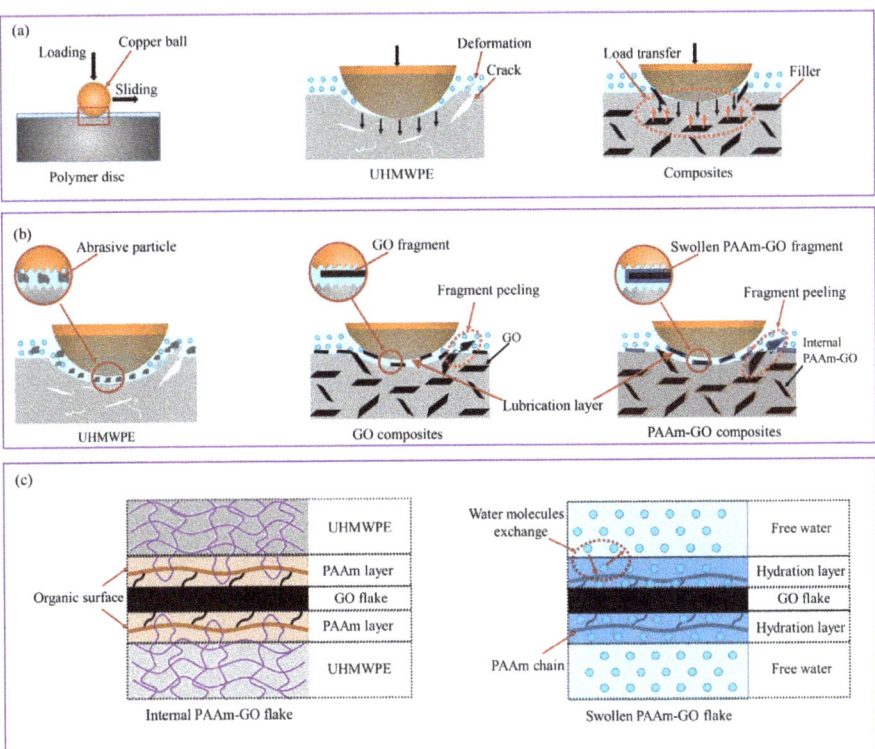

Figure 2. A comprehensive analysis of tribological mechanisms between GO and PAAm hydrogel. (**a**) The bearing statuses of UHMWPE and the composites; (**b**) the lubricating characteristics of the polymers; and (**c**) two states of PAAm-GO flakes [49], copyright (2021), with permission from Elsevier.

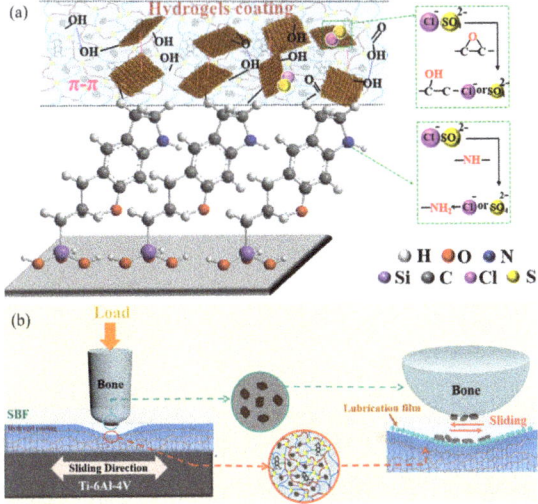

Figure 3. Schematic diagram of the mechanisms of GO-based hydrogel coating. (**a**) Anticorrosion mechanisms of PVA/PAA/GO/PDA samples; (**b**) biotribological mechanisms of PVA/PAA/GO/PDA samples [51], copyright (2023), with permission from Elsevier.

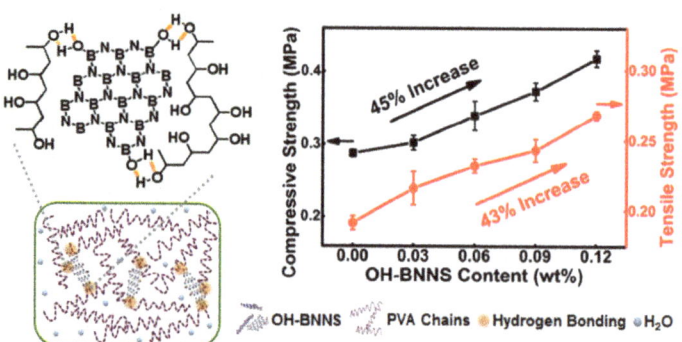

Figure 4. Structural characterization and mechanical responses of OH-BNNS/PVA composite hydrogels [54], copyright (2023), with permission from ACS.

Xue et al. [56] used surface-modified BN nanosheets as nanofillers to fabricate a novel PAA/BNNS-NH$_2$ composite hydrogel via hierarchical physical interactions. The addition of BNNS could enhance the mechanical properties of the PAA hydrogel, including offering a fracture stress of ~1311 kPa and toughness of ~4.7 MJ m^{-3}. Jiang et al. [57] also found that a dual crosslinked BNNS/PAA nanocomposite hydrogel could recover its mechanical strength even following severe structural breakdowns, such as after three consecutive cutting cycles. The functionalized material was prepared by Liu et al. [58] to effectively improve the mechanical modulus of polyurethane (PU) hydrogels from 1635 to 2776 kPa with only 0.066 wt. % BNNS loading, which could be highly useful in printed electronics. This enhancement can be primarily attributed to the robust hydrogen bonding interactions between OH-BNNS and the PU-based hydrogel matrix. Nevertheless, an excessive incorporation of OH-BNNS may lead to agglomeration and the introduction of more defects, which can compromise the material's integrity and facilitate stress concentration. Hu et al. [59] prepared ca-BNNS/PAAm nanocomposite hydrogels with high water retentivity and flexibility. They showed and elongation that exceeded 10,000% and a compressive strength of 8 MPa at 97% strain, and there was no obvious damage after the removal of the compression force. BNNSs, as a promising material, could provide excellent water retentivity and flexibility for the development of high-performance hydrogels.

A physically linked 3D f-BNNS/clay/PNIPAM ternary networks (TN) hydrogel was built by Tong et al. [60] using functionalized boron nitride nanosheets (f-BNNS) with H-grafted nitrogen/OH-grafted boron atoms. A soft polymer network embedded with 2D hard f-BNNS could improve the mechanical properties through effective load transfer and dissipated energy via the incorporation of a sacrificial non-covalent hydrogen bond. This high-toughness TN hydrogel might be used in various application fields, such as sensors, tissue engineering and flexible devices. Goncu et al. [61] found that the introduction of BN with different structures could play a dominant role in affecting the dynamic viscosity of the zero-shear point and the deformation rate, as well as the viscoelastic properties of the hybrid hydrogel. The lamellar structures of h-BN have been considered to be part of an effective method of joint injections for the treatment of osteoarthritis (OA).

2.3. Black Phosphorus

Black phosphorus (BP) is a graphite-like photoelectric material with an anisotropic lamellar structure and low interlayer interaction. Phosphorus atoms are deposited in a two-dimensional plane to form a folded honeycomb, which means it has great tensile and extrusion characteristics in the atomic plane, and it can be deformed under external forces to change its conductive ability [62]. BP's interlayer spacing (0.53 nm) is larger than that of GO (0.36 nm), which is conducive to the insertion and removal of ions, imparting broad application prospects in the fields of lithium-ion batteries and supercapacitors [63].

BP is prone to oxidation when exposed to air, forming phosphorus oxide with very low shear strength, which can achieve super-lubricity in a pure water environment [12]. The introduction of BP into carbon fiber (CF)/PTFE composites could help to form a lubrication film composed of phosphorus oxide and phosphoric acid at the contact interface, resulting in a reduced friction coefficient and the disappearance of adhesive wear [64,65]. Wu et al. [66] found that the formation of a highly mobile water layer on the oxidized BP/SiO$_2$ interface could impart a super-lubricity phenomenon, and the lubricious liquid water layer notably reduced the interfacial shear strength (~0.029 MPa).

However, single nanoparticles have been demonstrated to be insufficient when used as an additive to optimize the lubricating property of materials. By means of sol-gel, high-energy ball milling and mechanical stirring, BP could be combined with titanium dioxide (TiO$_2$) [67], silver nanoparticles [68], MoS$_2$ [69] or other nanoparticles, and used in oil-based lubricants [70] and water-based lubricants [71] to form a mixed lubrication additive. The synergistic effects of BP and the above particles further improved the anti-friction and anti-wear properties of friction materials. For example, BP powder was prepared by high-energy ball milling and then combined with TiO$_2$ to prepare BP/TiO$_2$ nanocomposites via a solvothermal reaction. The rolling tribology experiment (Figure 5) shows that the friction coefficient and wear rate of BP/TiO$_2$ nanocomposites are significantly lower than those of BP and TiO$_2$, which is due to the formation of a tribochemical reaction film on the surfaces of BP/TiO$_2$ nanocomposites, and the repair effects of TiO$_2$ on the worn area. The results show that the composite nanoparticles can repair the worn surface and form a complex tribochemical reaction film on the surface of the material, which contributes to the friction interface becoming smoother with the addition of nanoparticles, and the wear rate of the material surface being significantly reduced [69].

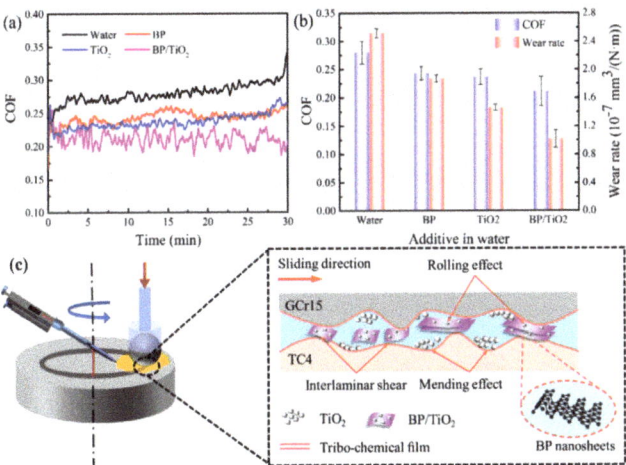

Figure 5. Friction experiments and synergy lubrication model of nanocomposite lubrication additives. (**a**) The curve of COFs and (**b**) wear rates of Ti–6Al–4V alloy discs of ultrapure water, BP water-based lubrication additive, TiO$_2$ lubrication additive, and BP/TiO$_2$ lubrication additive. (**c**) Collaborative lubrication mechanism of BP/TiO$_2$ composite nanoparticles [69], copyright (2022), with permission from Elsevier.

In addition, BP has excellent photothermal conversion efficiency and inherent photoacoustic properties, which endow it with potential applications in the biomedical field [72]. The encapsulated black phosphorus nanosheet (BPN) could be used to stably provide modest amounts of phosphorus for the enhancement of the compressive strength (up to 0.15 MPa) and compressive moduli (arrive at 0.9 MPa) of hydrogels. BPNs were successfully encapsulated within the hydrogel matrix to create a thicker structure with permeable holes.

This innovation results in a reinforced network architecture in the nanocomposite hydrogels. Additionally, the optimized porous configuration endows these hydrogels with enhanced water absorption capabilities, a feature that is particularly advantageous for facilitating the nutrient supply required for cellular growth and proliferation. Moreover, the phosphoric acid component contained in BP can promote the growth and reproduction of chondrocytes, which can be applied as bionic materials in the field of bone tissue engineering (Figure 6a) [73]. BP was mixed with chitosan (CS) for deposition on polyetheretherketone (PEEK) bone scaffold by 3D printing technology. Then, cell culturing in vitro and the growth of cells implanted into mice showed that this scaffold has good biological activity and can promote the expression of osteogenic genes [74]. The tests on the synthesized skeleton, including for its antibacterial property, in vitro osteosarcoma ablation, in vivo tumor ablation, in vitro cytocompatibility, in vitro osteogenic activity and in vivo bone regeneration, were performed to investigate its biocompatibility. Alkaline phosphatase was used as a marker of osteogenic differentiation to evaluate its capacity to promote bone growth. A novel scaffold system consisting of a 3D PEEK scaffold substrate and a BP/CS composite coating can effectively fill bone defects and provide the required mechanical support (Figure 6b).

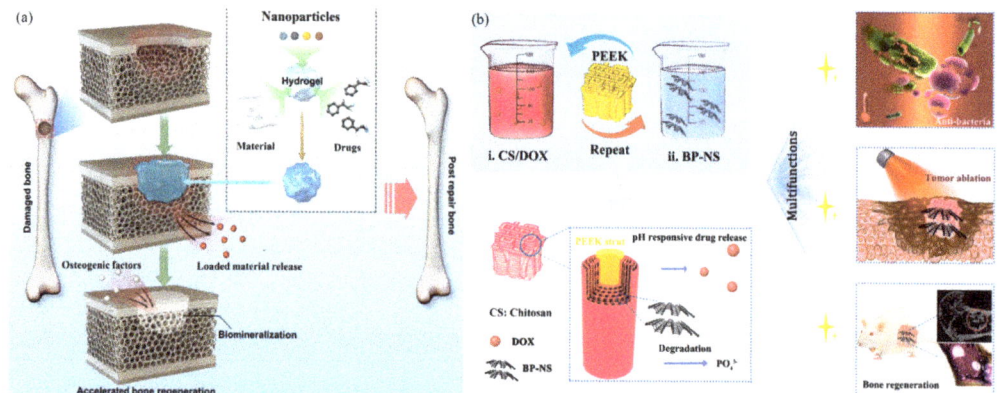

Figure 6. (a) Schematic illustration of BPN-containing hydrogel with the excellent mechanical behaviors required in bone tissue engineering [73], copyright (2022), with permission from Taylor & Francis. (b) Schematic of BP-NS/CS-DOX composite coating on a sulfonated PEEK scaffold with multiple functions [74], copyright (2022), with permission from Elsevier.

2.4. MXenes

In 2011, researchers from Drexel University obtained MXene materials by etching Al into ternary layered carbide Ti_3AlC_2 with hydrofluoric acid. These were composed of transition metal carbides, nitrides or carbon nitrides [75]. MXene has a two-dimensional graphene-like structure, excellent self-lubricating ability, and many groups on the surface, which means it has been the subject of much research in the fields of energy storage, catalysis, lubrication, and antibacterial and electromagnetic shielding [76]. Chhattal et al. [77] reviewed the effects of different synthesis methods on the properties of MXene (mechanics, lubrication and surface properties), and suggested that improving the synthesis of MXene 2D materials is an important way to effectively improve their properties. The tribological behavior of MXene at the micro and macro scales, as well as its research status as a coating, lubricating additive and composite reinforcing phase, is emphasized. It is considered that the dispersion and adsorption of MXene on the surface of the material greatly affects the lubrication effect of the contact area. The MXene interlayer has a weak force and is easy to peel off, leading to good self-lubricating properties. MXene is coated on the surface of stainless steel and silicon wafers to generate a dense $Ti_3C_2T_x$ friction film at the contact interface, which significantly reduces the friction coefficient and wear rate of the sliding interface [78,79]. Compared with traditional carbon-based lubrication fillers (graphite, graphene, carbon nanotubes), MXene has strong interface coupling properties [80].

Combining low-dimensional nanostructured materials with MXene can improve its dispersion performance. Das et al. [81] summarized the mechanical, magnetic and thermal stability and photoelectrical properties of composite nanomaterials synthesized from MXene and low-dimensional nanomaterials (quantum dot, magnetic nanoparticles, non-magnetic metal oxides and precious metals), as well as the application status of composite nanomaterials in the fields of sensors, new energy storage, catalysts, etc. MXene is mixed with metal nanoparticles and MoS_2 as a lubrication additive to enhance the mechanical strength of the physical adsorption film on the friction surface of the MXene layer, which imparts outstanding lubrication and anti-wear properties under high stress, and further provides a theoretical basis for the development of multi-functional lubrication materials [82,83]. The Cu^{2+} is loaded onto the surface of MXene nanosheets via electrostatic adsorption to prepare an MXene@Cu nanocomposite. The synergy of MXene nanosheets and Cu nanoparticles could lead to the formation of a stable protective film on the surface of the friction pair and thus provide excellent tribological performance [84]. Compared with MoS_2, tungsten-doped amorphous carbon (a-C:H:W) and hydrogen-free, more graphitelike amorphous carbon (a-C), MXene can significantly reduce the friction coefficient and wear rate of composites during the dry friction process [85]. However, due to the presence of the -OH, -O and -F functional groups on the surface, MXene has high surface energy, as well as poor dispersion and stability, in polar polymers or weakly polar polymers [86]. MXene was modified with tetradecylphosphonic acid and Poly[2(Perfluorooctyl)ethyl methacrylate] to improve its dispersibility in the lubrication oil, enhance its adsorption capacity on the substrate surface, and reduce the shear effect in the contact zone [5,87].

MXene has metal conductivity and colloidal processing properties, which can be used to enhance the friction and conductivity of hydrogels and enable broad application prospects in the field of temperature sensors and light-responsive soft robots [88,89]. Hydrogels can be used as drug carriers, coating drug molecules, reducing immune response and extending drug stability and storage life [90]. In order to realize the intelligent response of drug release, Yan et al. prepared a responsive nanofiber using MXene and hydrogel, which enhanced the dispersion of the hydrogel in a fiber membrane, and they controlled the drug release ability of the fiber membrane through light intensity [91]. MXene also has more active groups on its surface, meaning it can be used as a hydrogel crosslinking agent to optimize the structure, mechanical strength and tribolgical behavior of hydrogels. The tensile strength, toughness and elongation at break of the synthesized hydrogel with the addition of MXene are 0.251 MPa, 0.895 MJ m^{-3}, and 560.82%, respectively [92]. The formation of hydrogen bonds between MXene and the PVA molecular chains substantially fortifies the structure and hydration capacity of a hydrogel network, as depicted in Figure 7 [93]. The presence of reactive functional groups on the surface of MXene nanosheets, which share a graphene-like structure, allows for effective interaction with the hydrogel's molecular chains. This interaction not only reinforces the mechanical attributes of the hydrogels, but also enhances their lubricating properties, thereby contributing to a more robust and versatile material. Ye et al. added MXene and dopamine to a poly PEGDA–methylacrylic anhydride–gelatin hydrogel to synthesize a bionic heart patch with high electrical conductivity and excellent biocompatibility. The mechanical properties of the composite hydrogel containing MXene were enhanced by 60% [94]. However, using MXene as a single additive can only improve the mechanical properties of hydrogels in a limited way. MXene and nanocellulose were used to synthesize composite nanoadditives; here, nanocellulose could improve the dispersibility of MXene in hydrogels, and its synergy with MXene could promote the cross-linking of hydrogel networks [95].

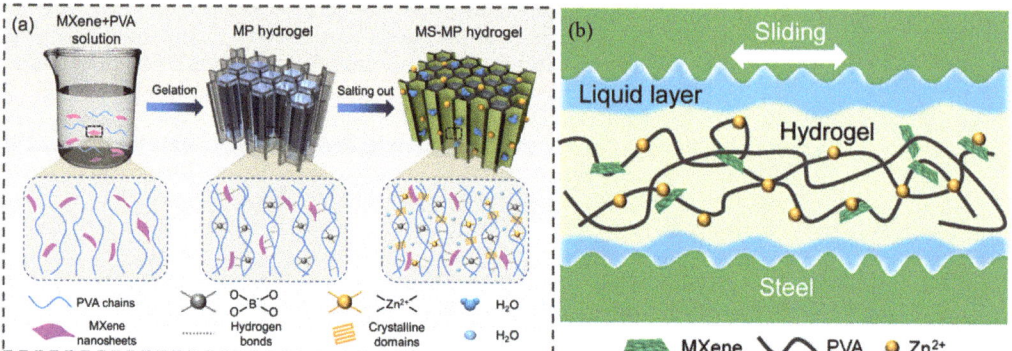

Figure 7. Schematic diagram of MXene-based hydrogels. (**a**) Multi-crosslinked and solid–liquid composite lubricating hydrogel; (**b**) the lubrication mechanism of hydrogels with MXene [93], copyright (2023), with permission from Elsevier.

A variety of synthesis methods have been proposed to improve the surface chemical properties of MXene, such as reactions between metals and metal halides, or selective etching in a mixture. Although it is theoretically possible to synthesize new MXene materials by changing M, X, n and Tx inserts, more oxidative impurities will be generated through etching, which will affect the microstructural and macroscopic properties of MXene, and lead to the production of toxic substances during synthesis processes, thus causing pollution to the human body and the environment [96,97]. The available uniform surface terminations of MXenes, including oxygen, imido group, sulfur, chlorine, selenium, bromine and tellurium, were also controlled using computational studies or various synthetic methods [98]. In addition, a universal 4D printing technology was used to prepare MXene-family hydrogels, such as Nb_2CT_x, $Ti_3C_2T_x$, and $Mo_2Ti_2C_3T_x$. The 4D-printed MXene hydrogel that was obtained could be used to create 3D porous architectures, with large specific surface areas, high electrical conductivities, and excellent mechanical properties [99].

Numerous studies have elucidated the considerable potential of MXene materials to enhance mechanical and tribological properties when employed as nanofillers in conjunction with other two-dimensional materials or nanoparticles. The unique elemental composition, surface terminations, and structural attributes of MXenes are the factors determining their mechanical characteristics [100]. The weak secondary interlayer bonding between adjacent layers of various combinations and forms of MXenes will result in a low shear resistance, which makes them promising for use in tribological applications [100]. Moreover, applying MXenes on different surfaces could decrease the interfacial adhesion, and their chemical reactivity at high pressures and temperatures might lead to the formation of a tribolayer in tribological contacts. The formed tribofilm tends to provide good substrate adhesion, as well as long-term low-friction and low-wear performance [101]. In addition, a variety of surface terminations on the outer surfaces of MXenes can be functionalized to improve their dispersion stability, oxidation resistance and compatibility with other matrix materials in composites, thus leading to an enhanced tribological performance [102].

2.5. Other 2D Materials

Recently, some other 2D nanoparticles with various functions have been under development [103]. Cadmium sulfide (CdS) photocatalytic nanoparticles were incorporated as a reinforced material into a P(AA-AM) composite hydrogel [104]. Increasing amounts of CdS could significantly enhance the mechanical strength of the hydrogel from 0.445 MPa to 1.014 MPa. The P(AA-AM)@CdS nanocomposite hydrogel also exhibited strong synergistic adsorption and photocatalytic degradation clearance effects with methylene blue. Thus, the introduction of CdS photocatalytic nanoparticles may enable the efficient enhancement of the mechanical properties of bifunctional hydrogel materials. Bioactive glass nanoparticles

(BG) with particle sizes of 12 and 25 nm were incorporated into Alginate–gelatin (Alg-Gel) hydrogel [105]. The nanocomposite hydrogel exhibited significantly enhanced stiffness and printability with increasing BG concentrations, as well as cellular proliferation and adhesion in the bioprinted constructs. The introduction of BG also did not significantly contribute to the highly porous structure and biodegradation of Alg-Gel hydrogel, which might ground its potential application in extrusion-based bioprinting. Dai et al. [106] used gold nanorods (AuNRs) to prepare a tough nanocomposite hydrogel with a designable gradient network structure via a facile post-photo regulation strategy. The photothermal effects of AuNRs could locally improve the typical yielding and forced elastic deformation of hydrogels with a kirigami structure by near-infrared light irradiation at room temperature, because the treated regions show better resistance to crack advancement. These tough hydrogels with programmable gradient structures and mechanics have many potential applications, such as in structural elements and biological devices. Wang et al. [107] prepared a novel hybrid hydrogel based on PVA, borax and poly-dopamine particles (PDAPs). With the increasing content of PDAPs, the hybrid hydrogels exhibited a higher tensile strength from 5.86 MPa to 16.71 MPa, and a better self-healing efficiency of 114%, after contact at room temperature for 10 min. This nanoparticle-induced strengthening might expand its potential applications, such as into electrical skin, tissue engineering, drug delivery, 3D printing and soft robots areas. Cellulose nanowafers (CNWs) were added by Du et al. [108] to the PAAM/Xanthan gum-Al^{3+} DN hydrogel. The prepared hydrogel showed a maximum stress of 0.14 MPa when the elongation at break was 707.1%. The improved mechanical properties and self-recovery were attributed to the combination between the PAAM and the CNWs. The developed hydrogel with high tensile strength showed great potential for use in various applications, such as in wearable sensors for the detection of human movement. Functionalized silicon nanoparticles (SiNPs) were used by Yang et al. [109] as the cross-linker to synthesize a novel organic–inorganic hybrid bilayer PNIPAm@PAAm hydrogel. The introduction of the doped SiNPs could considerably improve the rigidity of a PAAm-type hydrogel network, leading to excellent tensile and compressive strength (Figure 8). The low cost and excellent biocompatibility of the SiNPs could expand their potential use in future applications in smart hydrogel materials. Li et al. [110] used carboxyl-modified Fe_3O_4 nanoparticles as a photothermal agent to synthesize a carboxymethyl chitosans–Fe_3O_4–acrylamide (CMCS-Fe_3O_4-AM) hydrogel with good drug loading and antibacterial properties. The introduction of carboxyl-modified Fe_3O_4 could significantly change the structural density, mechanical strength, and photothermal properties of the nano-composite hydrogel. This research might provide a novel design for new hydrogels with good mechanical properties, controllable crosslink density and photothermal properties. Lu et al. [111] used TEMPO-oxidized cellulose nanofibers (TOCNs) as a filler to develop a novel TOCN/PAAM nanocomposite hydrogel. TOCNs with high strength and ultra-high aspect ratio could improve the energy dissipation capability (9.68 MJ m^{-3} at 60% strain), viscoelasticity (51.1 kPa) and self-recovery rate (about 93.2% after 30 min recovery) of polyacrylamide (PAAM) hydrogel. This nanocomposite hydrogel with good shape memory properties and excellent mechanical strength provides promising prospects for use in intelligent biomaterials used in soft actuators, biomedicine and sensory applications. Biocompatible micro/nanoparticles containing various ratios of Ca^{2+} and Mg^{2+} with sizes ranging from 1 to 8 μm were mixed with gellan gum (GG) solution to form a self-hardening multifunctional hydrogel [112]. The Ca^{2+}/Mg^{2+} particles could be efficiently bound to GG polymer chains for the enhancement of the macro-Young's modulus of the hydrogel from 2 kPa up to 100 kPa. This hydrogel with hydro-magnesite particles also exhibited a higher cell viability and greater hydroxyapatite production. This research opens new avenues for developing injectable reconstruction materials for use in biomedical applications related to bone regeneration.

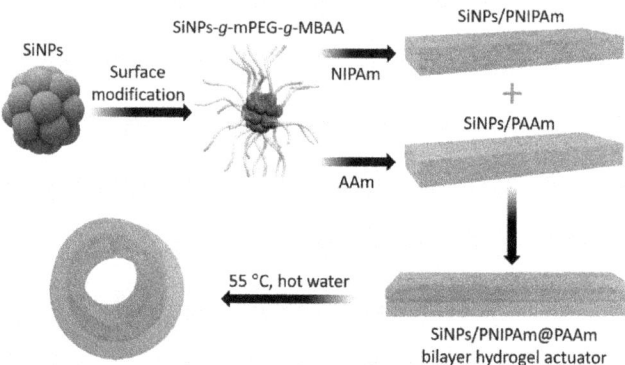

Figure 8. Synthetic procedure and structural illustration of the SiNPs/PNIPAm@PAAm bilayer hydrogel [109], copyright (2021), with permission from Elsevier.

It is here shown that 2D nanomaterials can interact with hydrogel polymer chains to synthesize nanocomposite hydrogels with excellent mechanical properties. Table 2 shows the mechanical and tribological properties of hydrogels reinforced with various 2D nanomaterials. Different types and mass fractions of these 2D nanomaterials affect the macroscopic properties of synthetic nanocomposite hydrogels. Zhang found that the mechanical properties of nanocomposite hydrogels began to decline when the mass fraction of GO was greater than 5.0 wt. %, mainly due to the polymerization inside the hydrogel, caused by an excessive amount of GO [42].

Table 2. Mechanical properties of hydrogels reinforced by 2D nanomaterials with optimal content.

2D Nanomaterials	Hydrogel	Maximum Mechanical Strength	Friction Coefficient	Concentration/Weight Percent
GO	PSBMA [23]	Compressive strength: 0.08 MPa~0.36 MPa Tensile strength: 50.7 kPa~151.9 kPa	0.006~0.03	0.005~0.025 wt. %
	PVA-PNIPA [39]	Compressive strength: 1.5 MPa~4.1 MPa		10~25 mg
	CS [43]	Tensile strength: 4.37 MPa~20.96 MPa Young's modulus: 0.122 to 0.364 MPa		0.1~0.5 wt. %
	PAAm [49]	Tensile strength: 10 MPa~50 MPa	0.05~0.12	0.2~2 wt. %
	PVA-PEG-HA [50]	Compressive strength: 1.95 MPa~4.79 MPa		1.5 wt. %
	PVA-PAA-PDA [51]	Compressive modulus: 1.12 MPa~2.53 MPa Young's modulus: 0.051 GPa~0.058 GPa	0.05~0.12	0.1~1 wt. %
	PVA [54]	Compressive strength: 0.29 MPa~0.42 MPa Tensile strength: 0.19 MPa~0.27 MPa	0.05~0.20	0~0.09 wt. %
h-BN	PAA [56]	Stiffness: 17.9 MPa, toughness: 10.5 MJ m^{-3}		0.1~1.0 mg mL^{-1}
	PU [58]	Young's modulus: 1632 kPa~2776 kPa		0.03~0.18 wt.%
	PAAm [59]	Compressive strength: ~8 MPa		0.1~2.5 mg mL^{-1}
	Clay- PNIPAM [60]	Compressive strength: 30 kPa~200 kPa Tensile strength: 17 kPa~40 kPa		0.04~0.32 wt. %
BP	PEA-GelMA [73]	Compressive strength: ~0.15 MPa Compressive moduli: 0.3 MPa~0.9 MPa		40 wt. %
	NS-CS coating [74]	Compressive moduli: 247.9 MPa~745.4 MPa Tensile moduli: 235 MPa~644 MPa		50 mg
MXene	oligo[poly(ethylene glycol) fumarate](OPF) [113]	Compression modulus: 497 kPa~734.5 kPa		0.1~1 mg mL^{-1}
	PAM [90]	Compressive strength: 400.6 kPa~819.4 kPa		0.0145~0.0436 wt. %
	PAA-PAM-TA [92]	Tensile strength: 0.251 ± 0.05 MPa		0.075 g
	PVA [93]		0.14~0.18	1~10 mg mL^{-1}
	Cryogel [94]	Compression modulus: 2.24 kPa~9.65 kPa		0.4~1.6 mg

3. Challenges and Perspectives

Hydrogels infused with 2D materials as nanofillers show immense potential utility in bioapplications, thanks to their improved mechanical and tribological characteristics. While significant advancements have been made in this field over the years, there remain a number of challenges that must be addressed to enable further progress in both scientific research and engineering applications:

- The dispersion stability of 2D materials in water-based solutions is governed by the surface energy, which significantly affects their internal interaction with polymer molecular chains of hydrogels [113]. Hence, the challenge lies in optimizing hydrogel systems to achieve a high load-bearing capacity and reduced friction at the lowest possible particle concentration. This aspect requires additional research if we are to realize the practical applications of these systems;
- The synthesis of 2D nanomaterials still involves some limitations. For example, the large-scale synthesis of 2D materials with precisely controlled nanostructures is difficult to realize. Advanced techniques and equipment for the preparation of 2D materials should be further developed.
- There is still a lack of extensive research on the optimization of basic parameters of 2D nanomaterials to be used as promising fillers for the feasible functional design of hydrogels. For example, the influence of some features of 2D materials, including the layer number, lateral size [114], types, and the concentration of functional groups, on the mechanical and tribological behaviors of hydrogels for specified bioapplications remains unclear. An in-depth investigation of the fundamental mechanisms and advanced techniques guiding the design and application of 2D materials as nanofillers in hydrogels is required in the future;
- The strengthening and lubrication mechanisms of 2D materials used as nanofillers of hydrogels still need to be further elucidated, and this should be based on a full understanding of the fundamental properties of both the 2D materials and the polymeric hydrogel matrix;
- The biocompatibility of hydrogels incorporating 2D materials as nanofillers necessitates a more rigorous and systematic investigation. This should encompass an extensive evaluation of multiple factors, including cytotoxicity, neurotoxicity, genotoxicity, and others, to ensure safety and efficacy over an extended period of service;
- Furthermore, the ongoing development of 2D nanomaterials opens up new avenues for enhancing the mechanical and tribological properties of nanocomposite hydrogels. For instance, the tribological attributes of nitride-MXenes have not been thoroughly investigated, despite their known superior mechanical properties [115–117]. Additionally, the potential synergistic effects of integrating diverse 2D nanomaterials into hydrogels are an area that is ripe for exploration, as there is insufficient research on this combined approach to offer comprehensive data and permit systematic studies. Moreover, incorporating novel "non-layered" 2D materials [118–121] as nanofillers could introduce distinctive characteristics to nanocomposite hydrogels, offering a new dimension of performance and functionality.

4. Conclusions

This review has underscored the significance of incorporating typical 2D nanomaterials in order to enhance the mechanical and tribological attributes of polymer hydrogels. It also highlights key challenges that must be addressed in future research. The objective is to offer a valuable resource for the design and advancement of hydrogels fortified with various functional 2D materials. We anticipate that these 2D nanomaterials will be pivotal in engineering composite hydrogels, and trust that the insights presented here will serve as a foundation for developing multifunctional hydrogels across a wide range of applications.

Author Contributions: Writing—original draft preparation, H.G. and Z.W.; funding acquisition, Y.L.; writing—review and editing, Y.L., J.Z. and Y.Q. All authors have read and agreed to the published version of the manuscript.

Funding: This work was financially supported by the National Natural Science Foundation of China (grant number 52005287), Beijing Institute of Technology Research Fund Program for Young Scholars, Tribology Science Fund of State Key Laboratory of Tribology in Advanced Equipment (SKLT) (No. SKLTKF21B14), and Young Elite Scientists Sponsorship Program by BAST (grant number BYESS2023288).

Data Availability Statement: The data presented in this study are available on request from the corresponding author.

Conflicts of Interest: Yi Qian was employed by Beijing Montagne Medical Device Co., Ltd. The remaining authors declare that the research was conducted in the absence of any commercial or financial relationships that could be construed as a potential conflict of interest.

References

1. Bonyadi, S.Z.; Demott, C.J.; Grunlan, M.A.; Dunn, A.C. Cartilage-like Tribological Performance of Charged Double Net-work Hydrogels. *J. Mech. Behav. Biomed. Mater.* **2021**, *114*, 104202. [CrossRef] [PubMed]
2. Feng, S.; Li, J.; Li, X.; Wen, S.; Liu, Y. Synergy of Phospholipid and Hyaluronan Based Super-Lubricated Hydrogels. *Appl. Mater. Today* **2022**, *27*, 101499. [CrossRef]
3. Zhao, X.; Xiong, D.; Liu, Y. Improving Surface Wettability and Lubrication of Polyetheretherketone (PEEK) by Combining with Polyvinyl Alcohol (PVA) Hydrogel. *J. Mech. Behav. Biomed. Mater.* **2018**, *82*, 27–34. [CrossRef] [PubMed]
4. Song, H.S.; Kwon, O.S.; Kim, J.H.; Conde, J.; Artzi, N. 3D hydrogel scaffold doped with 2D graphene materials for biosensors and bioelectronics. *Biosens. Bioelectron.* **2017**, *89*, 187–200. [CrossRef] [PubMed]
5. Chen, Y.; Song, J.; Wang, S.; Liu, W. Cationic Modified PVA Hydrogels Provide Low Friction and Excellent Mechanical Properties for Potential Cartilage and Orthopedic Applications. *Macromol. Biosci.* **2023**, *23*, 2200275. [CrossRef]
6. Lust, S.T.; Hoogland, D.; Norman, M.D.A.; Kerins, C.; Omar, J.; Jowett, G.M.; Yu, T.T.L.; Yan, Z.; Xu, J.Z.; Marciano, D.; et al. Selectively Cross-Linked Tetra-PEG Hydrogels Provide Control over Mechanical Strength with Minimal Impact on Diffusivity. *ACS Biomater. Sci. Eng.* **2021**, *7*, 4293–4304. [CrossRef] [PubMed]
7. Chen, S.; Huang, J.; Zhou, Z.; Chen, Q.; Hong, M.; Yang, S.; Fu, H. Highly Elastic Anti-Fatigue and Anti-Freezing Conductive Double Network Hydrogel for Human Body Sensors. *Ind. Eng. Chem. Res.* **2021**, *60*, 6162–6172. [CrossRef]
8. Wang, Z.; Meng, F.; Zhang, Y.; Guo, H. Low-Friction Hybrid Hydrogel with Excellent Mechanical Properties for Simulating Articular Cartilage Movement. *Langmuir* **2023**, *39*, 2368–2379. [CrossRef]
9. Li, Z.; Zhang, X.; Cheng, H.; Liu, J.; Shao, M.; Wei, M.; Evans, D.; Zhang, H.; Duan, X. Confined Synthesis of 2D Nanostructured Materials toward Electrocatalysis. *Adv. Energy Mater.* **2020**, *10*, 1900486. [CrossRef]
10. Kim, H.G.; Lee, H. Atomic Layer Deposition on 2D Materials. *Chem. Mater.* **2017**, *29*, 3809–3826. [CrossRef]
11. Chen, P.; Wu, Z.; Huang, Q.; Ji, S.; Weng, Y.; Wu, Z.; Ma, Z.; Chen, X.; Weng, M.; Fu, R.; et al. A quasi-2D material CePO4 and the self-lubrication in micro-arc oxidized coatings on Al alloy. *Tribol. Int.* **2019**, *138*, 157–165. [CrossRef]
12. Uzoma, P.C.; Hu, H.; Khadem, M.; Penkov, O.V. Tribology of 2D nanomaterials: A review. *Coatings* **2020**, *10*, 897. [CrossRef]
13. Rong, C.; Su, T.; Li, Z.; Chu, T.; Zhu, M.; Yan, Y.; Zhang, B.; Xuan, F. Elastic properties and tensile strength of 2D $Ti_3C_2T_x$ MXene monolayers. *Nat. Commun.* **2024**, *15*, 1566. [CrossRef] [PubMed]
14. Novoselov, K.S.; Fal'ko, V.I.; Colombo, L.; Gellert, P.R.; Schwab, M.G.; Kim, K. A roadmap for graphene. *Nature* **2012**, *490*, 192–200. [CrossRef] [PubMed]
15. Huang, S.; Mutyala, K.C.; Sumant, A.V.; Mochalin, V.N. Achieving superlubricity with 2D transition metal carbides (MXenes) and MXene/graphene coatings. *Mater. Today Adv.* **2021**, *9*, 100133. [CrossRef]
16. Rasul, M.G.; Kiziltas, A.; Arfaei, B.; Shahbazian-Yassar, R. 2D boron nitride nanosheets for polymer composite materials. *Npj 2D Mater. Appl.* **2021**, *5*, 56. [CrossRef]
17. Wu, H.; Yin, S.; Du, Y.; Wang, L.; Wang, H. An investigation on the lubrication effectiveness of MoS_2 and BN layered materials as oil additives using block-on-ring tests. *Tribol. Int.* **2020**, *151*, 106516. [CrossRef]
18. Tanjil, M.R.E.; Jeong, Y.; Yin, Z.; Panaccione, W.; Wang, C.M. Angstrom-Scale, Atomically Thin 2D Materials for Corrosion Mitigation and Passivation. *Coatings* **2019**, *9*, 133. [CrossRef]
19. Xia, M. 2D Materials-Coated Plasmonic Structures for SERS Applications. *Coatings* **2018**, *8*, 137. [CrossRef]
20. Liu, Y.; Li, J.; Li, J.; Yi, S.; Ge, X.; Zhang, X.; Luo, J. Shear-Induced Interfacial Structural Conversion Triggers Macroscale Superlubricity: From Black Phosphorus Nanoflakes to Phosphorus Oxide. *ACS Appl. Mater. Interfaces* **2021**, *13*, 31947–31956. [CrossRef]
21. Zheng, Y.; Hong, X.; Wang, J.; Feng, L.; Fan, T.; Guo, R.; Zhang, H. 2D Nanomaterials for Tissue Engineering and Regenerative Nanomedicines: Recent Advances and Future Challenges. *Adv. Healthc. Mater.* **2021**, *10*, 2001743. [CrossRef] [PubMed]

22. Marian, M.; Berman, D.; Nečas, D.; Emami, N.; Ruggiero, A.; Rosenkranz, A. Roadmap for 2D materials in biotribological/biomedical applications—A review. *Adv. Colloid Interface Sci.* **2022**, *307*, 102747. [CrossRef]
23. Wang, Z.; Li, J.; Jiang, L.; Xiao, S.; Liu, Y.; Luo, J. Zwitterionic Hydrogel Incorporated Graphene Oxide Nanosheets with Improved Strength and Lubricity. *Langmuir* **2019**, *35*, 11452–11462. [CrossRef] [PubMed]
24. Urade, A.R.; Lahiri, I.; Suresh, K.S. Graphene Properties, Synthesis and Applications: A Review. *J. Miner. Met. Mater. Soc. (TMS)* **2023**, *75*, 614–630. [CrossRef] [PubMed]
25. Jiang, H.; Zheng, L.; Liu, Z.; Wang, X. Two-dimensional materials: From mechanical properties to flexible mechanical sensors. *InfoMat* **2020**, *2*, 1077–1094. [CrossRef]
26. Suk, J.; Piner, R.; An, J.; Ruoff, R. Mechanical Properties of Monolayer Graphene Oxide. *ACS Nano* **2010**, *4*, 6557–6564. [CrossRef] [PubMed]
27. Falin, A.; Cai1, Q.; Santos, E.; Scullion, D.; Qian, D.; Zhang, R.; Yang, Z.; Huang, S.; Watanabe, K.; Taniguchi, T.; et al. Mechanical properties of atomically thin boron nitride and the role of interlayer interactions. *Nat. Commun.* **2017**, *8*, 15815. [CrossRef]
28. Jiang, J.; Park, H. Mechanical properties of single-layer black phosphorus. *J. Phys. D Appl. Phys.* **2014**, *47*, 385304. [CrossRef]
29. Li, L.; Yang, J. On mechanical behaviors of fewlayer black phosphoru. *Sci. Rep.* **2018**, *8*, 3227. [CrossRef]
30. Liu, Y.; Ge, X.; Li, J. Graphene lubrication. *Appl. Mater. Today* **2020**, *20*, 100662. [CrossRef]
31. Ge, X.; Chai, Z.; Shi, Q.; Liu, Y.; Wang, W. Graphene superlubricity: A review. *Friction* **2023**, *11*, 1953–1973. [CrossRef]
32. Mao, J.; Zhao, J.; Wang, W.; He, Y.; Luo, J. Influence of the micromorphology of reduced graphene oxide sheets on lubrication properties as a lubrication additive. *Tribol. Int.* **2018**, *119*, 614–621. [CrossRef]
33. Liang, H.; Xu, M.; Bu, Y.; Chen, B.; Zhang, Y.; Fu, Y.; Xu, X.; Zhang, J. Confined interlayer water enhances solid lubrication per-formances of graphene oxide films with optimized oxygen functional groups. *Appl. Surf. Sci.* **2019**, *485*, 64–69. [CrossRef]
34. Gupta, B.; Kumar, N.; Panda, K.; Dash, S.; Tyagi, A.K. Energy efficient reduced graphene oxide additives: Mechanism of effective lubrication and antiwear properties. *Sci. Rep.* **2016**, *6*, 18372. [CrossRef]
35. Chen, L.; Tu, N.; Wei, Q.; Liu, T.; Li, C.; Wang, W.; Li, J.; Lu, H. Inhibition of cold-welding and adhesive wear occurring on sur-face of the 6061 aluminum alloy by graphene oxide/polyethylene glycol composite water-based lubricant. *Surf. Inter-Face Anal.* **2021**, *54*, 218–230. [CrossRef]
36. Cheng, Y.H.; Cheng, S.J.; Chen, H.H.; Hsu, W.C. Development of injectable graphene oxide/laponite/gelatin hydrogel con-taining Wharton's jelly mesenchymal stem cells for treatment of oxidative stress-damaged cardiomyocytes. *Colloids Surf. B Biointerfaces* **2022**, *209*, 112150. [CrossRef]
37. Qiao, K.; Guo, S.; Zheng, Y.; Xu, X.; Meng, H.; Peng, J.; Fang, Z.; Xie, Y. Effects of graphene on the structure, properties, electro-response behaviors of GO/PAA composite hydrogels and influence of electro-mechanical coupling on BMSC differentiation. *Mater. Sci. Eng. C Mater. Biol. Appl.* **2018**, *93*, 853–863. [CrossRef]
38. Hou, Y.; Jin, M.; Liu, Y.; Jiang, N.; Zhang, L.; Zhu, S. Biomimetic construction of a lubricious hydrogel with robust mechanics via polymer chains interpenetration and entanglement for TMJ disc replacement. *Chem. Eng. J.* **2023**, *460*, 141731. [CrossRef]
39. Lei, Y.; Zhang, G.; Li, H. Thermal-responsive nanocomposite hydrogel based on graphene oxide-polyvinyl alcohol/poly (N-isopropylacrylamide). *IOP Conf. Ser. Mater. Sci. Eng.* **2017**, *274*, 012115. [CrossRef]
40. Li, G.; Zhao, Y.; Zhang, L.; Gao, M.; Kong, Y.; Yang, Y. Preparation of graphene oxide/polyacrylamide composite hydrogel and its effect on Schwann cells attachment and proliferation. *Colloids Surf. B Biointerfaces* **2016**, *143*, 547–556. [CrossRef] [PubMed]
41. Zhu, P.; Hu, M.; Deng, Y.; Wang, C. One-Pot Fabrication of a Novel Agar-Polyacrylamide/Graphene Oxide Nanocomposite Double Network Hydrogel with High Mechanical Properties. *Adv. Eng. Mater.* **2016**, *18*, 1799–1807. [CrossRef]
42. Zhang, Y.; Zhang, M.; Jiang, H.; Shi, J.; Li, F.; Xia, Y.; Zhang, G.; Li, H. Bio-inspired layered chitosan/graphene oxide nanocomposite hydrogels with high strength and pH-driven shape memory effect. *Carbohydr. Polym.* **2017**, *177*, 116–125. [CrossRef]
43. Nath, J.; Chowdhury, A.; Dolui, S.K. Chitosan/graphene oxide-based multifunctional pH-responsive hydrogel with significant mechanical strength, self-healing property, and shape memory effect. *Adv. Polym. Technol.* **2018**, *37*, 3665–3679. [CrossRef]
44. Zhang, S.; Gong, X.; Yu, R.; Xing, Q.; Xu, L.; Wang, Z. Self-lubrication and wear-resistance mechanism of graphene-modified coatings. *Ceram. Int.* **2020**, *46*, 15915–15924.
45. Wang, Z.; Hao, Z.; Yang, C.; Wang, H.; Huang, C.; Zhao, X.; Pan, Y. Ultra-sensitive and rapid screening of acute myocardial in-farction using 3D-affinity graphene biosensor. *Cell Rep. Phys. Sci.* **2022**, *3*, 100855. [CrossRef]
46. Yang, J.; Hu, D.; Li, W.; Jia, Y.; Li, P. Formation mechanism of zigzag patterned P(NIPAM-co-AA)/CuS composite microspheres by in situ biomimetic mineralization for morphology modulation. *RSC Adv.* **2021**, *11*, 37904. [CrossRef]
47. Wang, B.; Yuan, S.; Xin, W.; Chen, Y.; Fu, Q.; Li, L.; Jiao, Y. Synergic adhesive chemistry-based fabrication of BMP-2 immobilized silk fibroin hydrogel functionalized with hybrid nanomaterial to augment osteogenic differentiation of rBMSCs for bone defect repair. *Int. J. Biol. Macromol.* **2021**, *192*, 407–416. [CrossRef]
48. Meng, Y.; Ye, L.; Coates, P.; Twigg, P. In Situ Cross-Linking of Poly(vinyl alcohol)/Graphene Oxide–Polyethylene Glycol Nanocomposite Hydrogels as Artificial Cartilage Replacement: Intercalation Structure, Unconfined Compressive Behavior, and Biotribological Behaviors. *J. Phys. Chem. C* **2018**, *122*, 3157–3167. [CrossRef]
49. Wang, C.; Bai, X.; Guo, Z.; Dong, C.; Yuan, C. A strategy that combines a hydrogel and graphene oxide to improve the water-lubricated performance of ultrahigh molecular weight polyethylene. *Compos. Part A Appl. Sci. Manuf.* **2021**, *141*, 106207. [CrossRef]

50. Cao, J.; Meng, Y.; Zhao, X.; Ye, L. Dual-Anchoring Intercalation Structure and Enhanced Bioactivity of Poly(vinyl alcohol)/Graphene Oxide–Hydroxyapatite Nanocomposite Hydrogels as Artificial Cartilage Replacement. *Ind. Eng. Chem. Res.* **2020**, *59*, 20359–20370. [CrossRef]
51. Wang, C.; Zhu, K.; Gao, Y.; Han, S.; Ju, J.; Ren, T.; Zhao, X. Multifunctional GO-based hydrogel coating on Ti-6Al-4 V Alloy with enhanced bioactivity, anticorrosion and tribological properties against cortical bone. *Tribol. Int.* **2023**, *184*, 108423. [CrossRef]
52. Liu, L.; Xiao, L.; Li, M.; Zhang, X.; Chang, Y.; Shang, L.; Ao, Y. Effect of hexagonal boron nitride on high-performance polyether ether ketone composites. *Colloid Polym. Sci.* **2016**, *294*, 127–133. [CrossRef]
53. Belyaeva, L.; Ludwig, C.; Lai, Y.; Chou, C.; Shih, C. Uniform, Strain-Free, Large-Scale Graphene and h-BN Monolayers Enabled by Hydrogel Substrates. *Small* **2023**, *20*, 2307054. [CrossRef]
54. Jing, L.; Li, H.; Tay, Y.; Sun, B.; Tsang, S.H.; Cometto, O.; Lin, J.; Teo, E.; Tok, A. Biocompatible Hydroxylated Boron Nitride Nanosheets/Polyvinyl Alcohol Interpenetrating Hydrogels with Enhanced Mechanical and Thermal Responses. *ACS Nano* **2017**, *11*, 3742–3751. [CrossRef]
55. Yang, N.; Ji, H.; Jiang, X.; Qu, X.; Zhang, X.; Zhang, Y.; Liu, B. Preparation of Boron Nitride Nanoplatelets via Amino Acid Assisted Ball Milling: Towards Thermal Conductivity Application. *Nanomaterials* **2020**, *10*, 1652. [CrossRef]
56. Xue, S.; Wu, Y.; Guo, M.; Liu, D.; Zhang, T.; Lei, W. Fabrication of Poly(acrylic acid)/Boron Nitride Composite Hydrogels with Excellent Mechanical Properties and Rapid Self-Healing Through Hierarchically Physical Interactions. *Nanoscale Res. Lett.* **2018**, *13*, 393. [CrossRef]
57. Jiang, H.; Wang, Z.; Geng, H.; Song, X.; Zeng, H.; Zhi, C. Highly Flexible and Self-Healable Thermal Interface Material Based on Boron Nitride Nanosheets and a Dual Cross-linked Hydrogel. *ACS Appl. Mater. Interfaces* **2017**, *9*, 10078–10084. [CrossRef]
58. Liu, F.; Han, R.; Naficy, S.; Casillas, G.; Sun, X.; Huang, Z. Few-Layered Boron Nitride Nanosheets for Strengthening Polyurethane Hydrogels. *ACS Appl. Nano Mater.* **2021**, *4*, 7988–7994. [CrossRef]
59. Hu, X.; Liu, J.; He, Q.; Meng, Y.; Cao, L.; Sun, Y.; Chen, J.; Lu, F. Aqueous compatible boron nitride nanosheets for high-performance hydrogels. *Nanoscale* **2016**, *8*, 4260–4266. [CrossRef]
60. Tong, X.; Du, L.; Xu, Q. Tough, Adhesive and Self-Healing Conductive 3D Network Hydrogel of Physically Linked Functionalized-Boron Nitride/Clay/Poly(N-isopropylacrylamide). *J. Mater. Chem. A* **2018**, *6*, 3091–3099. [CrossRef]
61. Goncu, Y.; Ay, N. Boron Nitride's Morphological Role in the Design of Injectable Hyaluronic Acid Based Hybrid Artificial Synovial Fluid. *ACS Biomater. Sci. Eng.* **2023**, *9*, 6345–6356. [CrossRef] [PubMed]
62. Fan, L.; Zhang, X.; Liu, X.; Sun, B.; Li, L.; Zhao, Y. Responsive Hydrogel Microcarrier-Integrated Microneedles for Versatile and Controllable Drug Delivery. *Adv. Healthc. Mater.* **2021**, *10*, 2002249. [CrossRef]
63. Yang, G.; Wan, X.; Gu, Z.; Zeng, X.; Tang, J. Near infrared photothermal-responsive poly(vinyl alcohol)/black phosphorus com-posite hydrogels with excellent on-demand drug release capacity. *J. Mater. Chem. B* **2018**, *6*, 1622–1632. [CrossRef]
64. Lv, Y.; Wang, W.; Xie, G.; Luo, J. Self-Lubricating PTFE-Based Composites with Black Phosphorus Nanosheets. *Tribol. Lett.* **2018**, *66*, 61. [CrossRef]
65. Sun, X.; Yu, C.; Zhang, L.; Cao, J.; Kaleli, E.H.; Xie, G. Tribological and Antibacterial Properties of Polyetheretherketone Composites with Black Phosphorus Nanosheets. *Polymers* **2022**, *14*, 1242. [CrossRef]
66. Wu, S.; He, F.; Xie, G.; Bian, Z.; Ren, Y.; Liu, X.; Yang, H.; Guo, D.; Zhang, L.; Wen, S.; et al. Super-Slippery Degraded Black Phosphorus/Silicon Dioxide Interface. *ACS Appl. Mater. Interfaces* **2020**, *12*, 7717–7726. [CrossRef]
67. Luo, Z.; Yu, J.; Xu, Y.; Xi, H.; Cheng, G.; Yao, L.; Song, R.; Dearn, K.D. Surface characterization of steel/steel contact lubricated by PAO6 with novel black phosphorus nanocomposites. *Friction* **2020**, *9*, 723–733. [CrossRef]
68. Tang, G.; Su, F.; Xu, X.; Chu, P.K. 2D black phosphorus dotted with silver nanoparticles: An excellent lubricant additive for tribological applications. *Chem. Eng. J.* **2020**, *392*, 123631. [CrossRef]
69. Wang, W.; Gong, P.; Hou, T.; Wang, Q.; Gao, Y.; Wang, K. Tribological performances of BP/TiO$_2$ nanocomposites as water-based lubrication additives for titanium alloy plate cold rolling. *Wear* **2022**, *204278*, 494–495. [CrossRef]
70. Xu, Y.; Yu, J.; Dong, Y.; You, T.; Hu, X. Boundary Lubricating Properties of Black Phosphorus Nanosheets in Polyalphaolefin Oil. *J. Tribol.* **2019**, *141*, 072101. [CrossRef]
71. Wang, Q.; Hou, T.; Wang, W.; Zhang, G.; Gao, Y.; Wang, K. Tribological behavior of black phosphorus nanosheets as water-based lubrication additives. *Friction* **2021**, *10*, 374–387. [CrossRef]
72. Liu, W.; Zhu, Y.; Tao, Z.; Chen, Y.; Zhang, L.; Dong, A. Black Phosphorus-Based Conductive Hydrogels Assisted by Electrical Stimulus for Skin Tissue Engineering. *Adv. Healthc. Mater.* **2023**, *12*, 2301817. [CrossRef]
73. Du, C.; Huang, W. Progress and prospects of nanocomposite hydrogels in bone tissue engineering. *Nanocomposites* **2022**, *8*, 102–124. [CrossRef]
74. He, M.; Zhu, C.; Sun, D.; Liu, Z.; Du, M.; Huang, Y.; Huang, L.; Wang, J.; Liu, L.; Li, Y.; et al. Layer-by-layer assembled black phosphorus/chitosan composite coating for multi-functional PEEK bone scaffold. *Compos. Part B Eng.* **2022**, *246*, 110266. [CrossRef]
75. Malaki, M.; Varma, R.S. Mechanotribological Aspects of MXene-Reinforced Nanocomposites. *Adv. Mater.* **2020**, *32*, 2003154. [CrossRef]
76. Song, C.; Wang, T.; Sun, X.; Hu, Y.; Fan, L.; Guo, R. Lubrication performance of MXene/Brij30/H2O composite lamellar liquid crystal system. *Colloids Surf. A Physicochem. Eng. Asp.* **2022**, *641*, 128487. [CrossRef]

77. Chhattal, M.; Rosenkranz, A.; Zaki, S.; Ren, K.; Ghaffar, A.; Gong, Z.; Grützmacher, P.G. Unveiling the Tribological Potential of MXenes-Current Understanding and Future Perspectives. *Adv. Colloid Interface Sci.* **2023**, *321*, 103021. [CrossRef] [PubMed]
78. Marian, M.; Song, G.C.; Wang, B.; Fuenzalida, V.M.; Krauß, S.; Merle, B.; Tremmel, S.; Wartzack, S.; Yu, J.; Rosenkranz, A. Effective usage of 2D MXene nanosheets as solid lubricant-Influence of contact pressure and relative humidity. *Appl. Surf. Sci.* **2020**, *531*, 147311. [CrossRef]
79. Yin, X.; Jin, J.; Chen, X.; Rosenkranz, A.; Luo, J. Ultra-Wear-Resistant MXene-Based Composite Coating via in Situ Formed Nanostructured Tribofilm. *ACS Appl. Mater. Interfaces* **2019**, *11*, 32569–32576. [CrossRef] [PubMed]
80. Zhou, X.; Guo, Y.; Wang, D.; Xu, Q. Nano friction and adhesion properties on Ti_3C_2 and Nb_2C MXene studied by AFM. *Tribol. Int.* **2021**, *153*, 106646. [CrossRef]
81. Das, P.; Ganguly, S.; Rosenkranz, A.; Wang, B.; Yu, J.; Srinivasan, S.; Rajabzadeh, A.R. MXene/0D Nanocomposite Architectures: Design, Properties and Emerging Applications. *Mater. Today Nano* **2023**, *24*, 100428. [CrossRef]
82. Chen, Z.; Zhang, M.; Guo, Z.; Chen, H.; Yan, H.; Ren, F.; Jin, Y.; Sun, Z.; Ren, P. Synergistic effect of novel hyperbranched pol-ysiloxane and Ti_3C_2T MXene/MoS_2 hybrid filler towards desirable mechanical and tribological performance of bismaleimide composites. *Compos. Part B Eng.* **2023**, *248*, 110374. [CrossRef]
83. Guo, L.; Zhang, Y.; Zhang, G.; Wang, Q.; Wang, T. MXene-Al2O3 synergize to reduce friction and wear on epoxy-steel contacts lubricated with ultra-low sulfur diesel. *Tribol. Int.* **2021**, *153*, 106588. [CrossRef]
84. Cui, Y.; Xue, S.; Chen, X.; Bai, W.; Liu, S.; Ye, Q.; Zhou, F. Fabrication of two-dimensional MXene nanosheets loading Cu nano-particles as lubricant additives for friction and wear reduction. *Tribol. Int.* **2022**, *176*, 107934. [CrossRef]
85. Marian, M.; Feile, K.; Rothammer, B.; Bartz, M.; Wartzack, S.; Seynstahl, A.; Tremmel, S.; Krauß, S.; Merle, B.; Böhm, T.; et al. Ti_3C_2T solid lubricant coatings in rolling bearings with remarkable performance beyond state-of-the-art materials. *Appl. Mater. Today* **2021**, *25*, 101202. [CrossRef]
86. Zhou, C.; Li, Z.; Liu, S.; Ma, L.; Zhan, T.; Wang, J. Synthesis of MXene-Based Self-dispersing Additives for Enhanced Tribological Properties. *Tribol. Lett.* **2022**, *70*, 63. [CrossRef]
87. Guo, J.; Wu, P.; Zeng, C.; Wu, W.; Zhao, X.; Liu, G.; Zhou, F.; Liu, W. Fluoropolymer grafted $Ti_3C_2T_x$ MXene as an efficient lubri-cant additive for fluorine-containing lubricating oil. *Tribol. Int.* **2022**, *170*, 107500. [CrossRef]
88. Wang, Q.; Pan, X.; Wang, X.; Cao, S.; Chen, L.; Ma, X.; Huang, L.; Ni, Y. Fabrication strategies and application fields of novel 2D Ti_3C_2Tx (MXene) composite hydrogels: A mini-review. *Ceram. Int.* **2021**, *47*, 4398–4403. [CrossRef]
89. Xue, P.; Bisoyi, H.K.; Chen, Y.; Zeng, H.; Yang, J.; Yang, X.; Lv, P.; Zhang, X.; Priimagi, A.; Wang, L.; et al. Near-Infrared Light-Driven Shape-Morphing of Programmable Anisotropic Hydrogels Enabled by MXene Nanosheets. *Angew. Chem. Int. Ed. Engl.* **2021**, *60*, 3390–3396. [CrossRef]
90. Zhang, P.; Yang, X.J.; Li, P.; Zhao, Y.; Niu, Q.J. Fabrication of novel MXene (Ti_3C_2)/polyacrylamide nanocomposite hydrogels with enhanced mechanical and drug release properties. *Soft Matter* **2020**, *16*, 162–169. [CrossRef]
91. Yan, B.Y.; Cao, Z.K.; Hui, C.; Sun, T.C.; Xu, L.; Ramakrishna, S.; Yang, M.; Long, Y.Z.; Zhang, J. MXene@Hydrogel composite nanofibers with the photo-stimulus response and optical monitoring functions for on-demand drug release. *J. Colloid Interface Sci.* **2023**, *648*, 963–971. [CrossRef] [PubMed]
92. Qin, M.; Yuan, W.; Zhang, X.; Cheng, Y.; Xu, M.; Wei, Y.; Chen, W.; Huang, D. Preparation of PAA/PAM/MXene/TA hydrogel with antioxidant, healable ability as strain sensor. *Colloids Surf. B Biointerfaces* **2022**, *214*, 112482. [CrossRef] [PubMed]
93. Miao, X.; Li, Z.; Hou, K.; Gao, Q.; Huang, Y.; Wang, J.; Yang, S. Bioinspired multi-crosslinking and solid–liquid composite lubricating MXene/PVA hydrogel based on salting out effect. *Chem. Eng. J.* **2023**, *476*, 146848. [CrossRef]
94. Ye, G.; Wen, Z.; Wen, F.; Song, X.; Wang, L.; Li, C.; He, Y.; Prakash, S.; Qiu, X. Mussel-inspired conductive Ti_2C-cryogel promotes functional maturation of cardiomyocytes and enhances repair of myocardial infarction. *Theranostics* **2020**, *10*, 2047–2066. [CrossRef] [PubMed]
95. Quero, F.; Rosenkranz, A. Mechanical Performance of Binary and Ternary Hybrid MXene/Nanocellulose Hydro- and Aerogels—A Critical Review. *Adv Mater. Interfaces* **2021**, *8*, 2100952. [CrossRef]
96. Gogotsi, Y. The Future of MXenes. *Chem. Mater.* **2023**, *35*, 8767–8770. [CrossRef]
97. Naguib, M.; Barsoum, M.W.; Gogotsi, Y. Ten Years of Progress in the Synthesis and Development of MXenes. *Adv. Mater.* **2021**, *33*, 2103393. [CrossRef] [PubMed]
98. Anasori, B.; Gogotsi, Y. MXenes: Trends, Growth, and Future Directions. *Graphene 2D Mater.* **2022**, *7*, 75–79. [CrossRef]
99. Li, K.; Zhao, J.; Zhussupbekova, A.; Shuck, E.C.; Hughes, L.; Dong, Y.; Barwich, S.; Vaesen, S.; Shvets, V.I.; Möbius, M.; et al. 4D printing of MXene hydrogels for high-efficiency pseudocapacitive energy storage. *Nat. Commun.* **2022**, *13*, 6884. [CrossRef]
100. Wyatt, B.; Rosenkranz, A.; Anasori, B. 2D MXenes: Tunable Mechanical and Tribological Properties. *Adv. Mater.* **2021**, *33*, 2007973. [CrossRef]
101. Rosenkranz, A.; Righi, M.; Sumant, A.; Anasori, B.; Mochalin, V. Perspectives of 2D MXene Tribology. *Adv. Mater.* **2023**, *35*, 2207757. [CrossRef] [PubMed]
102. Parra-Munoz, N.; Soler, M.; Rosenkranz, A. Covalent functionalization of MXenes for tribological purposes—A critical review. *Adv. Colloid Interface Sci.* **2022**, *309*, 102792. [CrossRef] [PubMed]
103. Chakrabarty, A.; Maitra, U.; Das, A. Metal cholate hydrogels: Versatile supramolecular systems for nanoparticle embedded soft hybrid materials. *J. Mater. Chem.* **2012**, *22*, 18268. [CrossRef]

104. Chen, L.; He, C.; Yin, J.; Chen, S.; Zhao, W.; Zhao, C. Clearance of methylene blue by CdS enhanced composite hydrogel materials. *Environ. Technol.* **2020**, *43*, 355–366. [CrossRef] [PubMed]
105. Wei, L.; Li, Z.; Li, J.; Zhang, Y.; Yao, B.; Liu, Y.; Song, W.; Fu, X.; Wu, X.; Huang, S. An approach for mechanical property optimiza-tion of cell-laden alginate–gelatin composite bioink with bioactive glass nanoparticles. *J. Mater. Sci. Mater. Med.* **2020**, *31*, 103. [CrossRef]
106. Dai, C.; Zhang, X.; Du, C.; Frank, A.; Schmidt, H.; Zheng, Q.; Wu, Z. Photoregulated Gradient Structure and Programmable Me-chanical Performances of Tough Hydrogels with a Hydrogen-Bond Network. *ACS Appl. Mater. Interfaces* **2020**, *12*, 53376–53384. [CrossRef]
107. Wang, Z.; Yang, H.; Liang, H.; Xu, Y.; Zhou, J.; Peng, H.; Zhong, J.; Xi, W. Polydopamine particles reinforced poly(vinyl alcohol) hydrogel composites with fast self healing behavior. *Prog. Org. Coat.* **2020**, *143*, 105636. [CrossRef]
108. Du, Z.; Wang, Y.; Li, X. Preparation of Nanocellulose Whisker/Polyacrylamide/Xanthan Gum Double Network Conductive Hydrogels. *Coatings* **2023**, *13*, 843. [CrossRef]
109. Yang, B.; Zhang, S.; Wang, P.; Liu, C.; Zhu, Y. Robust and rapid responsive organic-inorganic hybrid bilayer hydrogel actuators with silicon nanoparticles as the cross-linker. *Polymer* **2021**, *228*, 123863. [CrossRef]
110. Li, S.; Wang, X.; Zhu, J.; Wang, Z.; Wang, L. Synthesis and characterization of photothermal antibacterial hydrogel with enhanced mechanical properties. *New J. Chem.* **2021**, *45*, 16804–16815. [CrossRef]
111. Lu, Y.; Han, J.; Ding, Q.; Yue, Y.; Xia, C.; Ge, S.; Le, Q.; Dou, X.; Sonne, C.; Lam, S. TEMPO-oxidized cellulose nano-fibers/polyacrylamide hybrid hydrogel with intrinsic self-recovery and shape memory properties. *Cellulose* **2021**, *28*, 1469–1488. [CrossRef]
112. Abalymov, A.; Lengert, E.; Van der Meeren, L.; Saveleva, M.; Ivanova, A.; Douglas, T.; Skirtach, A.; Volodkin, D.; Parakhon-skiy, B. The influence of Ca/Mg ratio on autogelation of hydrogel biomaterials with bioceramic compounds. *Biomater. Adv.* **2022**, *133*, 112632. [CrossRef] [PubMed]
113. Xiao, H.; Liu, S. 2D nanomaterials as lubricant additive: A review. *Mater. Des.* **2017**, *135*, 319–332. [CrossRef]
114. Liu, Y.; Yu, S.; Fan, Z.; Ge, X.; Wang, W. How does lateral size influence the tribological behaviors of graphene oxide as nanoadditive for water-based lubrication? *Carbon* **2024**, *219*, 118803. [CrossRef]
115. Zhang, N.; Hong, Y.; Yazdanparast, S.; Asle, M. Superior structural, elastic and electronic properties of 2D titanium nitride MXenes over carbide MXenes: A comprehensive first principles study. *2D Mater.* **2018**, *5*, 45004. [CrossRef]
116. Kurtoglu, M.; Naguib, M.; Gogotsi, Y.; Barsoum, M. First principles study of two-dimensional early transition metal car-bides. *MRS Commun.* **2012**, *2*, 133–137. [CrossRef]
117. Ronchi, R.; Lemos, H.; Nishihora, R.; Cuppari, M.; Santos, S. Tribology of polymer-based nanocomposites reinforced with 2D materials. *Mater. Today Commun.* **2023**, *34*, 105397. [CrossRef]
118. Lang, J.; Ding, B.; Zhang, S.; Su, H.; Ge, B.; Qi, L.; Gao, H.; Li, X.; Li, Q.; Wu, H. Scalable Synthesis of 2D Si Nanosheets. *Adv. Mater.* **2017**, *29*, 1701777. [CrossRef]
119. Liu, Y.; Yu, S.; Wang, W. Nanodiamond plates as macroscale solid lubricant: A "non-layered" two-dimension material. *Carbon* **2022**, *198*, 119–131. [CrossRef]
120. Liu, Y.; Yu, S.; Zhang, R.; Ge, X.; Wang, W. "Non-layered" two-dimensional nanodiamond plates as nanoadditives in water lubrication. *Wear* **2024**, *536–537*, 205174. [CrossRef]
121. Serles, P.; Arif, T.; Puthirath, A.; Yadav, S.; Wang, G.; Cui, T.; Balan, A.; Yadav, T.; Thibeorchews, P.; Chakingal, N.; et al. Friction of magnetene, a non–van der Waals 2D material. *Sci. Adv.* **2021**, *7*, 2041. [CrossRef] [PubMed]

Disclaimer/Publisher's Note: The statements, opinions and data contained in all publications are solely those of the individual author(s) and contributor(s) and not of MDPI and/or the editor(s). MDPI and/or the editor(s) disclaim responsibility for any injury to people or property resulting from any ideas, methods, instructions or products referred to in the content.

Article

Tribological Behavior of GTL Base Oil Improved by Ni-Fe Layered Double Hydroxide Nanosheets

Shuo Xiang [1], Xinghao Zhi [2], Hebin Bao [1,*], Yan He [1,*], Qinhui Zhang [1], Shigang Lin [1], Bo Hu [1], Senao Wang [1], Peng Lu [3], Xin Yang [1], Qiang Tian [4] and Xin Du [1]

1. Army Logistics Academy of PLA, Chongqing 401331, China; xslaplace@163.com (S.X.); 15223276757@163.com (Q.Z.)
2. Unit 326587, Jinzhou 121001, China; zxh052887@163.com
3. Unit 91967, Xingtai 054001, China
4. Chengdu Quality Supervision Station of PLA, Chengdu 610041, China
* Correspondence: baohb@vip.163.com (H.B.); 18623006256@163.com (Y.H.)

Abstract: The layered double hydroxide (LDH) has been practically applied in the field of tribology and materials science due to its unique physicochemical properties, weak bonding, flexible structural composition, and adjustable interlayer space. In this work, a series of ultrathin and flexible composition of Ni-Fe LDH samples were prepared via a cost-effective room-temperature co-precipitation process. Then, they were mechanically dispersed into GTL base oil and their lubricating performance were tested by a four-ball tribometer. It is found that the variation of Ni-Fe ratio of Ni-Fe LDH has a great influence on the improvement of lubricating performance of GTL base oil. At the same concentration (0.3 mg/mL), the Ni-Fe LDH with Ni/Fe ratio of 6 was demonstrated to exhibit the best lubricating performance and the AFC, WSD, the wear volume, surface roughness and average wear scar depth decreased 51.3%, 30.8%, 78.4%, 6.7% and 50.0%, respectively. SEM-EDS and X-ray photoelectron spectra illustrated that the tribo-chemical film consisting of iron oxides and NiO with better mechanical properties formed and slowly replaced the physical film, which resists scuffing and protect solid surface from severe collisions.

Keywords: gas to liquid; lubricating oil; lubricating performance; layered double hydroxide

Citation: Xiang, S.; Zhi, X.; Bao, H.; He, Y.; Zhang, Q.; Lin, S.; Hu, B.; Wang, S.; Lu, P.; Yang, X.; et al. Tribological Behavior of GTL Base Oil Improved by Ni-Fe Layered Double Hydroxide Nanosheets. *Lubricants* **2024**, *12*, 146. https://doi.org/10.3390/lubricants12050146

Received: 27 February 2024
Revised: 2 April 2024
Accepted: 10 April 2024
Published: 26 April 2024

Copyright: © 2024 by the authors. Licensee MDPI, Basel, Switzerland. This article is an open access article distributed under the terms and conditions of the Creative Commons Attribution (CC BY) license (https://creativecommons.org/licenses/by/4.0/).

1. Introduction

Friction and wear are widely used in the chemical industry, aerospace, machinery manufacturing, transportation, metallurgy, oceanography, bioengineering, renewable energy and daily life [1–5]. Along with the development of modern industry, friction and wear not only cause energy loss and mechanical failure, but also metal corrosion and environmental pollution [6,7]. To address these problems, lubricants are utilized extensively and have been proven as an efficient means of minimizing friction and wear in the fields of transportation, metal cutting, mining, construction, power generation, power transmission, and so on [8,9]. Modern commercial lubricants can be grouped as lubricating oils, lubricating greases, solid lubricants and compressed air or other gases, and the first one accounts for about 80% of them [10]. Lubricating oil is generally composed of two parts, a base oil and an additive. The former is the primary component of lubricating oil and typically makes up the majority of its composition (usually 70–95%), which determines the basic properties of lubricating oil [11].

With the progression of contemporary economy and technology, friction and wear not only culminate in energyipation and mechanical malfunction, also instigate metal corrosion and environmental contamination [6]. To address these predicaments, lubricants are extensively employed and have been substantiated as an efficacious means of minimizing friction and wear in the realms of transportation, metal cutting, mining, construction, power generation, power transmission, and so forth [8,9]. Modern commercial lubricants can be categorized as lubricating oils, lubricating greases, solid lubricants, and compressed air or other gases, wherein the former constitutes approximately 80% of the aforementioned [10]. Lubricating oil is generally comprised of two constituents, namely a base oil and an additive. The former serves as the principal component of lubricating oil and predominantly constitutes its composition (usually 70–95%), thereby determining the fundamental properties of the lubricating oil [11]. As a synthetic product produced from natural gas using the Fischer-Tropsch process, gas to liquid (GTL) base oil have excellent characteristics, including ultra-high VIs (VI > 140), originally no sulphur and nitrogen, very low evaporative losses and inappreciable aromatic content [12]. In light of this, GTL base oil is a promising and environmentally friendly alternative to traditional mineral base oil, which contains nitride and sulfide. Additives can compensate and improve the lubricating properties of base oils. Over the past decades, extensive research has been conducted on various forms of additives such as composites, nanoparticles, liquids and two-dimensional (2D) materials. Among them, 2D materials are expected to be an alternative to liquid lubricant additives due to their high specific surface area, chemical stability, and continuous formation of protective films on friction surfaces. Due to the characteristics of high specific surface area, remarkable chemical stability, and the ability to continuously form a protective film on the friction surface, a variety of two-dimensional (2D) materials have been extensively studied in the field of tribology, including graphene [13,14], black phosphorus (BP) [15,16], h-BN [17,18], MoS2 [19–21], WS2 [22,23], MXene [24–26], and layered double hydroxides (LDHs) [27,28]. As with other 2D materials, the synthesis strategies for layered double hydroxide nanosheets can be divided into two main categories: bottom-up direct synthesis processes and top-down exfoliation processes. Various bottom-up strategies have been developed for the direct preparation of LDHs from suitable precursors, including hydrothermal, co-precipitation, microemulsion, sol-gel, and reconstruction methods, through which LDHs with tunable dimensions, thicknesses, crystallinity, and shapes have been prepared. The top-down exfoliation process usually involves swelling/interpolation to enlarge the interlayer distance, followed by exfoliation of bulk LDHs. From the synthesis of the LDHs are indeed suitable candidates for tribological applications from the point of view of simplicity of process, flexibility of control and low cost [2,29]. More importantly, LDHs are environmentally friendly materials that meet the requirements of sustainable development, and have become potential environmentally friendly lubricant additives, replacing traditional lubricant additives, such as zinc dialkyl dithiophosphate (T202), isobutylene sulphide (T308), di-n-butyl phosphite (T304), chlorinated paraffin wax (T301), and molybdenum dithiocarbamate (MoDTC), etc.

In the present article, a series of ultrathin and flexible composition of Ni-Fe LDH samples were produced via a cost-effective room-temperature co-precipitation process. The obtained Ni-Fe LDH samples were characterized by various tools such as powder X-ray diffraction (XRD), Fourier transform infrared (FT-IR) spectroscopy, thermogravimetric/differential thermal analysis (TG/DTA). The anti-wear and friction-reducing properties of the as-synthesized Ni-Fe LDH samples in GTL base oil were systematically studied by a four-ball tribometer. White light interferometer (WLI) methods were applied to analyze the wear surface. The composition and microstructure of the tribo-chemical film were analyzed by scanning electron microscope, energy dispersion spectrum (SEM-EDS), and X-ray photoelectron spectroscopy (XPS). The lubrication mechanism of GTL base oil improved by Ni-Fe LDH was presented. The lubrication mechanism of Ni-Fe LDH in GTL base oil was further discussed.

2. Materials and Methods

2.1. Materials

Sodium bicarbonate ($NaHCO_3$), nickel nitrate hexahydrate ($Ni(NO_3)_2 \cdot 6H_2O$), iron nitrate nonahydrate ($Fe(NO_3)_3 \cdot 9H_2O$), methyl alcohol (CH_3OH), methyl alcohol (CH_3CH_2OH), and petroleum ether (90–120 °C) were of analytical grade and purchased from Shanghai Aladdin Biochemical Technology Co., Ltd. (Shanghai, China). The Risella X 430 base oil

(GTL 430) supplied by Shell was used as fluid lubricant to formulate dispersions (Typical characteristics showed in Table 1). All chemicals were used as received without further purification and distilled water was used during all experiments.

Table 1. Typical characteristics of the Risella X 430 base oil (GTL 430).

Test Description	Result	Method
Kinematic viscosity (mm^2/s) (40 °C)	44.23	ASTM D445
Kinematic viscosity (mm^2/s) (100 °C)	7.62	ASTM D445
Viscosity Index	140	ASTM D2270
Appearance	Clear to bright	Visual
Colour saybolt	30	ASTM D156
Colour (ASTM) (Quantitative)	0.50	ASTM D1500
Density (kg/m^3) (15 °C)	827.70	ASTM D4052
Refractive index (20 °C)	1.46	ASTM D1218
Pour point (°C)	−45	ASTM D6749
Flash point (°C) (PMcc)	234	ASTM D93

2.2. Sample Preparation

Ni(NO$_3$)$_2$·6H$_2$O, urea, Fe(NO$_3$)$_3$·9H$_2$O and TEA were added into a 200 mL solution using deionized water such that their typical concentrations were 7.5 mM, 2.5 mM, 17.5 mM and 10 mM. A magnetic stir bar was added to the flask and the resulting solution was stirred at 300 rpm for 24 h at room temperature. The reaction mixture was heated to reflux at 100 °C with continuous stirring in a silicone oil bath for 48 h. Subsequently, the system was cooled to room temperature in the air. The dispersion was centrifuged at 3000 rpm for 10 to 15 min at room temperature to separate the precipitate from the solvent. The precipitate was then washed three times with deionized water by shaking and then centrifuging to separate. The clean precipitate was then redispersed in isopropanol with an additional two washing steps. The as-produced platelets were subsequently tip-sonicated for 1 h using a Fischer Scientific Sonic Dismembrator Ultrasonic Processor at 40% power at a frequency of 37 kHz (with Fisher Scientific Isotemp refrigerated bath circulator cooling system held at 5 °C). Further modification of the solid lubricant is necessary for stable dispersion in lubricating oils. Ni-Fe LDH powders were added to an ethanol solution containing 15 wt.% oleic acid and sonicated for 30 min. The resulting mixture was then dried to obtain the modified powder. The 0.wt.%, 0.1 wt.%, 0.2 wt.%, 0.3 wt.%, 0.4 wt.%, 0.5 wt.% modified powders were then added to GTL 430 and dispersed for 30 min using an ultrasonic immersion probe sonicator.

2.3. Characterization of Ni-Fe LDH

The as synthesized Ni-Fe LDH powders were characterized by using different analysis techniques. The crystal structure of Ni-Fe LDH powders was recorded using a powder X-ray diffraction (XRD, Bruker D8 ADVANCE, Karlsruhe, Germany) with Cu Kα radiation in the 2θ range from 5° to 90° and scan step 0: 02°. The infrared spectroscopy of Ni-Fe LDH powders was performed on Fourier transform infrared spectroscopy (FT-IR PerkinElmer Spectrum, Shelton, CT, USA) at a resolution of 4 cm^{-1}. Thermal stability of Ni-Fe LDH powders was evaluated using thermogravimetric analysis instrument (TG/DSC, SDT-Q600, TA Instruments, New Castle, DE, USA) under pure N$_2$ flow (100 mL min^{-1}) to study the weight changes with temperature. The samples were heated from room temperature to 850 °C at a heating rate of 10 °C/min.

2.4. Tribological Experiments and Evaluation

The friction and wear properties of the lubricant were examined on an SHM-1A four-ball tester (Jinan Shunmao test instrument Co., Ltd., Jinan, China) with a rotational speed of 1200 rpm, pressure loads of 200 N, 300 N, 400 N, 500 N, and 600 N, and a test time of 1 h at room temperature. Before friction test, the steel balls were successively ultrasonicated in acetone and anhydrous ethanol to remove a wide variety of contaminants on the surface, then washed with deionized water and dried in a nitrogen atmosphere. For each friction test, 100 mL of lubricant was added to the contact area of the ball friction pair to ensure that the friction pair was lubricated throughout the sliding process.

In order to obtain accurate and reliable tribological data, each set of experiments was repeated three times and the coefficient of friction of the friction tester was recorded with an accuracy of ±0.01. All the tests were carried out in humid air at room temperature and relative humidity of 30–50%. The parameters of the tested steel balls are shown in Table 2.

Table 2. Experimental parameters and basic characteristics of the steel balls used.

Parameter	GTL 430	0.1 mg/mL Ni-Fe LDH	0.2 mg/mL Ni-Fe LDH	0.3 mg/mL Ni-Fe LDH	0.4 mg/mL Ni-Fe LDH	0.5 mg/mL Ni-Fe LDH
Speed	1200 rpm	1200 rpm	1200 rpm	1200 rpm	1200 rpm	1200 rpm
Load	200 N	-	-	-	-	-
	300 N	-	-	-	-	-
	400 N	-	-	-	-	-
	500 N	-	-	-	-	-
	600 N	600 N	600 N	600 N	600 N	600 N
Temperature	RT	RT	RT	RT	RT	RT
Test Duration	60 min	60 min	60 min	60 min	60 min	60 min
Component	Elastic modulus (MPa)	Poisson ratio	Diameter	Rockwell	Surface roughness	
GCr15	2.085×10^5	0.3	12.7 mm	60 ± 1	0.256 μm	

To explore the friction mechanism of the Ni-Fe LDH, the morphology of the worn surface was observed using a JEOL JSM-6610LV scanning electron microscope (JEOL, Tokyo, Japan) and a white light interferometer (WLI) with a Contour GT-X 3D optical profiler (Bruker, Karlsruhe, Germany), and the elemental distribution and composition of the worn surface was identified and quantified using an Oxford X-Max 20 mm² energy dispersive X-ray spectrometer (Oxford Instruments, Oxford, UK). In order to further determine the elemental composition and chemical state of the ternary films on the worn surfaces, X-ray photoelectron spectroscopy replication (XPS) tests were carried out using an ESCALAB 250Xi X-ray photoelectron spectrometer (Bruker, Karlsruhe, Germany) to probe the deposition of the ternary films.

3. Results and Discussion

3.1. Material Characterization

As can be seen from the XRD pattern of the Ni-Fe LDH powders with various Ni/Fe ratio (Figure 1a,b), the characteristic diffraction peaks of the as-prepared samples at 11.3°, 22.7°, 34.4°, 38.9°, and 60.1° corresponded to the (003) (006), (012) (015), and (110) crystal planes of NiFe-LDHs (PDF#40–0215), respectively, indicating that these samples have typical hydrotalcite-like crystal structure [27]. Most noteworthy, the XRD patterns of the samples with smaller Ni-Fe ratio (2:1, 3:1) displayed sharper and narrower peaks, implying higher degrees of crystallinity increase [2]. All the characteristic peaks of the LDH structure appeared in the XRD pattern of the samples can be well indexed to the characteristic Ni-Fe LDH structure (JCPDS no. 40-0215), conforming the successful preparation of NiFe-LDHs.

The FT-IR spectra of the Ni-Fe LDHs are shown in Figure 1c. The infrared bands around 3500 cm^{-1} corresponds to the OH-stretching vibration and the one at 1627 cm^{-1} is attributed to the H-O-H deformation, which are the evidence of the hydration of the LDH samples. The peak located at 1341 cm^{-1}, which originates from the ν stretching vibration of the NO_3 groups in the LDH interlayer. Bands under 1000 cm^{-1} can be ascribed to bonds between O and metallic atoms forming the hydroxide layers.

Figure 2 displays the TG and DTG curves. The DTG curve exhibited endothermic peaks at a temperature lower than 100 °C, which could be related to the release of absorbed H_2O molecules. The endothermic peaks at the temperature range of 200–500 °C are believed to result from the dissipation of -OH groups in the Ni-Fe LDH. Accordingly, assuming that the dissipation of interlayer H$_2$O molecules occurred at the temperature range of 100–200 °C, the H$_2$O content in the interlayer was calculated from the weight loss of 5.4% which occurred between 100 °C and 200 °C.

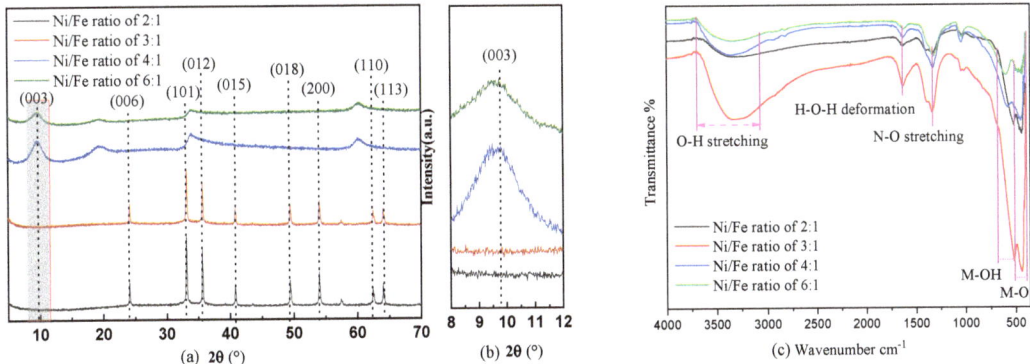

Figure 1. The powder X-ray patterns of the four as-prepared Ni-Fe LDH samples (**a**,**b**). FTIR measurement of the four synthesized Ni-Fe LDH samples (**c**).

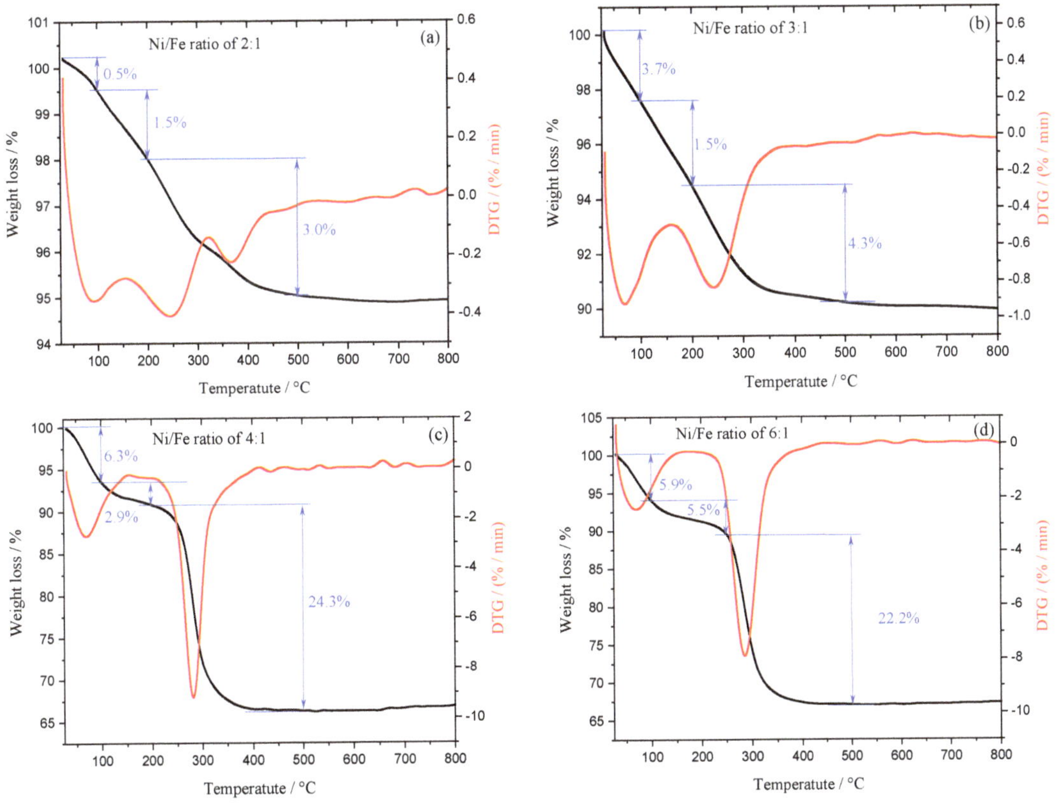

Figure 2. TG and DTG curves of Ni-Fe LDH samples. Ni/Fe ratios of 2 (**a**), 3 (**b**), 4 (**c**) and 6 (**d**).

3.2. Friction and Wear Performance

Figure 3 provided the variations of the friction coefficient (COF), the average friction coefficient (AFC) and the wear scar diameter (WSD) of GTL base oil with load, as well as the variations of COF, AFC and WSD of GTL base oil with Ni-Fe ratio and concentration. The lubricating performance of GTL base oil was tested under 200–600 N in Figure 3a,b. All the COF curves of GTL base oil remained stable under the loads of 200, 300,400, 500, and 600 N. As the load increases, the AFC decreases and then increases, and the WSD gradually increases from 0.54 mm to 0.91 mm. This is because, the increase in load will result in the lubrication state in the metal-to-metal contact zone transition from boundary lubrication to mixed lubrication according to the Stribeck curve. In order to study the tribological performance of Ni-Fe LDHs, the variations of COF, AFC and WSD of GTL base oil with different Ni-Fe ratio and concentration are shown in Figure 3c–j. It can be seen that as the Ni-Fe ratio increases from 2:1 to 6:1, the AFC and WSD of GTL base oil with Ni-Fe LDHs presents similar variations. As the concentration of Ni-Fe LDHs increased from 0 mg/mL to 0.3 mg/mL, the AFC and WSD decreases, while the concentration of Ni-Fe LDHs increased from 0.4 mg/mL to 0.5 mg/mL, the AFC and WSD increases. At the same concentration (0.3 mg/mL), the AFC and WSD of Ni-Fe LDH (Ni/Fe ratio of 6) are the smallest, reducing by 51.3% and 30.8%, respectively, which shows the best lubricating performance.

In order to better understand the wear characteristic of GTL base oil before and after enhanced with 0.3 mg/mL Ni-Fe LDH with Ni/Fe ratio of 6, wear surface morphology was observed by WLI. The steel ball with wear scars was firstly cleaned by ultrasonic with ethanol. A wear scar with the wear volume of 3,076,504 μm^3, surface roughness (Ra) value of 3.0 μm and the average wear scar depth of 38 μm were formed for GTL based oil as seen in Figure 4a–c. It was obvious that the wear volume, the surface roughness and the average depth of the wear marks on the steel balls differed when using GTL base oil + 0.3 mg/mL Ni-Fe LDH (Ni/Fe ratio of 6). The obtained results showed that GTL based oil + 0.3 mg/mL Ni-Fe LDH with Ni/Fe ratio of 6 which had smaller AFC and WSD displayed the wear scar with lower wear volume, surface roughness and average wear scar depth (se Figure 4d–f). In a comparison, the wear volume, surface roughness and average wear scar depth decreased 78.4%, 6.7% and 50.0%, respectively. It was believed that the effects of Ni-Fe LDH alleviated adhesion.

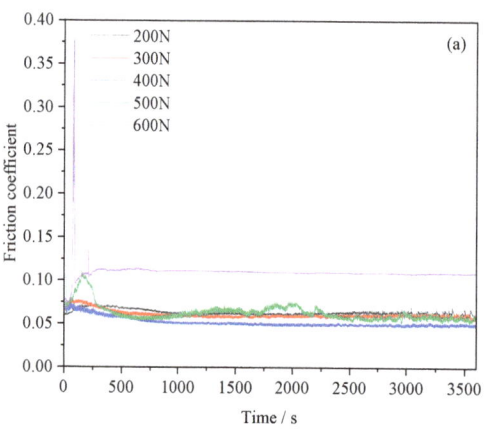
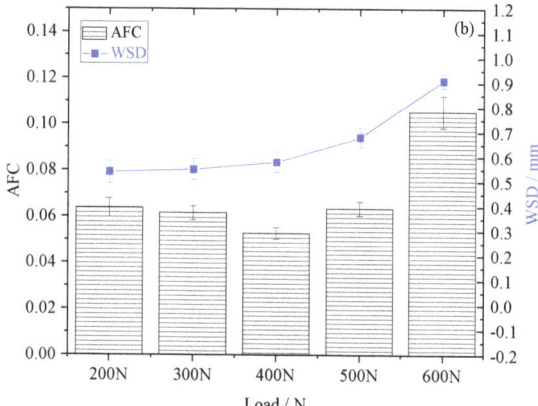

Figure 3. Cont.

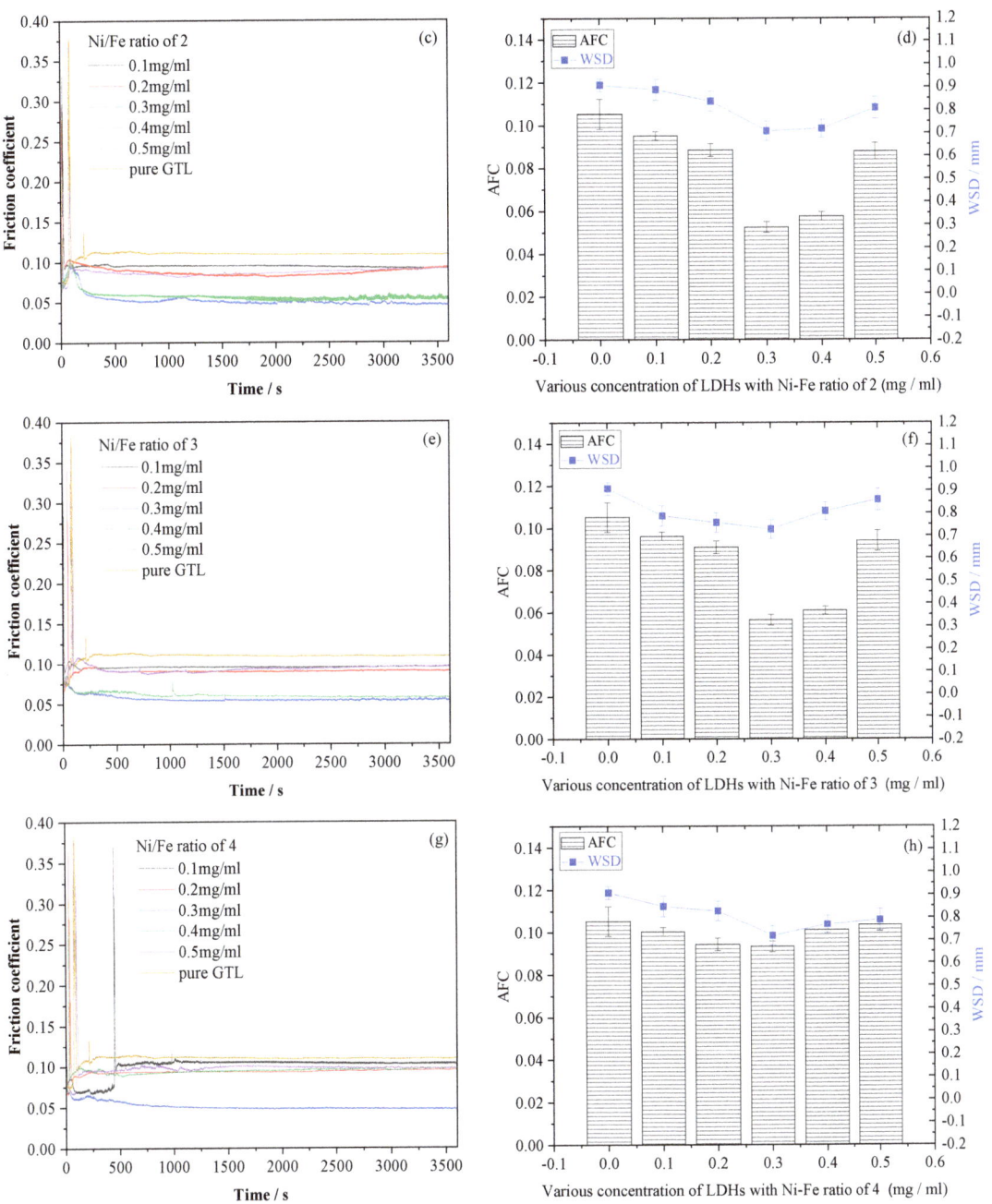

Figure 3. Cont.

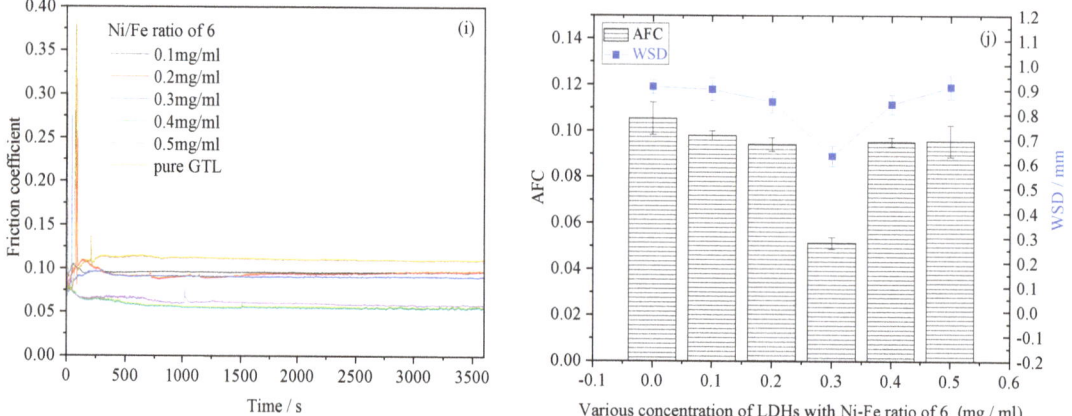

Figure 3. Tribological performance of Ni-Fe LDHs. (**a,b**) variations of COF, AFC and WSD of GTL base oil with load (60 min, RT, 1200 rpm), (**c–j**) variations of COF, AFC and WSD of GTL base oil with Ni-Fe ratio and concentration (600 N, 60 min, RT, 1200 rpm).

3.3. Worn Surface Analysis

In order to investigate the lubrication effect of Ni-Fe LDHs as lubricant additives on the contact area of steel balls during the friction test, the morphology of the wear marks was observed using an optical microscope and a scanning electron microscope equipped with an EDS detector to detect the distribution of the corresponding elements on the wear mark surface after the friction test. As shown in Figure 5, deep grooves and flaking pits were observed on the wear surfaces lubricated with GTL base oil. The elements C, O and Fe were detected on the surface of the wear marks, with O coming from air and C from thermal decomposition of GTL base oil.

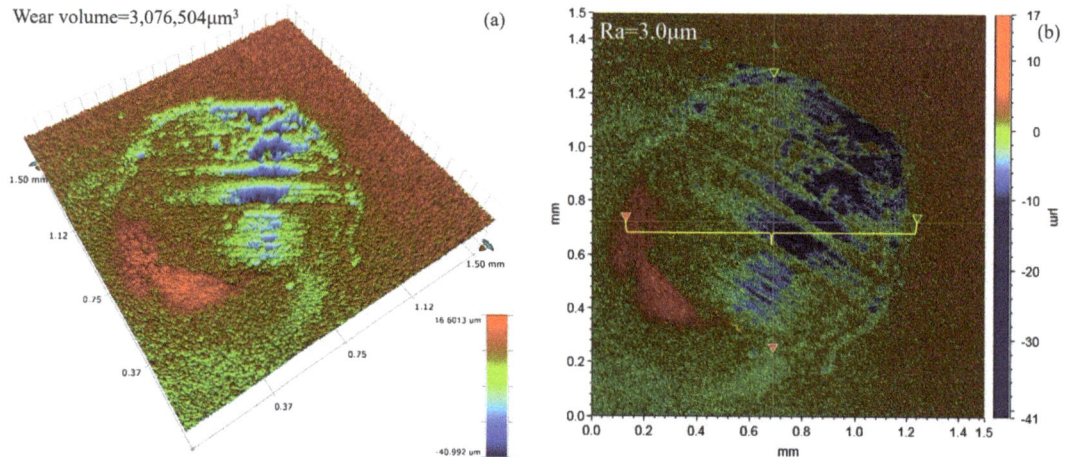

Figure 4. *Cont.*

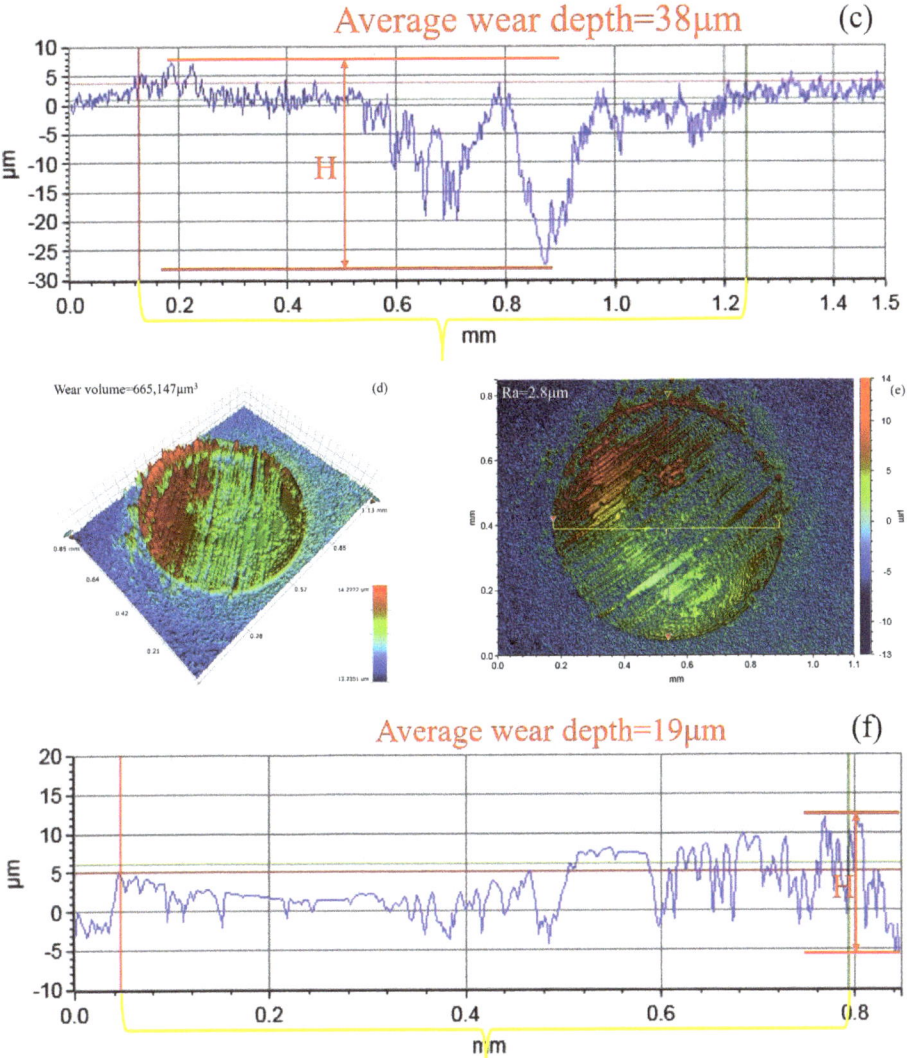

Figure 4. 3D morphology and average depth of steel ball wear scar surfaces, (**a–c**) lubricated by GTL base oil, (**d–f**) lubricated by GTL base oil + 0.3 mg/ml Ni−Fe LDH with Ni/Fe ratio of 6.

After that, the morphology and the corresponding element distribution on the wear scar surface lubricated by GTL base oil + 0.3 mg/mL Ni-Fe LDH with Ni/Fe ratio of 6 is shown in Figure 6 to prove the lubricating effect of Ni-Fe LDHs on the two sliding surfaces. It can be readily found that the worn surface looks very smooth with little visible wear after adding the Ni-Fe LDHs into GTL base oil. The C, O, Fe, and Ni four kinds of elements were detected on the wear scar surface. In addition, after adding Ni-Fe LDHs, elemental O is obviously increased compared with the GTL base oil, which certifies that the wear scar surface of the steel ball produced a tribofilm after the addition of Ni-Fe LDHs. The detected Ni elements were primarily from the Ni-Fe LDHs. Additionally, Ni observed on the wear scar surface, indicating that the adsorption of Ni-Fe LDHs occured on the surface of the bottom balls from tests.

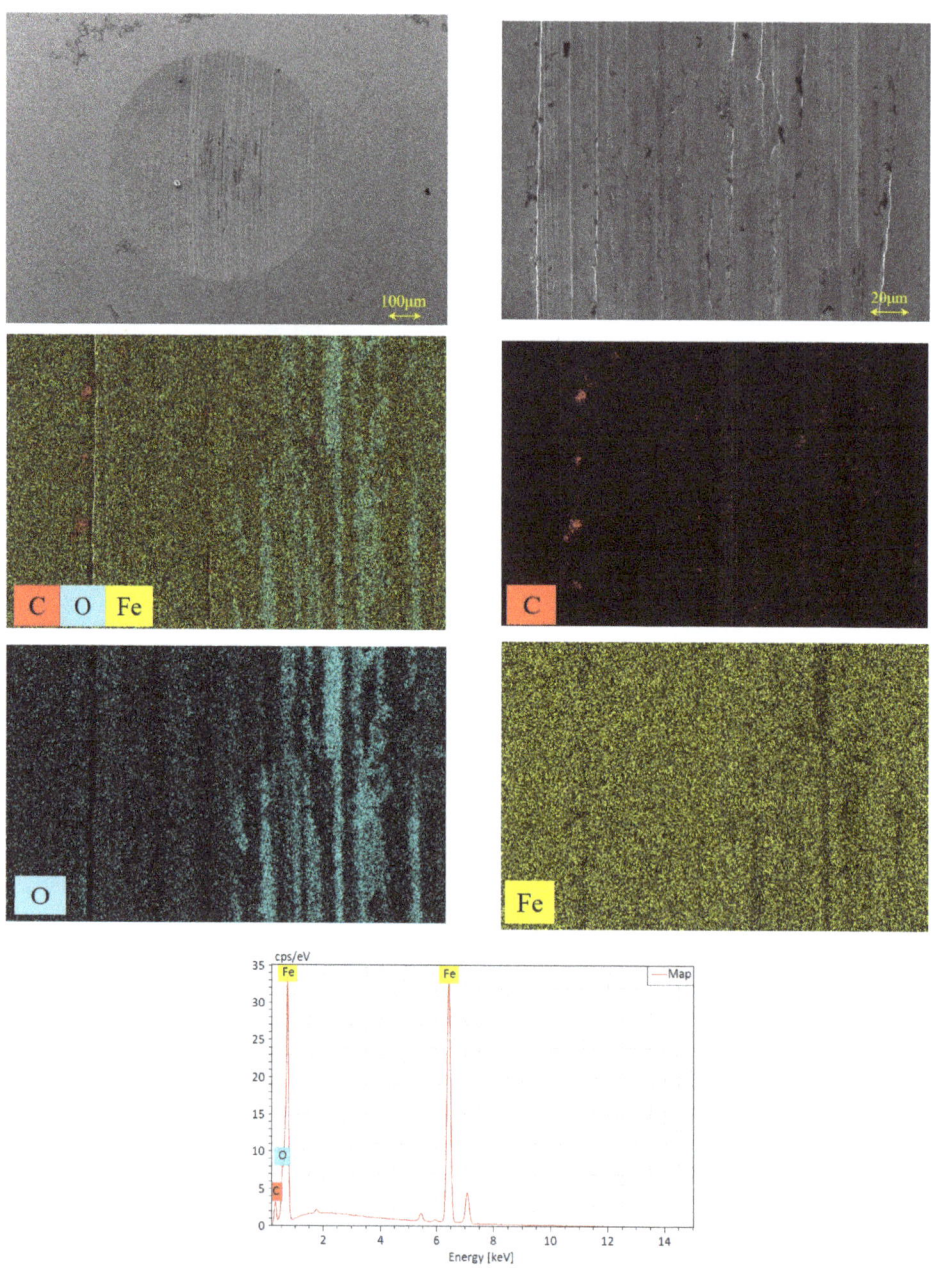

Figure 5. SEM images and EDS mapping of worn surface of steel ball lubricated by GTL base oil.

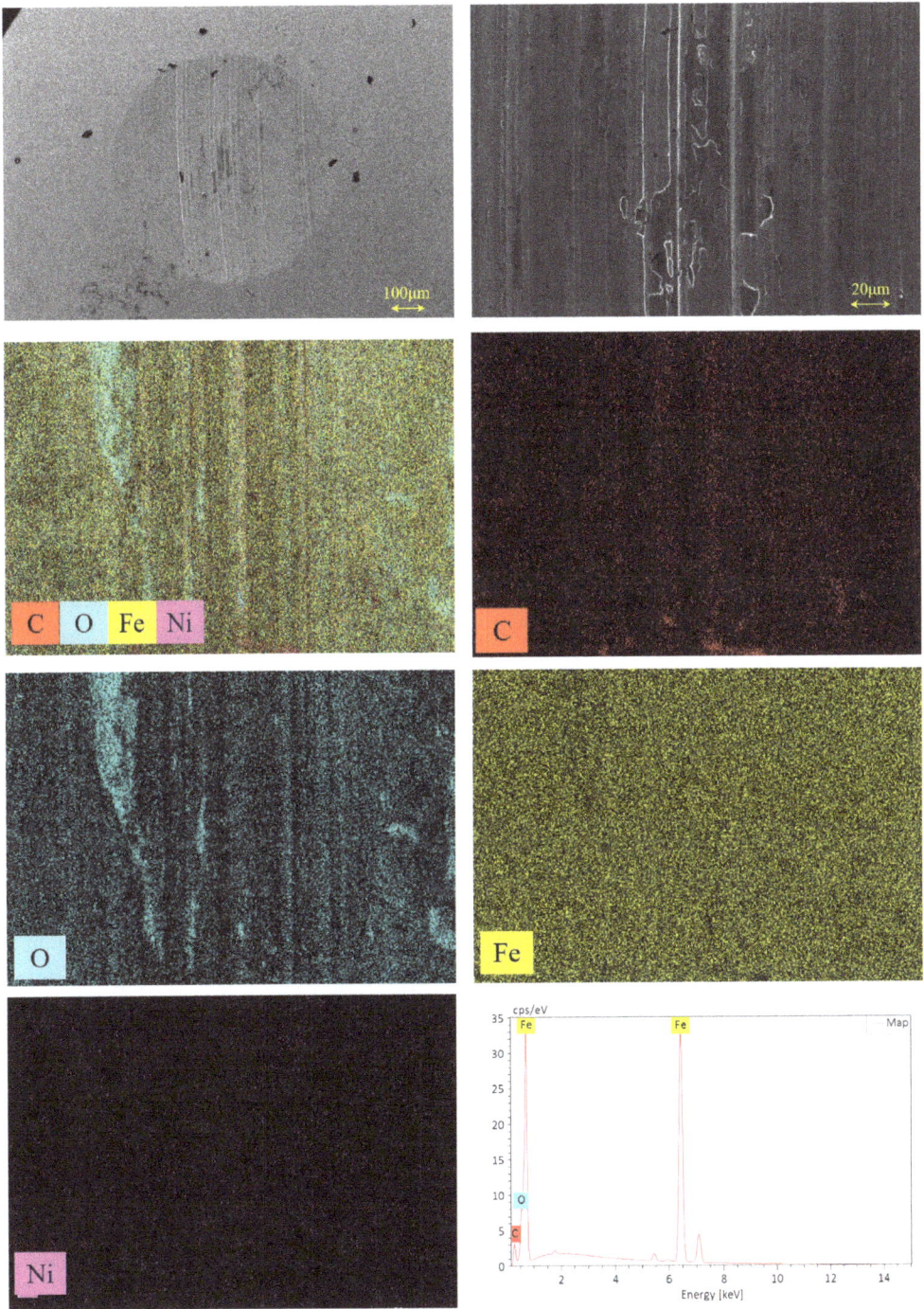

Figure 6. SEM images and EDS mapping of worn surface of steel ball lubricated by GTL base oil + 0.3 mg/mL Ni-Fe LDH with Ni/Fe ratio of 6.

To further clarify the composition of the tribochemical or adsorption film, the XPS was used to characterize the contact area on the steel ball worn surface. Figure 7 presents that the survey spectra of the steel ball worn surface and the high resolution spectra of C 1s, Fe 2p, Ni 2p, O 1s. According to Figure 7a, the peaks of C1s, Fe2p, and O1s were founded on the survey spectrum lubricated by GTL base oil. While the peaks of C1s, Fe2p, and O1s and Ni 2p were founded on the survey spectrum lubricated by GTL base oil + 0.3 mg/mL Ni-Fe LDH with Ni/Fe ratio of 6. According to Figure 7b, the peaks of element C near 284.63 and 288.78 eV respectively assigned to C=C and C=O, suggesting that some carbon-containing material on the worn surface originates from thermal cracking and coke deposition of the GTL base oil. In Figure 7c, the subpeak at a higher binding energy (725.13 eV) is attributed to Fe_3O_4, while the lower subpeak (710.83 eV) is contributed by Fe_2O_3. The Ni2p spectrum (Figure 7d) with subpeak at 854.88 eV suggests that Ni exists as NiO on the rubbing surface by tribochemical reaction. In Figure 7e, the peaks near 529.63 eV and 531.43 eV of the O1s spectrum lubricated by GTL base oil + 0.3 mg/mL Ni-Fe LDH with Ni/Fe ratio of 6 attributed to the NiO and iron oxides, respectively. While the peak near 531.88 eV of the O1s spectrum lubricated by GTL base oil attributed to the iron oxides. Therefore, it is justified to conclude that the tribofilm deposited on the worn surface of the steel ball uniformly and densely, which was primarily made up of of iron oxides and NiO, proving that Ni-Fe LDHs indeed benefited to enhance the tribological behavior of GTL base oil.

3.4. Lubrication Mechanism of Ni-Fe LDH

In this paper, the four-sphere point contact model in contact mechanics is used to investigate the contact stress pattern of a point contact object under pressure in order to assess its tribological behaviour (Figure 8a).

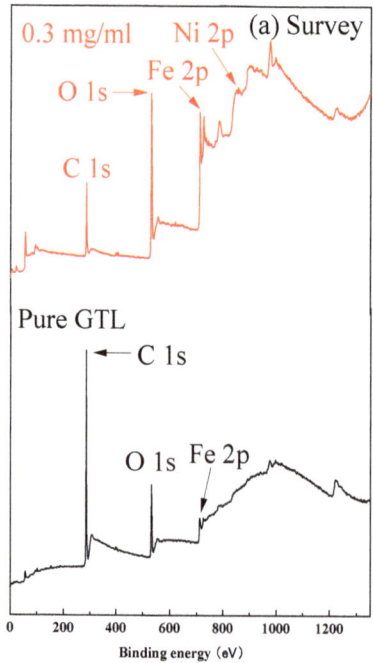
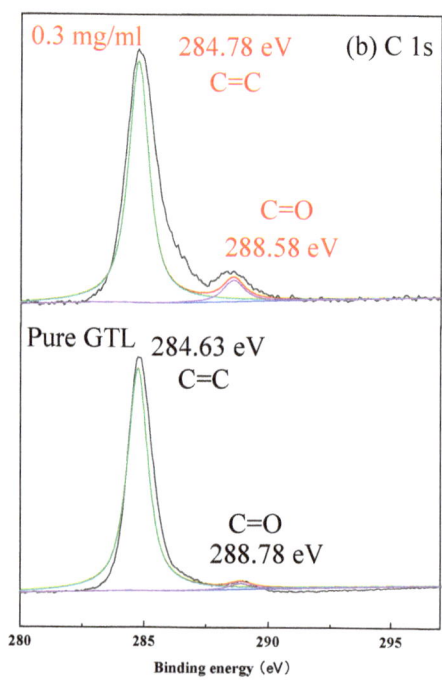

Figure 7. Cont.

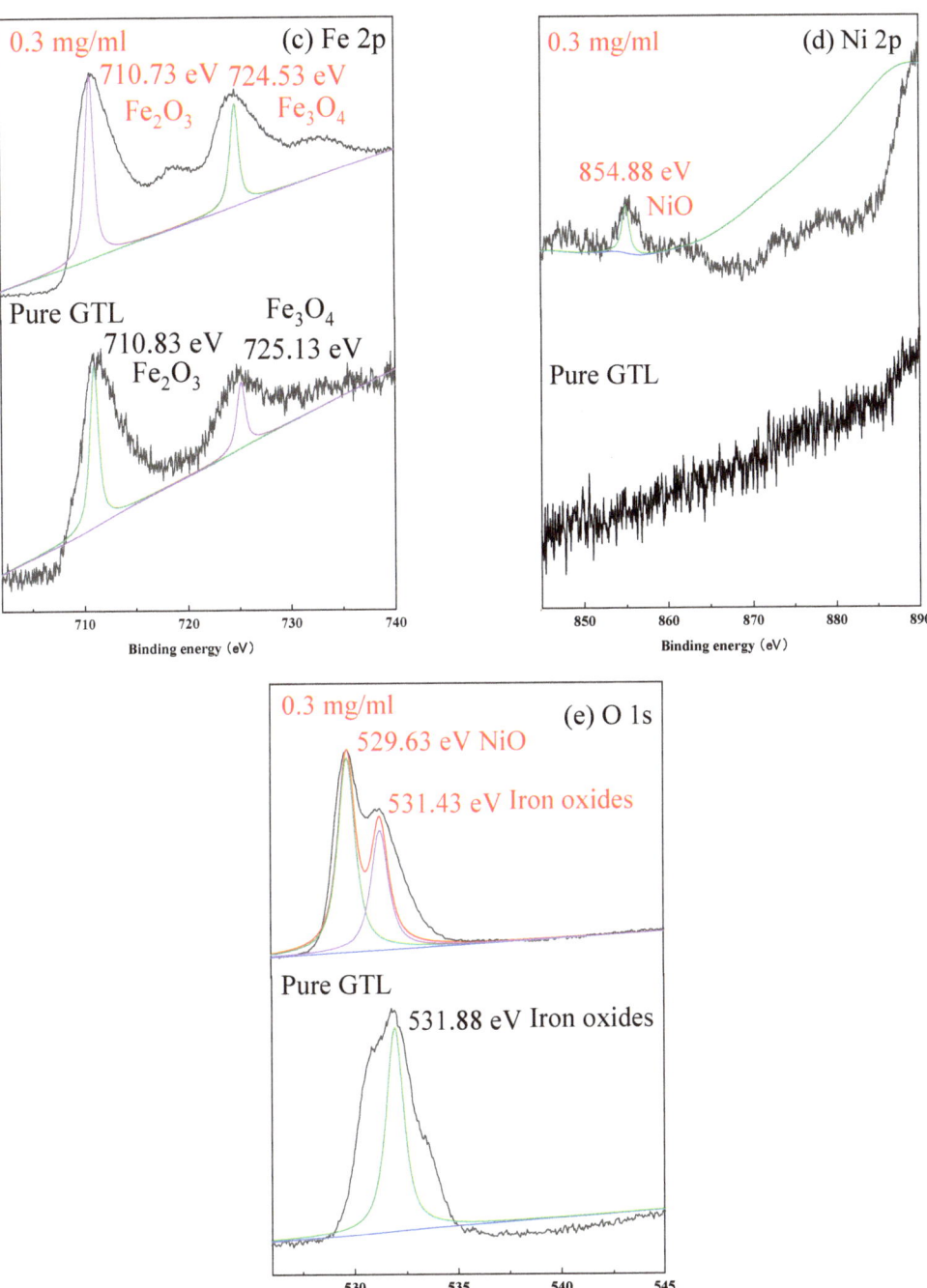

Figure 7. XPS spectrum of steel ball wear spot surface lubricated by GTL base oil and GTL base oil + 0.3 mg/ml Ni−Fe LDH with Ni/Fe ratio of 6. (**a**) XPS survey spectra, (**b**) C 1s, (**c**) Fe 2p, (**d**) Ni 2p, (**e**) O 1s.

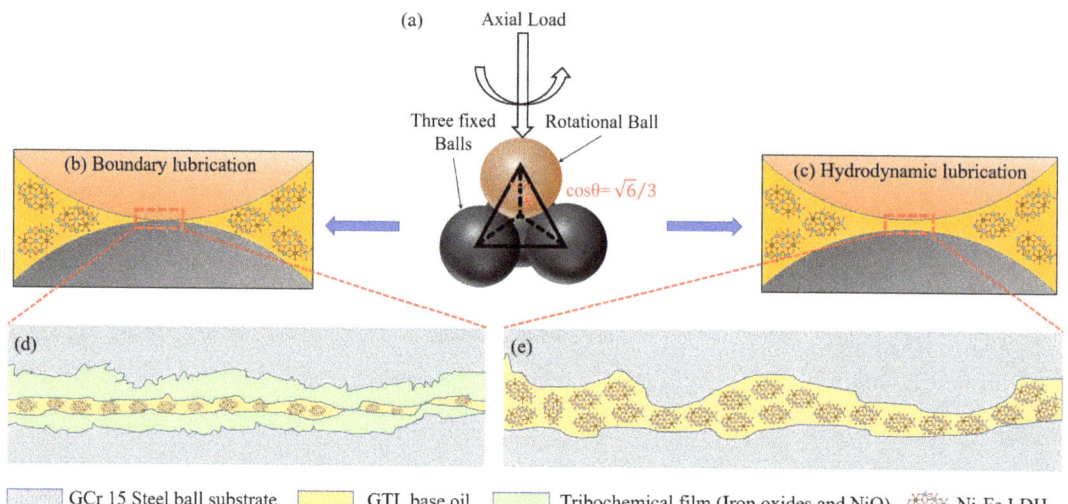

Figure 8. Schematic diagrams of lubrication mechanism for GTL base oil enhanced with Ni-Fe LDH with Ni/Fe ratio of 6: (**a**) schematic illustration of the four-sphere point contact model, (**b**,**d**) boundary lubrication between two sliding surfaces of a solid rough peaks and ridges, (**c**,**e**) hydrodynamic lubrication in the area of fluid contact.

The rotating upper ball was located on a spindle, whereas the lower three stationary balls were fixed in an oil cup. Based on Hertz contact theory, the maximum contact pressure at the center of the circular contact area is given by:

$$q_{max} = \frac{3p}{2\pi a^2} \quad (1)$$

$$a = \left(\frac{3}{4} \times \frac{pR'}{E^*}\right)^{\frac{1}{3}} \quad (2)$$

Herein, q_{max} is the maximum contact stress between the steel balls, p is the effective load, a is the radius of the steel balls contact area, E^* is the equivalent Young's modulus, R' is the comprehensive radius. p, E^*, and R' are defined as:

$$w = 3p\cos\theta \quad (3)$$

$$\frac{1}{R'} = \frac{1}{R_{above}} + \frac{1}{R_{below}} \quad (4)$$

$$E^* = \left(\frac{1-v_{above}^2}{E_{above}} + \frac{1-v_{below}^2}{E_{below}}\right)^{-1} \quad (5)$$

In this model contact, the polygon of the steel ball friction pair is a regular triangular pyramid with four ball centers as vertices, according to which $\cos\theta = \sqrt{6}/3$, and the effective load (p) of every steel ball is calculated by the equation (Equation (3)), where w is the total load, R_{above} is the radius of the rotating upper ball, R_{below} are the radii of the lower three stationary balls, v_{above} and E_{above} are Poisson's ratio and Young's modulus of the rotating upper ball, respectively, v_{below} and E_{below} are Poisson's ratio and Young's modulus of the lower three stationary balls, respectively.

The effective load of each ball is estimated to be 245 N for the total applied load of 600 N and the maximum contact pressure calculated by the Hertz contact stress formula (Equations (1) and (2)) was 3.51 GPa.

For further understanding of the lubrication state between friction pairs and to investigate the tribological properties of Ni-Fe LDH in GTL base oil, the thickness of the lubricating film (h_{min}) was calculated from the Hamrock-Dowson equation (Equation (6)) [30–32]:

$$h_{min} = 3.63 \frac{G^{0.49} U^{0.68} R'}{W^{0.073}} \left(1 - 0.61 e^{-0.68k}\right) \quad (6)$$

where $G = \alpha E'$, $U = \eta_0 u / E' R'$, $W = p / E' R'^2$, $k = 1.03 (R_y/R_x)^{0.64} = 1.03$ were dimensionless material parameter, dimensionless speed parameter, dimensionless load parameter, ellipticity parameter, respectively. α and η_0 were the viscosity-pressure coefficient and the dynamic viscosity at 25 °C of GTL base oil + 0.3 mg/mL Ni-Fe LDH with Ni/Fe ratio of 6, respectively. u (0.461 m/s) was the relative sliding velocity of the two friction pairs, k was the ellipticity parameter, $E' = 2E^*$ was the effective modulus of elasticity.

Here, the minimum oil film thickness (h_{min}) between the two friction pairs was 1.306 μm. The lubrication state under the four ball experiment could be subsequently classified in accordance with the relationship between the film thickness and surface roughness as follows equation (Equation (7)):

$$\lambda = \frac{h_{min}}{\sqrt{\sigma_1^2 + \sigma_2^2}} \quad (7)$$

where σ_1 (2.8 μm) and σ_2 (2.8 μm) are the surface roughness of the worn area of the upper rotating ball and the lower stationary ball, respectively. This showed that λ is about 2.56 for GTL base oil + 0.3 mg/mL Ni-Fe LDH with Ni/Fe ratio of 6, indicating that the contact area is in a mixed lubrication state which is an intermediary condition between boundary lubrication (Figure 8b) and hydrodynamic lubrication (Figure 8c) [32,33].

Boundary lubrication occurs mainly between two sliding surfaces of a solid rough peaks and ridges (Figure 8d). During the collision close to the cusp, the obvious microscopic fold structure is firstly destroyed under high contact pressure. At the same time, a large number of ultrathin Ni-Fe LDH nanosheets are formed. To some extent, the possibility of direct contact of Ni-Fe LDH nanosheets with sliding solid surfaces is greatly increased. During violent friction, the ultrathin Ni-Fe LDH nanosheets physically hinder the direct collision of the pointed surface in the middle of the contact zone [33]. As sliding proceeds, the physically adsorbed film ruptures under harsh conditions. At this point, the gradual generation of heat, plastic deformation and defects on the worn surface also provide an environment for the subsequent ternary chemical reaction [29]. The ternary chemical reaction between the Ni-Fe LDH additive and the surface of the sliding steel ball generates a new relatively dense ternary protective film, which has better mechanical properties, and can greatly prevent scuffing and protect the solid surfaces from severe collisions.

At the same time, hydrodynamic lubrication occurs mainly in the area of fluid contact (Figure 8e). The dynamic pressure effect caused by the relatively rapid movement of the friction partner also produces a film of hydrodynamic lubrication between the two friction partners. In hydrodynamic lubrication, this film completely separates the concave and convex peaks, thus reducing friction and wear between the balls.

4. Conclusions

In this paper, some kinds of Ni-Fe LDH powders with various Ni/Fe ratio were synthesized via a cost-effective room-temperature co-precipitation process, and its lubricating performance as an oil-based additive in steel–steel contact was studied. It shows that the Ni/Fe ratio of Ni-Fe LDH had a remarkable influence on the lubricating performance improvement of GTL base oil. At the same concentration (0.3 mg/mL), the Ni-Fe LDH with Ni/Fe ratio of 6 was demonstrated to exhibit the best lubricating performance and the AFC, WSD, the wear volume, surface roughness and average wear scar depth decreased 51.3%, 30.8%78.4%, 6.7% and 50.0%, respectively. The lubrication mechanisms are summarized as follows. (1) As a result of the benefit of nanoscale size and layered structure, the microstructure of Ni-Fe LDH is broken down under the action of high applied pressure at the initial stage, producing a mass of ultrathin Ni-Fe LDH nanosheets, which form a physical adsorption film on the worn surface of the steel ball, polishing and mending the microbulges of worn surface and the defects resulted from harsh friction. (2) As friction proceeds, the physical film described above ruptures under harsh conditions. As a result of heating, plastic deformation and peeling of the worn surface of the steel ball, a new triple chemical film consisting of iron and nickel oxides with better mechanical properties is produced and gradually replaces the physical adsorption film, thus resisting scratches and protecting the solid surface from extremely severe impacts.

Author Contributions: Investigation, S.X., X.Z., H.B., Q.Z. and Y.H.; resources, S.X. and Q.Z.; methodology and validation, S.X., X.Y. and Q.T.; visualization and formal analysis, S.X., P.L. and B.H.; supervision, S.X., Q.Z., X.D. and Y.H.; writing—original draft preparation, S.X., S.L., Q.Z. and S.W.; writing—review and editing, S.X., X.Z., Q.Z., X.Y., P.L., Q.T., X.D. and Y.H. All authors have read and agreed to the published version of the manuscript.

Funding: This research was funded by Science and Technology Research Program of Chongqing Municipal Education Commission, grant number KJZD-K202212905 and Natural Science Foundation of Chongqing, grant number cstc2019jcyj-msxmX0453.

Data Availability Statement: The original contributions presented in the study are included in the article, further inquiries can be directed to the corresponding author.

Conflicts of Interest: The authors declare no conflicts of interest.

References

1. Duan, L.; Li, J.; Duan, H. Nanomaterials for lubricating oil application: A review. *Friction* **2023**, *11*, 647–684. [CrossRef]
2. Jiang, H.; Hou, X.; Qian, Y.; Liu, H.; Ahmed Ali, M.K.; Dearn, K.D. A tribological behavior assessment of steel contacting interface lubricated by engine oil introducing layered structural nanomaterials functionalized by oleic acid. *Wear* **2023**, *524–525*, 204675. [CrossRef]
3. Rawat, S.S.; Harsha, A.P.; Khatri, O.P. Tribological Investigations of Two-Dimensional Nanostructured Lamellar Materials as Additives to Castor-Oil-Derived Lithium Grease. *J. Tribol.* **2022**, *144*, 091902. [CrossRef]
4. Singh, A.; Chauhan, P.; Mamatha, T.G. A review on tribological performance of lubricants with nanoparticles additives. *Mater. Today Proc.* **2020**, *25*, 586–591. [CrossRef]
5. Zhou, C.; Li, Z.; Liu, S.; Zhan, T.; Li, W.; Wang, J. Layered double hydroxides for tribological application: Recent advances and future prospective. *Appl. Clay Sci.* **2022**, *221*, 106466. [CrossRef]
6. Luo, J.; Liu, M.; Ma, L. Origin of friction and the new frictionless technology—Superlubricity: Advancements and future outlook. *Nano Energy* **2021**, *86*, 106092. [CrossRef]
7. Chen, W.; Feng, Y.; Wan, Y.; Zhang, L.; Yang, D.; Gao, X.; Yu, Q.; Wang, D. Investigation on anti-wear and corrosion-resistance behavior of steel-steel friction pair enhanced by ionic liquid additives under conductive conditions. *Tribol. Int.* **2023**, *177*, 108002. [CrossRef]
8. Bowden, F.P.; Tabor, D. *The Friction and Lubrication of Solids*; Oxford University Press: Oxford, UK, 2001.
9. Li, S.; Bhushan, B. Lubricating performance and mechanisms of Mg/Al-, Zn/Al-, and Zn/Mg/Al-layered double hydroxide nanoparticles as lubricant additives. *Appl. Surf. Sci.* **2016**, *378*, 308–319. [CrossRef]
10. Gong, H.; Yu, C.; Zhang, L.; Xie, G.; Guo, D.; Luo, J. Intelligent lubricating materials: A review. *Compos. Part B Eng.* **2020**, *202*, 108450. [CrossRef]
11. Xia, L.; Long, J.; Zhao, Y.; Wu, Z.; Dai, Z.; Wang, L. Molecular Dynamics Simulation on the Aggregation of Lubricant Oxidation Products. *Tribol. Lett.* **2018**, *66*, 104. [CrossRef]
12. Sulima, S.I.; Bakun, V.G.; Chistyakova, N.S.; Larina, M.V.; Yakovenko, R.E.; Savost'yanov, A.P. Prospects for Technologies in the Production of Synthetic Base Stocks for Engine Oils (A Review). *Pet. Chem.* **2021**, *61*, 1178–1189. [CrossRef]
13. Paul, G.; Hirani, H.; Kuila, T.; Murmu, N.C. Nanolubricants dispersed with graphene and its derivatives: An assessment and review of the tribological performance. *Nanoscale* **2019**, *11*, 3458–3483. [CrossRef] [PubMed]
14. Jin, B.; Chen, G.; He, Y.; Zhang, C.; Luo, J. Lubrication properties of graphene under harsh working conditions. *Mater. Today Adv.* **2023**, *18*, 100369. [CrossRef]
15. Wang, Q.; Hou, T.; Wang, W.; Zhang, G.; Gao, Y.; Wang, K. Tribological behavior of black phosphorus nanosheets as water-based lubrication additives. *Friction* **2022**, *10*, 374–387. [CrossRef]
16. Tang, G.; Wu, Z.; Su, F.; Wang, H.; Xu, X.; Li, Q.; Ma, G.; Chu, P.K. Macroscale Superlubricity on Engineering Steel in the Presence of Black Phosphorus. *Nano Lett.* **2021**, *21*, 5308–5315. [CrossRef] [PubMed]
17. Chen, H.; Xiao, G.; Chen, Z.; Yi, M.; Zhang, J.; Li, Z.; Xu, C. Hexagonal boron nitride (h-BN) nanosheets as lubricant additive to 5CB liquid crystal for friction and wear reduction. *Mater. Lett.* **2022**, *307*, 131007. [CrossRef]

18. An, L.; Yu, Y.; Bai, C.; Bai, Y.; Zhang, B.; Gao, K.; Wang, X.; Lai, Z.; Zhang, J. Simultaneous production and functionalization of hexagonal boron nitride nanosheets by solvent-free mechanical exfoliation for superlubricant water-based lubricant additives. *NPJ 2d Mater. Appl.* **2019**, *3*, 28. [CrossRef]
19. Xiang, S.; Long, X.; Zhang, Q.; Ma, P.; Yang, X.; Xu, H.; Lu, P.; Su, P.; Yang, W.; He, Y. Enhancing Lubricating performance of Calcium Sulfonate Complex Grease Dispersed with Two-Dimensional MoS_2 Nanosheets. *Lubricants* **2023**, *11*, 336. [CrossRef]
20. Liu, C.; Meng, Y.; Tian, Y. Potential-Controlled Boundary Lubrication Using MoS_2 Additives in Diethyl Succinate. *Tribol. Lett.* **2020**, *68*, 72. [CrossRef]
21. Wu, H.; Yin, S.; Du, Y.; Wang, L.; Wang, H. An investigation on the lubrication effectiveness of MoS_2 and BN layered materials as oil additives using block-on-ring tests. *Tribol. Int.* **2020**, *151*, 106516. [CrossRef]
22. Lu, Z.; Cao, Z.; Hu, E.; Hu, K.; Hu, X. Preparation and tribological properties of WS_2 and WS_2/TiO_2 nanoparticles. *Tribol. Int.* **2019**, *130*, 308–316. [CrossRef]
23. Wang, C.; Zhang, X.; Jia, W.; Deng, Q.; Leng, Y. Preparation and Tribological Properties of Modified Field's Alloy Nanoparticles as Additives in Liquid Poly-alfa-olefin Solution. *J. Tribol.* **2019**, *141*, 1. [CrossRef]
24. Zhou, C.; Li, Z.; Liu, S.; Ma, L.; Zhan, T.; Wang, J. Synthesis of MXene-Based Self-dispersing Additives for Enhanced Tribological Properties. *Tribol. Lett.* **2022**, *70*, 63. [CrossRef]
25. Miao, X.; Li, Z.; Liu, S.; Wang, J.; Yang, S. MXenes in tribology: Current status and perspectives. *Adv. Powder Mater.* **2023**, *2*, 100092. [CrossRef]
26. Boidi, G.; de Queiróz, J.C.F.; Profito, F.J.; Rosenkranz, A. Ti_3C_2Tx MXene Nanosheets as Lubricant Additives to Lower Friction under High Loads, Sliding Ratios, and Elevated Temperatures. *ACS Appl. Nano Mater.* **2023**, *6*, 729–737. [CrossRef]
27. Wang, K.; Wu, H.; Wang, H.; Liu, Y. Superior extreme pressure properties of different layer LDH nanoplatelets used as boundary lubricants. *Appl. Surf. Sci.* **2020**, *530*, 147203. [CrossRef]
28. Pancrecious, J.K.; Gopika, P.S.; Suja, P.; Ulaeto, S.B.; Gowd, E.B.; Rajan, T.P.D. Role of layered double hydroxide in enhancing wear and corrosion performance of self-lubricating hydrophobic Ni-B composite coatings on aluminium alloy. *Colloids Surf. A Physicochem. Eng. Asp.* **2022**, *634*, 128017. [CrossRef]
29. Wang, H.; Wang, Y.; Liu, Y.; Zhao, J.; Li, J.; Wang, Q.; Luo, J. Tribological behavior of layered double hydroxides with various chemical compositions and morphologies as grease additives. *Friction* **2021**, *9*, 952–962. [CrossRef]
30. Wei, X.; Li, W.; Fan, X.; Zhu, M. MoS_2-functionalized attapulgite hybrid toward high-performance thickener of lubricating grease. *Tribol. Int.* **2023**, *179*, 108135. [CrossRef]
31. Hamrock, B.J.; Dowson, D. Isothermal Elastohydrodynamic Lubrication of Point Contacts: Part II—Ellipticity Parameter Results. *J. Lubr. Technol.* **1976**, *98*, 375–381. [CrossRef]
32. Liu, Y.; Li, J.; Li, J.; Yi, S.; Ge, X.; Zhang, X.; Luo, J. Shear-Induced Interfacial Structural Conversion Triggers Macroscale Superlubricity: From Black Phosphorus Nanoflakes to Phosphorus Oxide. *ACS Appl. Mater. Interfaces* **2021**, *13*, 31947–31956. [CrossRef] [PubMed]
33. Wang, D.; Yang, J.; Wei, P.; Pu, W. A mixed EHL model of grease lubrication considering surface roughness and the study of friction behavior. *Tribol. Int.* **2021**, *154*, 106710. [CrossRef]

Disclaimer/Publisher's Note: The statements, opinions and data contained in all publications are solely those of the individual author(s) and contributor(s) and not of MDPI and/or the editor(s). MDPI and/or the editor(s) disclaim responsibility for any injury to people or property resulting from any ideas, methods, instructions or products referred to in the content.

Article

Study on the Lubricating Characteristics of Graphene Lubricants

Yi Dong [1], Biao Ma [1], Cenbo Xiong [1,*], Yong Liu [2] and Qin Zhao [1]

[1] School of Mechanical Engineering, Beijing Institute of Technology, Beijing 100081, China; dongyi0219@163.com (Y.D.); mabiao@bit.edu.cn (B.M.); yudewuhou@outlook.com (Q.Z.)
[2] School of Energy and Power Engineering, North University of China, Taiyuan 030051, China; yongliu_epe@nuc.edu.cn
* Correspondence: xiongcenbo@bit.edu.cn

Abstract: Graphene is considered a good lubricant additive. The lubricating properties of graphene lubricant at different concentrations and temperatures are studied via a four-ball friction and wear-testing machine. The results show that the coefficient of friction (COF) and wear scar diameter (WSD) of the steel ball with 0.035 wt% graphene lubricant decreased by 40.8% and 50.4%, respectively. Finally, through surface analysis, the following lubrication mechanism is proposed: as the added graphene particles can easily fill and cover the pores of the friction surface, the contact pressure of the rough peak is reduced, resulting in a lower COF and smoother surface. Although the COF increases with temperature, graphene lubricants still exhibit good lubrication effects.

Keywords: graphene lubricant; friction and wear properties; roughness; lubrication characteristics

Citation: Dong, Y.; Ma, B.; Xiong, C.; Liu, Y.; Zhao, Q. Study on the Lubricating Characteristics of Graphene Lubricants. *Lubricants* **2023**, *11*, 506. https://doi.org/10.3390/lubricants11120506

Received: 6 November 2023
Revised: 25 November 2023
Accepted: 27 November 2023
Published: 30 November 2023

Copyright: © 2023 by the authors. Licensee MDPI, Basel, Switzerland. This article is an open access article distributed under the terms and conditions of the Creative Commons Attribution (CC BY) license (https://creativecommons.org/licenses/by/4.0/).

1. Introduction

In mechanical systems, frictional components serve as integral parts of the execution process, and they directly affect the system performance. During the working process, friction components are often operated in harsh environments, such as high-temperature, high-speed, and high-pressure conditions. Consequently, they are inevitably accompanied by a large amount of frictional heat and energy loss. It should be noted that energy loss due to friction is one of the typical forms of global energy consumption; more exactly, about 1/3 of the world's primary energy comes from friction consumption, and almost 1/2 of the power of transportation equipment is consumed in friction [1]. Subsequently, the resulting wear is the main cause of mechanical equipment failure. Generally, there are four ways to reduce friction and wear, including optimizing the bearing structure, improving the working conditions, and developing new friction materials and better lubricants [2].

Lubricants are indispensable because of their unique and effective anti-friction effects, especially for oil-lubricated bearings. Since the base oil does not have integral properties that enable it to withstand different working environments, it must be mixed with additives to improve its working performance [3]. Accordingly, the research of lubricant additives is promising regarding improvements in the anti-friction and anti-wear properties of the base oil [4]. Additives are one or more compounds added into lubricants in order to improve their properties [5]. The reasonable use of additives is crucial to ensure the quality of lubricants. Recently, research on various types of additives has become an attractive area. Nano additives have emerged with the development of the atomic friction model [6]. Nanoparticles have a large specific surface area and are easily adsorbed on the contact surface. Since atoms in the same atomic layer are bound by covalent bonds, a single-layer structure is formed with high modulus and high strength, avoiding direct contact with the friction pair [7]. Several excellent studies about nano additives in lubricants are available, such as metal nanoparticles [6–9], carbon element nanoparticles [10,11], oxide nanoparticles [12,13], various inorganic compound nanoparticles [14,15], polymer nanoparticles [16,17] and composite nanoparticles [18,19], etc.

Many researchers so far have focused on graphene additives. They have several advantages, such as the chemical structure of ultra-thin glass sheets, kinetic and physical properties, and a better fluid self-lubricating transmission performance [20]. However, the deficiency of dispersibility has hindered their development in lubricating oil. Ka et al. [21] explored the tribological properties of graphene as an oil additive. After thorough research work [22], graphene was found to be prone to agglomerating in acidic lubricating oil and found to be unstable in alkaline solvent [23,24]. As for the synthesis methods used for graphene lubricants, there are three strategies employed to improve their dispersion stability [25,26], namely physical modification, chemical structure modification and microstructure. Su et al. [27] observed the lubricated wear scar surface after adding 0.25% graphite nanoparticles to vegetable-based oil. Graphite nanoparticles can be stably adsorbed on the friction surface and form a physical adsorption film on it. Flakes are formed due to precipitation and agglomeration. Since the modification effect can significantly improve the dispersibility of graphene, CI et al. [28] prepared fluorinated reduced graphene oxide nanosheets using a gas fluorination method to improve the load-bearing capacity and wear resistance of lubricants. Lau et al. [29] investigated the suspension stability of GBC and GSF particles (0.05 wt%) dispersed in a low-viscosity polyol ester lubricating oil and their tribological performance. It should be noted that, although graphene lubricants have already been proposed and applied [30], graphene's aggregation and precipitation phenomena in different lubricants, as well as its mechanism of friction and wear, still need to be further studied.

In the extensive applications of bearings, especially in some important devices such as heavy-duty vehicles, trains and vessels, they require regular lubrication to maintain exceptional performance. A lower and more stable coefficient of friction (COF) can improve the working performance and efficiency of bearings. In addition, a lower wear rate can significantly extend the practical life of bearings and devices. In this work, the friction and wear characteristics of graphene lubricants with different concentrations and temperatures are experimentally explored. The lubrication and wear mechanism are further proposed to illuminate the tribochemical interaction between carbon surfaces. The results are expected to provide some basic knowledge for improving the design of nanoparticle applications in lubricants.

2. Lubricant Preparation

The graphene lubricants were prepared using a constant-temperature magnetic stirring ultrasonic method. The lubricant preparation process is shown in Figure 1. Based on our previous study [31], oleic acid and stearic acid were selected as the graphene modifier ①, and the corresponding ratio was 9:10. Firstly, the mixture of oleic acid–stearic acid was placed in a beaker and heated to 80 °C; secondly, the modifier was obtained after stirring the mixture for 30 min; thirdly, a certain amount of graphene powder was weighed to prepare graphene lubricants of different concentrations. Subsequently, the Class I base oil 500SN without additives was added to the beaker as the lubricating liquid, and heated to 80 °C in a water bath with a thermostatic magnetic stirrer; then, the weighed graphene flakes were added into the beaker, and the magnetic stirring was started at a speed of 1200 rpm. The modifier was then added into the beaker at a constant speed through a micropipette and stirred for 60 min. After that, the solution obtained above was slowly moved into the ultrasonic two-dimensional material stripper for ultrasonic treatment for 30 min. Finally, a homogeneously dispersed and stable graphene lubricant was obtained [31].

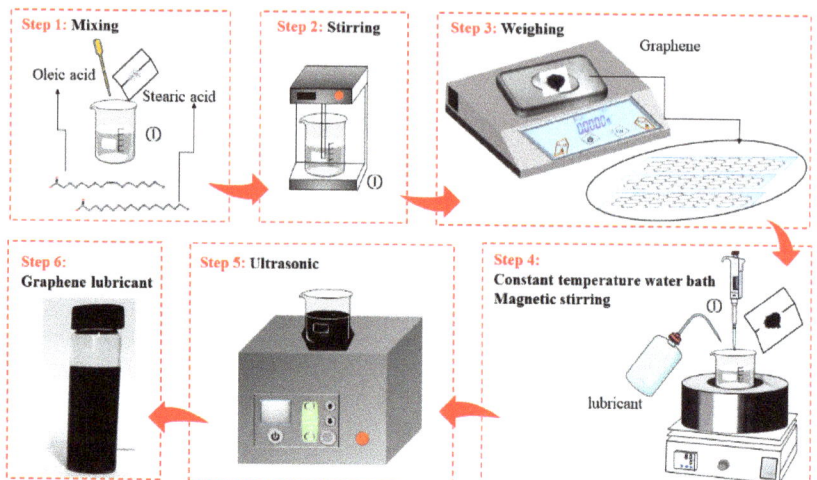

Figure 1. Flow chart of the overall preparation of graphene lubricant.

3. Friction and Wear Experiment

In this paper, we aim to explore the lubricating properties of graphene lubricants at different concentrations and temperatures experimentally. For this purpose, we would like to first introduce the experimental method and conditions that we used in this section.

3.1. Experimental Method

The experimental study was carried out on a lever-type four-ball friction and wear-testing machine (MR-S10G, Jinan Hengke Test Equipment Co., Ltd., Shandong, China, as shown in Figure 2), which is composed of a spindle drive system, a weight lever system, an oil box and a heater, a friction force measurement system, an electrical control system, a strong current system, etc., all of which are installed on the frame with the base as the main body. The spindle drive system is responsible for transmitting the rotational speed. A lever system is used to adjust the applied load. In addition, a measurement system is used to collect the coefficient of friction (COF). The COF is inversely proportional to the axial test force and proportional to the friction torque value. The obtained instantaneous COF is used to characterize the change in the whole experimental process, and the average COF is used to compare and analyze the globe friction characteristics under different conditions. The friction test was conducted using ASTM D5183-95 (1999) standard, from SH/T0762-2005.

The friction samples were four steel balls with a diameter of 12.7 mm. During the test, three steel balls were clamped in 10 mL of running-in oil; the other one was used as the upper steel ball to make three-point contact with the three steel balls, as shown in Figure 2. In this study, a long grinding test was carried out; the spindle speed was 1200 r/min, the test time was 60 min, and the oil pool was lubricated. Before the test, the required lubricating oil was sonicated in an ultrasonic machine for 30 min to ensure a good dispersion of the graphene in the base oil. After the test, petroleum ether was used to clean the test steel ball several times, and place it in a cool place to air dry. After the friction test, the surface morphology was observed via a scanning electron microscope (SEM), and the wear scar WSD of the worn surface could be obtained to evaluate the lubricating effect of the lubricant.

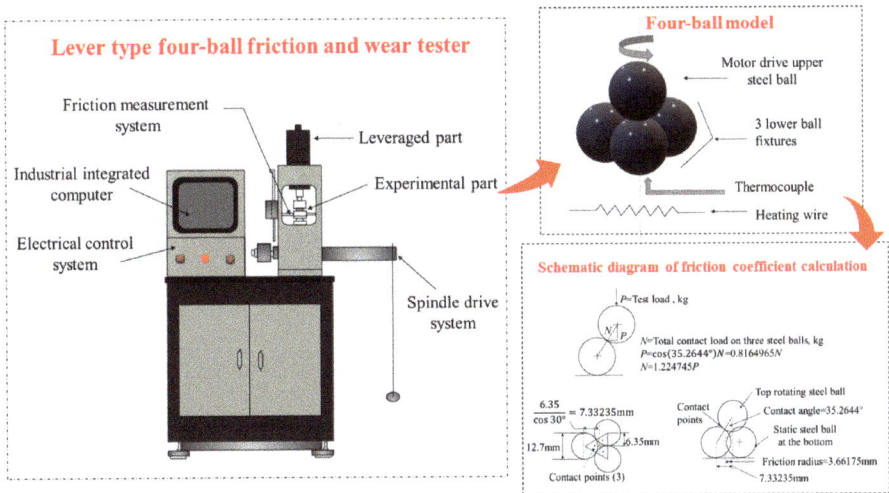

Figure 2. Structure and principle diagram of lever type-four-ball friction and wear testing machine.

3.2. Experimental Conditions

In this test, a 60 min long grinding test is carried out under the lubricated status. The specific experimental conditions are presented in Table 1. Since the effects of temperature and concentration are concerned in this study, four different temperature values and five different concentration values are chosen to compare the results. There are 20 groups of experiments, each with new steel balls. In addition, each experiment is repeated three times. Before the test, the graphene lubricant is sonicated in an ultrasonic machine for 30 min to ensure a good dispersion effect. After the test, the tested steel balls are cleaned several times with the petroleum ether and then dried in the shade.

Table 1. Experimental conditions of friction and wear tests.

Factors	Values
Load/N	392
Rotating speed/rpm	1200
Temperature/°C	25, 100, 150, 200
Concentration/wt%	0, 0.030, 0.035, 0.040, 0.050

4. Results and Discussion

In order to study the tribological behavior of the friction pair within lubricants, friction experiments were carried out under the same load of 392 N. In this section, the COF, surface morphology and 3D topography of the balls were tested to reveal the effects of temperature and concentration.

4.1. Tribological Characteristics

Figure 3 shows a set of typical COFs under different temperature and concentration conditions in one testing cycle. It can be observed that the COF curves display a running-in stage at the beginning of the test, which lasts about 1000 s. The COF in pure oil is almost higher than all of the other concentrations, except for the running-in stage in Figure 3d. During the running-in stage, the COF dramatically increases, decreases or fluctuates with time due to the change in friction surface topography. After the running-in stage, the COF becomes relatively stable. Since the variation trends of COF in one testing cycle may lead to incorrect conclusions, it is hard to tell the effects of concentration on the COF under different temperature conditions. In order to obtain more reliable data and reduce the occasionality, the experiments are repeated for three times.

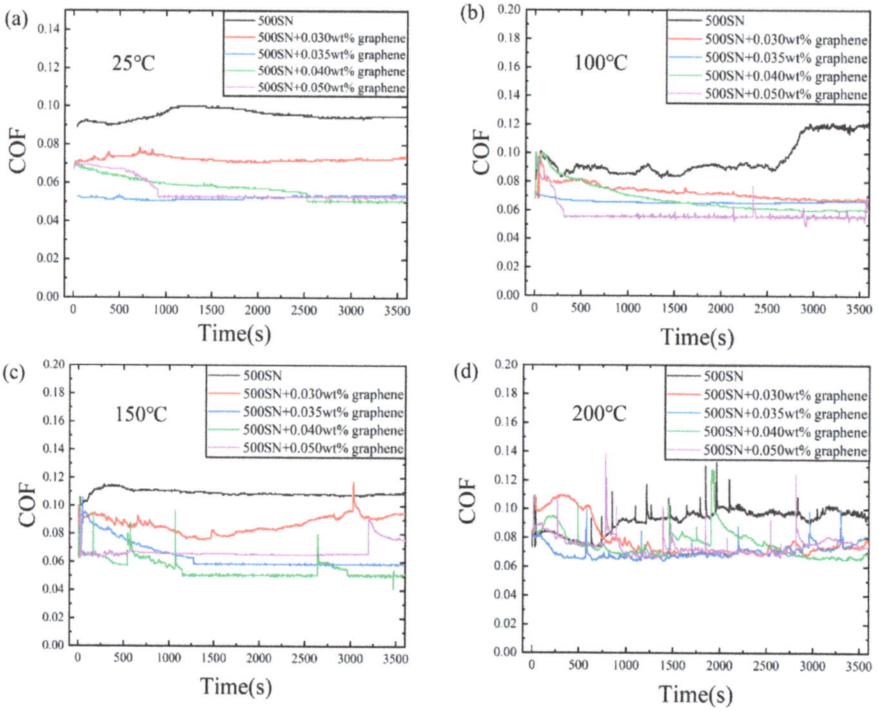

Figure 3. Instantaneous COF with different concentrations and temperatures: (**a**) 25 °C, (**b**) 100 °C, (**c**) 150 °C, (**d**) 200 °C.

Even though the COF undergoes fluctuations during some of the tests, it is stable most of the time. According to some relevant references, the average COF is commonly used to interpret the friction characteristics [6,32,33]. Therefore, the average COF for the whole run is also adopted here to show the effects of temperature and concentration. In Table 2, the time-average values of the COF during three tests are listed in the columns Test 1, Test 2 and Test 3. $\bar{f}$ is the mean value of the time-average COFs and σ denotes the standard deviation of the time-average COFs. The standard deviation of the test results is located in the range of 0.0003 to 0.0065 under 20 groups of working conditions. In this way, it is sufficient to prove the reproducibility of the experiment.

Figure 4 shows the mean COFs and the error bars under different temperature and concentration conditions. In terms of graphene concentration, it can be seen that not only are the mean COFs of the pure oil much higher than those of the graphene lubricants, but also their standard deviations are much larger than those of the graphene lubricants. It means that graphene lubricants make the average COF more stable than the pure oil. With the given temperature, the general trend is that, when the graphene concentration increases, the mean COF firstly decreases rapidly and then increases slowly. When the temperature is 25 °C, 0.035 wt% graphene lubricant has the lowest mean COF of 0.05170, which is 40.8% lower than the pure lubricant. In terms of temperature, the mean COF generally grows as the temperature increases. With a higher concentration of 0.035 wt% to 0.05 wt% and a higher temperature of 150 °C and 200 °C, the mean COFs are very close to each other.

Table 2. The average COF values of repeated experiments.

Temperature (°C)	Concentrations (wt%)	Test 1	Test 2	Test 3	$\bar{f}$	σ
25	0	0.08007	0.09248	0.08931	0.08729	0.00645
	0.03	0.06319	0.07008	0.06858	0.06728	0.00362
	0.035	0.05263	0.05061	0.05187	0.05170	0.00102
	0.04	0.05561	0.05431	0.05477	0.05490	0.00066
	0.05	0.05910	0.05613	0.05749	0.05757	0.00149
100	0	0.09601	0.09411	0.10243	0.09752	0.00436
	0.03	0.07248	0.07818	0.0749	0.07519	0.00286
	0.035	0.05495	0.05530	0.05565	0.05530	0.00035
	0.04	0.05742	0.06222	0.06424	0.06129	0.00350
	0.05	0.06405	0.06537	0.06595	0.06512	0.00097
150	0	0.09721	0.10414	0.10555	0.10230	0.00446
	0.03	0.07827	0.08466	0.08112	0.08135	0.00320
	0.035	0.06726	0.06852	0.06918	0.06832	0.00098
	0.04	0.06810	0.07198	0.07089	0.07032	0.00200
	0.05	0.06909	0.06781	0.06781	0.06824	0.00074
200	0	0.10927	0.11823	0.10941	0.11230	0.00513
	0.03	0.08701	0.08598	0.08841	0.08713	0.00122
	0.035	0.07164	0.07364	0.07213	0.07247	0.00104
	0.04	0.07537	0.07453	0.07249	0.07413	0.00148
	0.05	0.07653	0.07766	0.07615	0.07678	0.00079

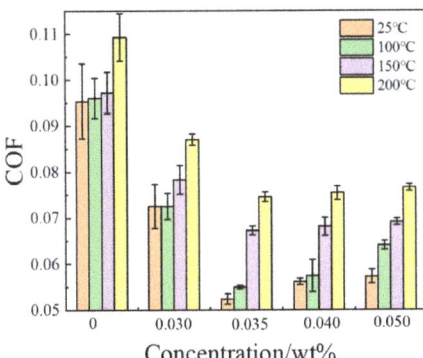

Figure 4. Average COFs with variation in temperature and concentrations.

4.2. Wear Properties

4.2.1. Surface Topography

After the tribological test, the wear extent of the upper steel ball is evaluated via SEM. The corresponding WSD is listed in Table 3 and plotted in Figure 5. It is notable that, although the WSD increases with temperature, the change in graphene concentration has little effect on the WSD at the same temperature. At 25 °C, the WSD of the steel ball surface with 0.035 wt% graphene lubricant is the lowest, which is 50.4% lower than that of the steel ball with pure oil.

Table 3. WSD under different conditions (μm).

Concentration/wt%	Temperature (°C)			
	25	100	150	200
0	835.22	1024.92	1050.06	1483.91
0.030	705.65	947.54	1109.02	1159.07
0.035	414.50	988.30	1094.76	1338.59
0.040	467.75	977.10	1131.56	1369.45
0.050	595.66	870.73	1114.49	1282.68

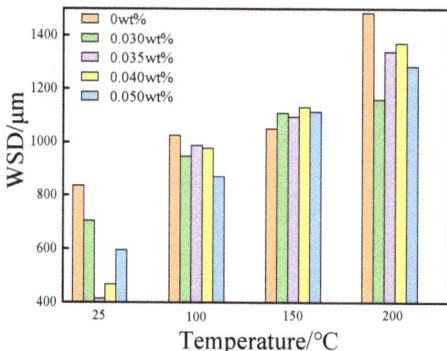

Figure 5. WSD under different conditions.

Figure 6 shows the wear morphology of the steel ball with different concentrations of graphene lubricant. It can be clearly seen in Figure 6a that the worn surface of the lubricant has many deep wear marks, and that the obvious grooves are large. After adding graphene, the WSD of the steel ball becomes smaller and the wear scar becomes shallower, contributing to a smoother surface. Accordingly, the worn surface is effectively separated by the friction reaction film during the friction process. The graphene lubricant film effectively inhibits the asperity summits shearing and the adhesive wear on the friction surface.

The transferred graphene particles are stacked on the surface, as shown by the red dashed circles in Figure 6b,c. There are two-body abrasive wear and three-body abrasive wear on the worn surfaces. In two-body wear, hard abrasive particles are fixed on the surface layer on one side of the friction pair. Furrowing occurs under the effect of shear force, and obvious scratches or grooves are produced on the worn surface. External abrasive particles move between the two friction surfaces. In three-body wear, hard abrasive particles are captured by two relatively moving surfaces, but they are still in a loose state. The damage is manifested by frictional irregular bite marks, pits, and a small amount of scratches and grooves on the subsurface.

Figure 7 depicts the morphology of adhesive wear on the steel surface when exposed to a high temperature. During the initial running-in process, the wear surface undergoes shear fracture, leading to the detachment of the sheared material and the formation of wear debris, resulting in adhesive wear. As the temperature increases, the width of the wear scar gradually expands, exacerbating the severity of wear. Specifically, at 200 °C as shown in Figure 7a, the wear scar exhibits a pronounced irregular shape, accompanied by the significant accumulation of graphene. When the COF is high and unstable, most of the graphene layers are worn or removed from the wear track, indicating the presence of graphite flakes near the track. This suggests that the graphene undergoes structural disorder due to modifications during the sliding process. The adhesion between surfaces leads to shear fracture and the subsequent detachment of material, giving rise to adhesive wear.

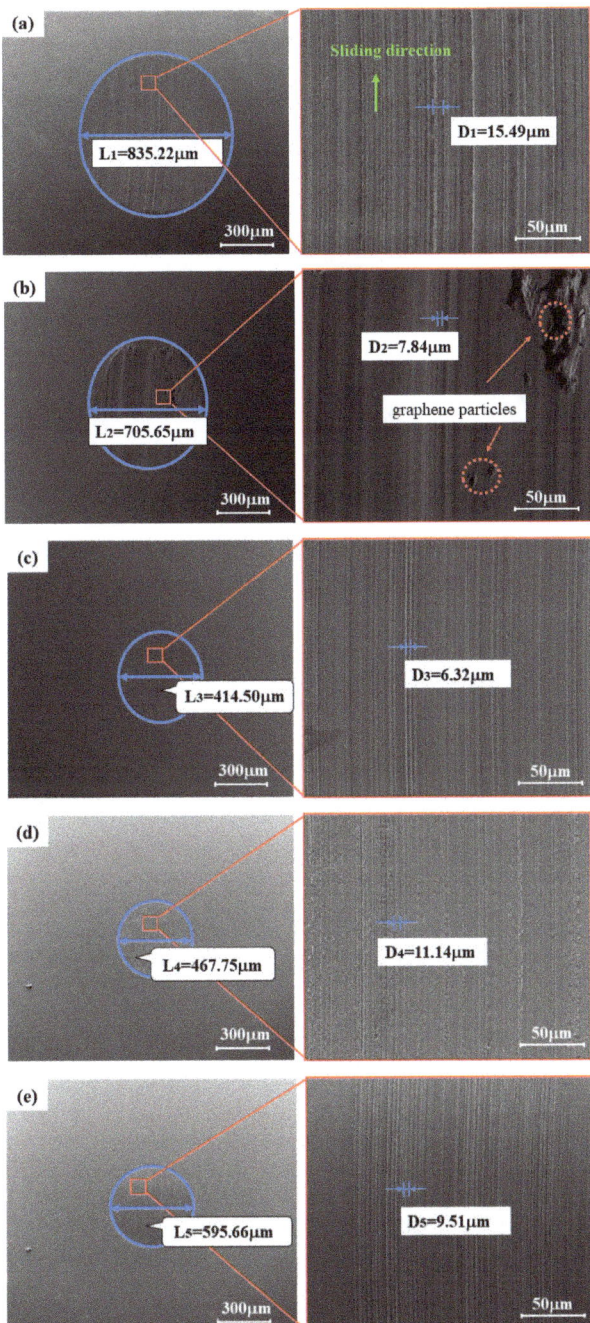

Figure 6. SEM images of a rigid ball with different graphene lubricants. (**a**) Pure oil; (**b**) 0.030 wt% graphene lubricant; (**c**) 0.035 wt% graphene lubricant; (**d**) 0.040 wt% graphene lubricant; (**e**) 0.050 wt% graphene lubricant.

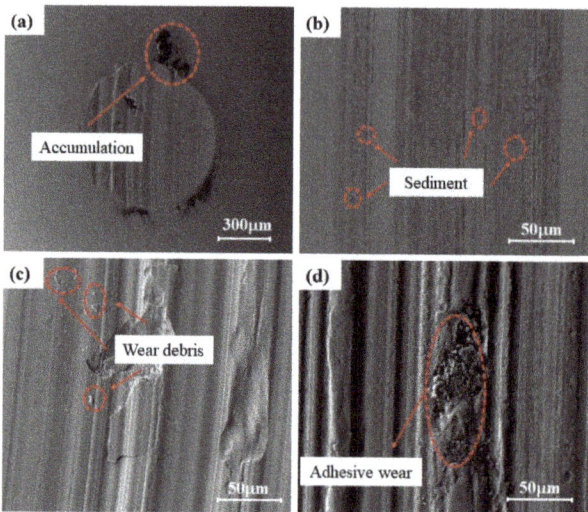

Figure 7. Adhesive wear surface morphology under a high temperature (200 °C): (**a**) Graphene accumulation, (**b**) Surface sediment, (**c**) Surface debris, (**d**) Adhesive wear.

The increase in temperature also causes a decline in the anti-wear and anti-friction properties of graphene lubricants and weakens their resistance to high temperatures. As the temperature increases, the oxidation reaction of the lubricant intensifies. Graphene plays a catalytic role in lubricating oil oxidation as a solid additive and accelerates its oxidation. Lubricants undergo polymerization and shear fragmentation with graphene that result in contamination and deterioration. This effect leads to an increase in lubricant viscosity, a decrease in the lubrication performance and severe wear. Consequently, the lubrication state between friction pairs evolves from elastohydrodynamic lubrication to mixed lubrication. Hence, it is imperative to enhance the high-temperature resistance of graphene lubricants.

In summary, the above analysis demonstrates that the thermal stability of the graphene oil film surpasses that of the base oil film, owing to the inhibited oxidation of graphite at high temperatures. This finding aligns with the exceptional tribological performance exhibited by the prepared coating under high-temperature conditions, as depicted in Figure 8. The energy spectrum of the worn surface when subjected to pure oil and graphene lubricating oil illustrates the effective integration of graphene into the friction interface, its ability to fill pits, and its wear-reducing properties.

As the temperature of the friction surface increases, the mechanical activation energy increases. As the local temperature and stress on the surface of the friction pair continue to increase, some nanoparticles firstly interact. The surface micro asperities of the friction pair produce nanometers and nanometers with a low COF, forming a chemical film with a low shear strength. In addition, nano-oxidation and slow oxidation form dense nano-films, whose COF is also very low. They form a chemical reaction film with nanometers and nanometers, and form an extreme pressure anti-wear repair layer together with the physical adsorption film, and the repair layer changes friction. The contact method of the secondary surface inhibits the furrowing, adhesion and fatigue wear of the friction pair surface, and improves the bearing capacity and anti-wear and anti-friction performance of the friction pair. Under higher loads, the friction pair is in a mixed lubrication state of fluid lubrication and boundary lubrication.

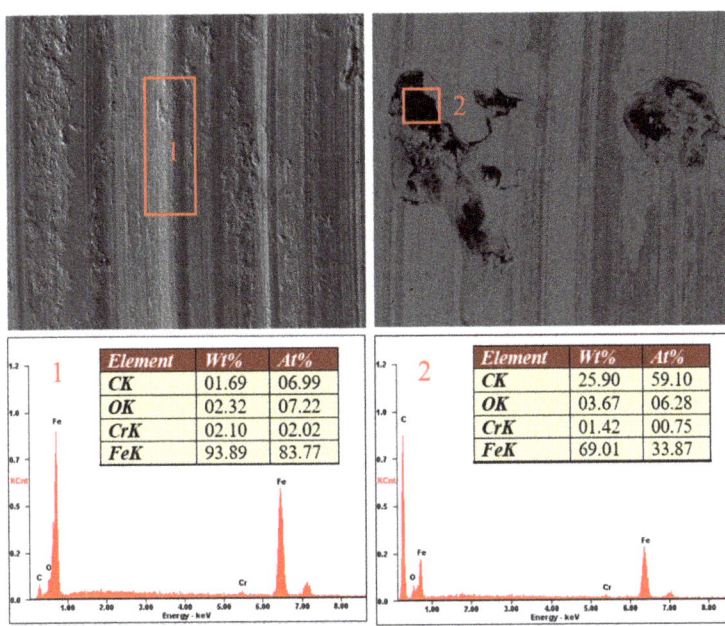

Figure 8. EDS of rigid sphere under the action of pure oil and graphene lubricants.

The addition of nanoparticles improves the anti-friction performance of the material, but does not improve the anti-wear performance of the material. This is because nanoparticles increase internal friction. In the later stage of violent friction, due to long-term friction, the increase in the surface temperature of the friction pair will reactivate the nanoparticles deposited on the wear marks and pits of the friction pair, and transfer to the surrounding environment along the surface with the friction. Under this action, the tribochemical reaction will eventually repair and smooth the worn surface.

Consequently, the increase in temperature leads to a gradual decrease in the anti-wear and anti-friction abilities of the graphene lubricant. Nevertheless, the graphene lubricant film still has a better thermal stability than the lubricant film, which is attributed to the oxidation inhibition of graphite at high temperatures. This is also consistent with the excellent tribological behavior of the as-prepared coatings at high temperatures.

4.2.2. The 3D Topography

To further compare the tribological properties of different concentrations of graphene lubricant at different temperatures, Figure 9 shows the three-dimensional image and surface roughness of the worn surface obtained using a white light interferometer. The wear depth data are shown in Figure 10. With the increase in the graphene concentration, the Ra and Rq of the worn surface first decrease and then increase, which is consistent with the change in WSD. As the surface roughness of the material increases, the contact area of the rough surface increases, leading to an increase in the COF. At different temperatures, the graphene lubricant has smaller wear marks (smaller wear volume and wear depth) than the pure oil. Adding the graphene, the morphology of the wear surface of the matrix changes obviously. Compared with pure oil, the addition of graphene can effectively reduce parallel grooves in the sliding direction, resulting in a smoother wear morphology. However, there are some gaps on the worn surface due to the abrasive wear effect of graphene as a hard particle.

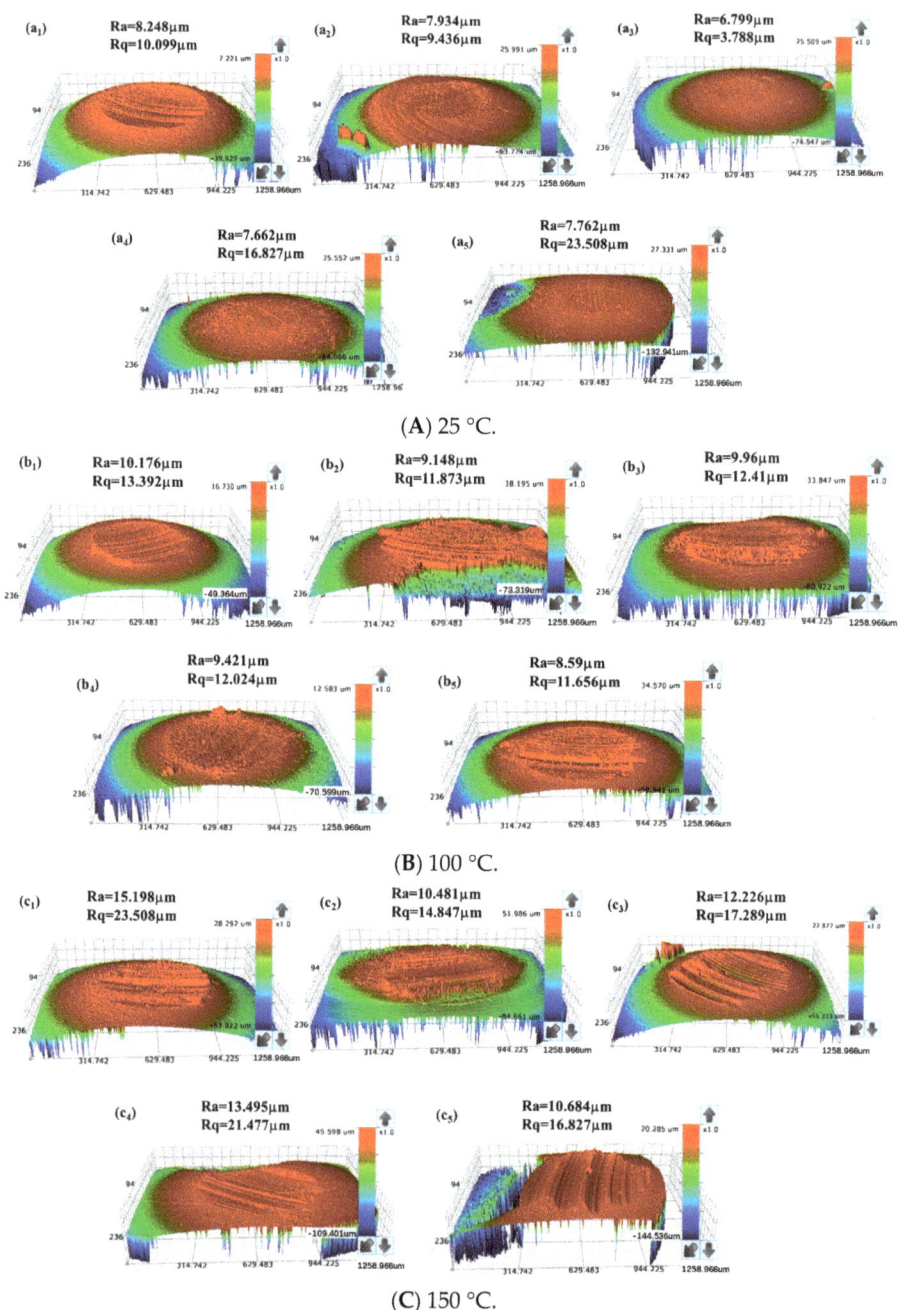

Figure 9. The 3D surface profiles and surface roughness with different graphene lubricants. (**A**) 25 °C, (**B**) 100 °C, (**C**) 150 °C.

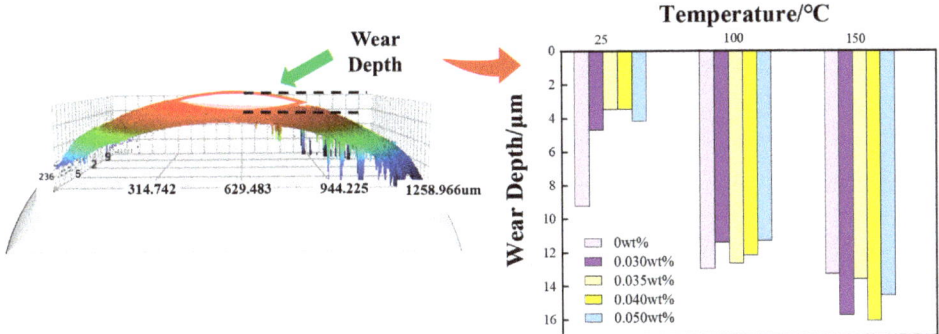

Figure 10. Surface wear depth with different graphene lubricants.

As the temperature increases, the number of asperities on the worn surface increases, and the morphology basically shows a sharp peak state. From Figure 10, it can be seen that, when the surface roughness is small, the local film thickness between the rough surfaces is thinner, and the fluid dynamic pressure is higher. The colors in the left figure represent the height of the ball, where red color represents the highest. Most of the normal loads are borne by the lubricating oil film, while the normal loads borne by the asperity are smaller. Therefore, the contact pressure of the asperity is lower, and the number of asperity contacts is smaller, as well as the actual contact area. Due to the small contact area of the asperity at this time, most of the friction heat is generated by the shear oil film. When the surface roughness is large, the local film thickness between rough surfaces increases, and the hydrodynamic pressure effect decreases. The normal load borne by the lubricating oil film decreases, while the normal load borne by the asperity increases. Therefore, the contact pressure of the asperity is higher, and the number of asperity contacts is larger, as well as the actual contact area. This is due to the high frictional heat generated by shear asperity contact at this moment.

4.3. Discussion

According to the experimental results of the COF and wear in Sections 3.1 and 3.2, the mechanisms can be assumed. Figure 11 shows a schematic drawing of the lubricating characteristics of graphene lubricants under different conditions, including the effects of graphene concentration and temperature. It is known that the microscopic topography of the contact surface consists of a large number of asperity summits and pits. On the surface of the asperity summits, the graphene particles and polar molecules in the lubricant act together as a physical adsorption film. At the onset of friction, since the density of the particles is larger than that of the lubricant, homogeneously dispersed graphene particles are rapidly deposited in the pits under the action of hydrodynamic pressure. When the mass fraction of graphene increases, the COF tends to be stable; this is because the lubricant film formed is sufficient to withstand the shear stress under the test conditions. When the graphene lubricant concentration continues to increase, it causes graphene to accumulate on the surface of the friction pair, so that the oil film of the lubricant cannot be formed normally, resulting in an increase in the COF. Therefore, further increasing the additive amount will not further improve the tribological properties. This could be the reason that the graphene lubricant with a 0.035% concentration has a good performance regarding the tested steel–steel friction pairs.

Graphene nanoparticles and fine iron abrasive particles will drop into the wear scars and fill the pits after lubrication to reduce wear. After adding a certain amount of graphene to the base oil, the oil film can still play an important role in the friction and wear process. At low concentrations, graphene particles play a certain lubricating role and can repair rough surfaces. At an appropriate concentration, the graphene particles have the best lubricating effect, and the rough surface can be repaired to be smoother. However, graphene particles at high concentrations will agglomerate in a large area, and the agglomerated particles will act as wear debris to intensify wear.

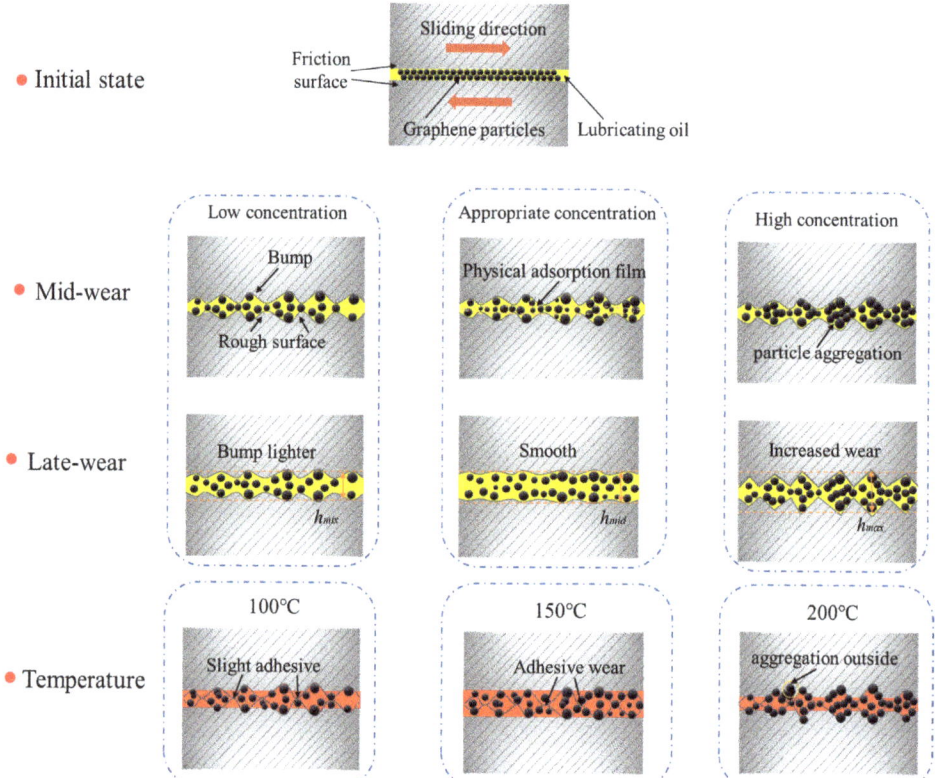

Figure 11. Schematic drawing of graphene lubricant under different conditions.

The effect of temperature on the friction performance is the change in the lubrication state, from hydrodynamic lubrication to boundary lubrication and even dry friction. This causes changes in the surface structure during the friction process; that is, there are friction changes between the friction surface and the surrounding medium, such as diffusion, adsorption or desorption between surface atoms or molecules, surface structure changes and phase transitions. The phenomenon in Figure 3d is caused by the increase in the thermal motion of atoms with increased temperature, resulting in a transition of the lubricating state. The transformation from hydrodynamic lubrication to boundary lubrication or even dry friction causes changes in the surface layer structure during the friction process. Additionally, with the increase in temperature, the adhesive wear of the worn surface intensifies, and some graphene particles agglomerate around the wear scar.

5. Conclusions

The results of the friction and wear experiments show that graphene lubricants have excellent anti-wear and anti-friction abilities. Compared with the pure oil, the COF and WSD of steel balls under the 0.035 wt% graphene lubricant decreased by 40.8% and 50.4%. After micro three-dimensional surface analysis, the lubrication characteristics were attributed to the synergistic effect of graphene particles on the wear surface to protect and repair the friction surface. The adsorbed lubricating film inhibits furrowing, adhesion, and fatigue wear on the friction pair's surface, improving the bearing capacity, as well as the anti-wear and friction-reducing effects. It shows that graphene has broad application prospects as a lubricant additive.

Author Contributions: Conceptualization, Y.D. and Y.L.; methodology, Y.D.; software, Q.Z.; validation, Y.D., C.X. and Y.L.; formal analysis, Y.D.; investigation, Y.D.; resources, B.M. and C.X.; data curation, Y.D.; writing—original draft preparation, Y.D.; writing—review and editing, C.X.; visualization, Y.D.; supervision, B.M. and C.X.; project administration, B.M. and C.X.; funding acquisition, B.M. and C.X. All authors have read and agreed to the published version of the manuscript.

Funding: This research was funded by the National Natural Science Foundations of China, grant number [52175037 and 51805289] and the APC was funded by [51805289].

Data Availability Statement: Data are contained within the article.

Conflicts of Interest: The authors declare no conflict of interest.

References

1. Zhao, J.; Huang, Y.; He, Y.; Shi, Y. Nanolubricant additives: A review. *Friction* **2021**, *9*, 891–917. [CrossRef]
2. Li, Y.; Yang, R.; Hao, Q.; Lei, W. Tribological Properties of the Functionalized Graphene/Montmorillonite Nanosheets as a Lubricant Additive. *Tribol. Lett.* **2021**, *69*, 117. [CrossRef]
3. Cyriac, F.; Yi, T.X.; Poornachary, S.K.; Chow, P.S. Boundary lubrication performance of polymeric and organic friction modifiers in the presence of an anti-wear additive. *Tribol. Int.* **2022**, *165*, 107256. [CrossRef]
4. Yin, Y.; Lei, H.; Song, J.; Zhao, G.; Ding, Q. Molecular Dynamics Simulation on the Tribological Properties of Polytetrafluoroethylene Reinforced with Modified Graphene. *Tribology* **2022**, *42*, 598–608. [CrossRef]
5. Zhao, X.; Tian, C.; Hao, L.; Xu, H.; Dong, J. Tribology and Rheology of Polypropylene Grease with MoS_2 and ZDDP Additives at Low Temperatures. *Lubricants* **2023**, *11*, 464. [CrossRef]
6. Ali, M.K.A.; Hou, X. Exploring the lubrication mechanism of CeO_2 nanoparticles dispersed in engine oil by bis(2-ethylhexyl) phosphate as a novel anti-wear additive. *Tribol. Int.* **2022**, *165*, 107321. [CrossRef]
7. Ares, P.; Novoselov, K.S. Recent advances in graphene and other 2D materials. *Nano Mater. Sci.* **2022**, *4*, 3–9. [CrossRef]
8. Li, Z.; Gao, C.; Zhao, H. Porous biomass-derived carbon modified by Cu, N co-doping and Cu nanoparticles as high-efficient electrocatalyst for oxygen reduction reaction and zinc-air battery. *J. Alloys Compd. Interdiscip. J. Mater. Sci. Solid-State Chem. Phys.* **2022**, *897*, 163175. [CrossRef]
9. Zhang, K.Y.; Yin, Y.G.; Zhang, G.T.; Ding, S.G.; Chen, Q. Tribological Properties of FeS/Cu Copper-Based Self Lubricating Bearing Materials Prepared by Mechanical Alloying. *Tribol. Trans.* **2020**, *63*, 197–204. [CrossRef]
10. Hu, Z.; Chen, J.; Pan, P.; Liu, C.; Zeng, J.; Ou, Y.; Qi, X.; Liang, T. Porous N-doped $Mo_2C@C$ nanoparticles for high-performance hydrogen evolution reaction. *Int. J. Hydrogen Energy* **2022**, *47*, 4641–4652. [CrossRef]
11. Huai, W.J.; Zhang, C.H.; Wen, S.Z. Graphite-based solid lubricant for high-temperature lubrication. *Friction* **2021**, *9*, 1660–1672. [CrossRef]
12. Jamel, R.S.; Al-Murad, M.A.; Alkhalidi, E.F. The efficacy of reinforcement of glass fibers and ZrO_2 nanoparticles on the mechanical properties of autopolymerizing provisional restorations (PMMA). *Saudi Dent. J.* **2023**, *35*, 707–713. [CrossRef]
13. Qi, H.; Zhang, G.; Zheng, Z.; Yu, J.; Hu, C. Tribological properties of polyimide composites reinforced with fibers rubbing against Al_2O_3. *Friction* **2020**, *9*, 301–314. [CrossRef]
14. Wang, F.J. Spherical-shaped CuS modified carbon nitride nanosheet for efficient capture of elemental mercury from flue gas at low temperature. *J. Hazard. Mater.* **2021**, *415*, 125692. [CrossRef]
15. Qian, Y.T. Facile synthesis of sub-10 nm ZnS/ZnO nanoflakes for high-performance flexible triboelectric nanogenerators. *Nano Energy* **2021**, *88*, 106256. [CrossRef]
16. Soetaredjo, F.E.; Santoso, S.P.; Waworuntu, G.; Darsono, F.L. Cellulose Nanocrystal (CNC) Capsules from Oil Palm Empty Fruit Bunches (OPEFB). *Biointerface Res. Appl. Chem.* **2022**, *12*, 2013–2021. [CrossRef]
17. Liu, J.P.; Zhang, H.R.; Yan, Q.L. Anti-sintering behavior and combustion process of aluminum nano particles coated with PTFE:A molecular dynamics study. *Def. Technol.* **2023**, *24*, 46–57. [CrossRef]
18. Su, Y.; Li, Y.F.; Gong, S.G.; Song, Y.H.; Li, B.; Wu, X.L.; Zhang, J.P.; Liu, D.T.; Shao, C.L.; Sun, H.Z. Graphene wrapped $TiO_2@MoSe_2$ nano-microspheres with sandwich structure for high-performance sodium-ion hybrid capacitor. *Appl. Surf. Sci. J. Devoted Prop. Interfaces Relat. Synth. Behav. Mater.* **2023**, *610*, 155494. [CrossRef]
19. Wang, G.; Liu, X.B.; Zhu, G.X.; Zhu, Y.; Liu, Y.F.; Zhang, L.; Wang, J.L. Tribological study of $Ti_3SiC_2/Cu_5Si/TiC$ reinforced Co-based coatings on SUS304 steel by laser cladding. *Surf. Coat. Technol.* **2022**, *432*, 128064. [CrossRef]
20. Liu, L.; Zhou, M.; Mo, Y.; Bai, P.; Wei, Q.; Jin, L.; You, S.; Wang, M.; Li, L.; Chen, X.; et al. Synergistic lubricating effect of graphene/ionic liquid composite material used as an additive. *Friction* **2021**, *9*, 1568–1579. [CrossRef]
21. Kaleli, E.H.; Demirtas, S. Experimental investigation of the effect of tribological performance of reduced graphene oxide additive added into engine oil on gasoline engine wear. *Lubr. Sci.* **2023**, *35*, 118–143. [CrossRef]

22. Sun, S.; Ru, G.; Qi, W.; Liu, W. Molecular dynamics study of the robust superlubricity in penta-graphene van der Waals layered structures. *Tribol. Int.* **2023**, *177*, 107988. [CrossRef]
23. Zhang, Y.; Tai, X.; Zhou, J.; Zhai, T.; Xu, L.; Diao, C.; Xie, X.; Hou, C.; Sun, X.; Zhang, X.; et al. Enhanced high-temperature thermal conductivity of the reduced graphene oxide@SiO$_2$ composites synthesised by liquid phase deposition. *Ceram. Int.* **2022**, *48*, 8481–8488. [CrossRef]
24. Wang, H.; Bai, Q.; Chen, S.; Dou, Y.; Guo, W. Nanoscale mechanism of suppression of friction and wear of the diamond substrate by graphene. *Mater. Today Commun.* **2022**, *33*, 104894. [CrossRef]
25. Chen, G.Y.; Zhao, J.; He, Y.Y.; Luo, J.B. Synthesis and Structure Control of Graphene Lubricant Additives. *Tribology* **2021**, *41*, 758–772. [CrossRef]
26. Saufi, M.A.; Mamat, H. Comparison of dispersion techniques of graphene nanoparticles in polyester oil. *Mater. Today Proc.* **2022**, *66*, 2747–2751. [CrossRef]
27. Sarath, P.S.; Reghunath, R.; Thomas, S.; Haponiuk, J.T.; George, S.C. An investigation on the tribological and mechanical properties of silicone rubber/graphite composites. *J. Compos. Mater.* **2021**, *55*, 002199832110316. [CrossRef]
28. Ci, X.; Zhao, W.; Luo, J.; Wu, Y.; Ge, T.; Xue, Q.; Gao, X.; Fang, Z. How the fluorographene replaced graphene as nanoadditive for improving tribological performances of GTL-8 based lubricant oil. *Tribology* **2021**, *9*, 488–501. [CrossRef]
29. Lau, G.A.; Neves, G.O.; Salvaro, D.B.; Binder, C.; Klein, A.N.; de Mello, J.D. Stability and Tribological Performance of Nanostructured 2D Turbostratic Graphite and Functionalised Graphene as Low-Viscosity Oil Additives. *Lubricants* **2023**, *11*, 155. [CrossRef]
30. Zhang, C.; Zhang, X.; Zhang, W.; Zhao, Z.; Fan, X. Functionalized Graphene from Electrochemical Exfoliation of Graphite toward Improving Lubrication Function of Base Oil. *Lubricants* **2023**, *11*, 166. [CrossRef]
31. Liu, Y.; Dong, Y.; Zhang, Y.; Liu, S.; Bai, Y. Effect of different preparation processes on tribological properties of graphene. *Nanomater. Nanotechnol.* **2020**, *10*, 184798042094665. [CrossRef]
32. Jiang, Z.; Yang, G.; Zhang, Y.; Gao, C.; Ma, J.; Zhang, S.; Zhang, P. Facile method preparation of oil-soluble tungsten disulfide nanosheets and their tribological properties over a wide temperature range. *Tribol. Int.* **2019**, *135*, 287–295. [CrossRef]
33. Han, Z.; Gan, C.; Li, X.; Feng, P.; Ma, X.; Fan, X.; Zhu, M. Electrochemical preparation of modified-graphene additive towards lubrication requirement. *Tribol. Int.* **2021**, *161*, 107057. [CrossRef]

Disclaimer/Publisher's Note: The statements, opinions and data contained in all publications are solely those of the individual author(s) and contributor(s) and not of MDPI and/or the editor(s). MDPI and/or the editor(s) disclaim responsibility for any injury to people or property resulting from any ideas, methods, instructions or products referred to in the content.

Article

Effect of Argon Flow Rate on Tribological Properties of Rare Earth Ce Doped MoS$_2$ Based Composite Coatings by Magnetron Sputtering

Changling Tian [1], Haichao Cai [1], Yujun Xue [1,2,3,*], Lulu Pei [1,2,3] and Yongjian Yu [1,2,3]

1. School of Mechatronics Engineering, Henan University of Science and Technology, No. 48 Xiyuan Street, Luoyang 471003, China
2. Henan Key Laboratory for Machinery Design and Transmission System, Henan University of Science and Technology, No. 48 Xiyuan Street, Luoyang 471003, China
3. Longmen Laboratory, No. 1 Keji Street, Luoyang 471000, China
* Correspondence: yjxue@haust.edu.cn; Tel.: +86-379-64278961

Citation: Tian, C.; Cai, H.; Xue, Y.; Pei, L.; Yu, Y. Effect of Argon Flow Rate on Tribological Properties of Rare Earth Ce Doped MoS$_2$ Based Composite Coatings by Magnetron Sputtering. *Lubricants* 2023, 11, 432. https://doi.org/10.3390/lubricants11100432

Received: 30 July 2023
Revised: 8 September 2023
Accepted: 29 September 2023
Published: 7 October 2023

Copyright: © 2023 by the authors. Licensee MDPI, Basel, Switzerland. This article is an open access article distributed under the terms and conditions of the Creative Commons Attribution (CC BY) license (https://creativecommons.org/licenses/by/4.0/).

Abstract: Exploring the doping components of the coating is of great significance for improving the tribological properties of the MoS$_2$-based coating. The optimization of magnetron sputtering process parameters can also improve the coating quality. In this paper, the effects of working gas flow rate on the microstructure in a vacuum chamber, nano-hardness, and tribological properties of Ce-Ti/MoS$_2$ coatings were studied using DC and RF unbalanced co-sputtering technology. It is found that the coating structure was coarse and porous when the Ar flow rate was excessive (70 sccm), significantly affecting the mechanical properties; there are pit defects on the surface of the coating when the flow rate is just minor (30 sccm), and the coating easily falls off during the friction process. When the flow rate is 40~60 sccm, the coating grows uniformly, the hardness reaches 7.85 GPa at 50 sccm, and the wear rate is only 4.42×10^{-7} mm^3 N^{-1} m^{-1} at 60 sccm. The coating doped with Ce and Ti is an approximate amorphous structure. Under appropriate gas flow rate conditions, the friction induces a transfer film with a layered structure, and the MoS$_2$ (002) crystal plane orientation is arranged in parallel at the edge of the wear debris, effectively reducing the shear force during sliding and reducing wear. Based on rare earth doping, this study improves the tribological properties by optimizing the working gas parameters, which plays a reference role in preparing high-quality MoS$_2$-based coatings.

Keywords: MoS$_2$-based coating; rare element doping; tribology performance; working gas flow

1. Introduction

Transition metal dichalcogenides (TMDs) materials have a sandwich-layered structure. The adjacent two layers are connected by weak van der Waals, which reduces the shear strength between the layers [1,2]. It can be expressed as MX$_2$ (M = Mo, W; X = S, Se). Among them, MoS$_2$ coating can be used for long-term and stable use under harsh conditions such as high load, high-low temperature, and high vacuum, and is widely used in critical moving parts in ground and space environments. However, the mechanical properties of pure MoS$_2$ coatings are poor, and the loose structure is easy to combine with water and oxygen to form MoO$_3$ [3]. Therefore, MoS$_2$ coatings have significant performance differences in different application environments.

To solve the problem of low bearing capacity and limited application range of pure MoS$_2$ coating, component doping of MoS$_2$-based coating is a typical improvement method. At present, various metal and non-metal elements, such as C, N, Cu, Si, Ta, Ag, Zr, and Ti [4–11], as well as compounds such as TiN and Al$_2$O$_3$ [12,13], have been studied for doping MoS$_2$ coatings to improve wear resistance. Among them, metal doping can induce the growth of MoS$_2$ coating with (002) preferred orientation, and the coating structure is more compact, improving the film's wear resistance. On the other hand, due to the preferential oxidation of some metal elements, the oxidation of MoS_2 on the contact surface and the destruction of the layered structure are prevented, and the oxidation resistance of the MoS_2 coating is improved [14].

In addition, as a highly anisotropic material, MoS_2 is very sensitive to the preparation process parameters. The process parameters of MoS_2 composite coating prepared by unbalanced magnetron sputtering, such as deposition pressure, sputtering target power, workpiece bias, and gas flow rate, can affect the crystal orientation and coating density of MoS_2 and then change its tribological properties, mechanical properties, and catalytic properties [15–17]. Therefore, improving the process is also essential to improving the performance of MoS_2-based coatings. Kokalj et al. [18] studied the structural changes of MoS_x films under pulse frequency and other parameters. They found that low pulse frequency shortened the Mo-S bond length, which could significantly improve the hardness and elastic stiffness of the film. Tillmann et al. [19] used HiPIMS to prepare MoS_x films and studied the influence of bias voltage and deposition pressure parameters. It was found that a low deposition rate was beneficial to the growth of the (002) basal plane, while high kinetic energy caused by high bias voltage promoted the growth of the (100) basal plane of the coating. These studies have analyzed the influence of parameters on the coating structure and application performance from the perspective of sputtering particle kinetic energy. In the chamber of PVD coating, in addition to sputtering power, the pressure and flow rate of working gas are also the main factors affecting the energy of sputtering particles. At present, the research on gas flow rate focuses on the process of reactive magnetron sputtering, involving the influence of the flow rate ratio of reactive gas and argon on the reaction products of the coating [20,21]. At the same time, there are few studies on the flow rate of a single working gas.

Nowadays, the research of doped metals in surface engineering has been widely involved in rare earth elements. They have the effect of regulating optical and magnetic properties [22–24]. In addition to the thermal barrier function [25,26], their oxidation products are also used to regulate grain growth and slow down sliding wear [27]. For MoS_2-based coatings, there are few studies on the improvement of tribological properties of rare earth multi-doped coatings, and the influence of process improvement, especially working gas flow, is also rare. Therefore, in this paper, binary doped Ce-Ti/MoS_2 coatings were prepared by co-sputtering Ce-Ti alloy and MoS_2 using unbalanced magnetron sputtering equipment. The effects of argon flow rate on the structure, mechanical properties, and tribological properties of the coating were studied, which provided ideas for improving the preparation process of MoS_2-based coating and the wear resistance of the coating.

2. Materials and Methods

2.1. Coatings Preparation

Ce-Ti/MoS_2 composite coatings were deposited by an unbalanced magnetron sputtering system (JGP045CA, SKY, Shenyang, China). MoS_2, Cr, and Ce-Ti alloy(1:1) targets were used as sputtering targets. Monocrystalline silicon was used to test composite coatings' mechanical properties and section images; 9Cr18 steel was used to test the friction, wear properties, and surface morphology of composite coatings. The 9Cr18 steel is polished and cleaned to ensure a surface roughness of Ra $\leq$ 0.2 μm. Before sputtering, a 10 min Cr transition layer was deposited on the substrate before Ce-Ti/MoS_2 composite coatings to improve the adhesion of Ce-Ti/MoS_2 composite coatings. Then, Ce-Ti/MoS_2 composite coatings were deposited at different Argon flow rates by co-sputtering the Ce-Ti alloy target and MoS_2 target. The specific process parameters are shown in Table 1.

Table 1. Preparation parameters of each sample.

Argon Flow Rate (sccm)	MoS$_2$ Target Power (W)	Ce-Ti Target Power (W)	Cr Target Power (W)	Deposition Time of Composite Film (min)
30, 40, 50, 60, 70	250	70	100	60

2.2. Characterization Analysis

FESEM (field emission scanning electron microscopy, Sigma300, ZEISS, Jena, Germany) was used to observe the microstructure of coatings and analyze element distribution by EDS (Energy Dispersive Spectrometer) components. The crystal structure of the coatings was analyzed by GIXRD (grazing incidence X-ray diffractometer, Rigaku Smartlab 3 kW, Tokyo, Japan) with CuKα radiation. The composition of coatings was obtained by XPS (X-ray photoelectron spectroscopy, PHI-Vesoprobe 5000 III, Thermo Fischer, Waltham, MA, USA), the vacuum degree of the analysis room was 4×10^{-9} Pa with Al kα radiation (hv = 1486.6 eV). Working voltage and working current were 14.6 kV and 13.5 mA, respectively. The binding energies were corrected according to the standard of C1s peak at 284.8 eV. The morphology and structure of wear debris were observed by HRTEM (high-resolution transmission electron microscopy, Talos F200, FEI, Hillsboro, OR, USA). The samples used for HRTEM measurements were obtained by mechanical scraping on a Cu grid fitted with a carbon membrane. The microhardness and elastic modulus of Ce-Ti/MoS$_2$ composite coatings were analyzed using nanoindentation (iNano, KLA, Ann Arbor, MI, USA). Berkovich indenter was selected to test the single-point hardness on the monocrystalline silicon. To avoid the test error, five different positions were tested, the average value of the test results was taken, the test load was 10 mN, and the maximum indentation depth was set to be no more than 1/10 of the coating's thickness.

2.3. Tribological Tests

The friction and wear properties of Ce-Ti/MoS$_2$ composite coatings were tested on a friction and wear testing machine in the atmospheric environment. The friction mode was circular sliding friction under dry friction. Sliding wear tests used 2.4 N normal loads, 1000 r/min and 9Cr18 ball (ϕ 6 mm) as the counterpart. The sliding time was set at 20 min in an environment with a humidity of 50%RH.

A three-dimensional white light interferometer (Rtec, San Jose, CA, USA) was used to measure the cross-section morphology of the wear marks, and the cross-sectional area was obtained by integration. The wear volume was obtained by multiplying the total length of the wear mark, and the wear rate (W) was calculated according to the formula W = V/(F $\times$ L). In the formula, W is the wear rate (mm^3 N^{-1} m^{-1}), V is the wear volume (mm^3), F is the applied normal load (N), and L is the total friction stroke (m). To reduce the error, the average wear rate of 5 times was calculated as an index to measure the wear resistance of the coating.

3. Results and Discussion

3.1. Morphology and Microstructure

Figure 1 shows the surface and cross-section images of Ce-Ti/MoS$_2$ composite films at different Ar flows. Different from the morphology of pure MoS$_2$ [28], a porous and fiber-like surface in the report, the Ce-Ti/MoS$_2$ coating exhibits cauliflower-like; there are pores between multiple uplifts. The cross-section morphology in Figure 1a is porous sponge-like, and signs of columnar growth perpendicular to the substrate can be observed; there are also many sputtering defects on the surface. In addition to the surface roughness and large pores, the agglomeration size is also uneven. The surface of Figure 1c,d is smooth, and the grain size is small, which plays the role of fine grain strengthening. When the argon flow rate increases to 70 sccm, large particles appear on the surface of the Figure 1e coating, indicating that the sputtering particle energy is low and the coating is difficult to grow densely.

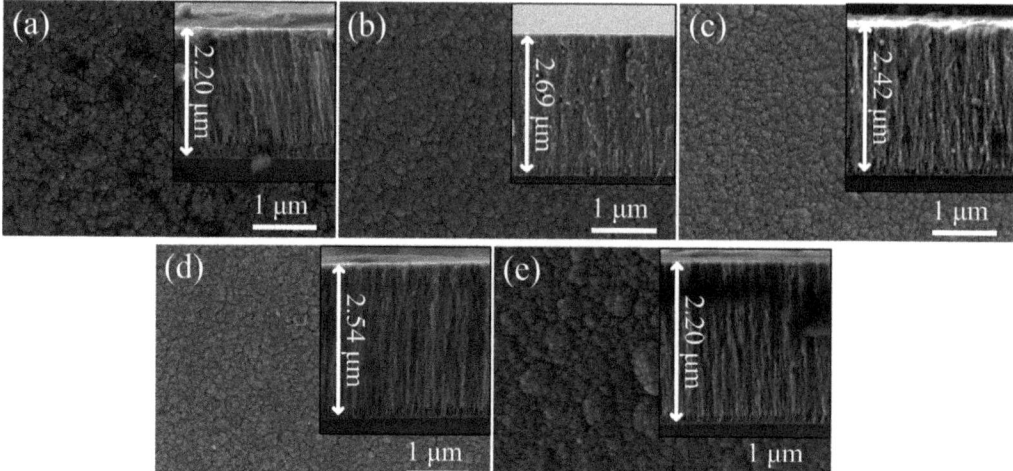

Figure 1. Surface and Cross-section FESEM micrographs of the composite coatings. (**a**) 30 sccm, (**b**) 40 sccm, (**c**) 50 sccm, (**d**) 60 sccm, (**e**) 70 sccm.

The content of doped elements and oxygen was counted in Table 2. It was observed that the change in the percentage of doped metal atoms caused by the change in the Ar flow rate was not obvious, and the sputtering efficiency of the Ti element was significantly higher than that of the Ce element. It is worth noting that the presence of O element was also found in the Ce-Ti/MoS$_2$ coating, and the content was relatively low when the flow rate was 60 sccm. The analysis shows that the oxygen content of the coating is mainly related to the content of the doped metal elements and the growth structure of the coating. The Ti element doped in the composite coating will generate oxide TiO$_2$ [29]. Ce, as a rare earth element with very active chemical properties, makes it easy for O to generate oxide CeO$_2$ and Ce$_2$O$_3$ [30]. On the other hand, the Mo element in the MoS$_2$-based coating is also easily oxidized to MoO$_3$ in the atmosphere. At 60 sccm, the coating structure is dense, which can effectively curb the further reaction of H$_2$O and O$_2$ in the atmospheric environment with the coating, so the content of the O element is relatively reduced. In addition, the coating thickness also shows a trend of thickening first and then thinning with the increase of the Ar flow rate, indicating that an appropriate Ar flow rate (40~60 sccm) can significantly improve the deposition efficiency.

Table 2. Chemical composition and thickness of coatings.

Argon Flow Rate/sccm Pressure/Pa	Ti/at.%	Ce/at.%	O/at.%	Thickness/μm
30	6.24	2.32	18.16	2.20
40	6.18	2.20	12.39	2.69
50	7.21	2.32	11.67	2.42
60	6.93	2.26	7.25	2.54
70	6.06	2.48	12.47	2.20

According to the morphology and element content, the change in the Ar flow rate mainly affects the energy of sputtering particles and finally makes the quality of the finished coating different [31]. The analysis of the sputtering process in the vacuum chamber shows that the increase of the Ar flow rate increases the ionization amount of argon, the voltage will decrease under constant sputtering power, and the initial kinetic energy of the emitted particles will decrease. This phenomenon will cause some target components cannot be attached to the substrate, reducing the deposition efficiency and making the coating structure coarse and porous. On the other hand, when the working gas flow rate is too low, the particles with too much kinetic energy will collide and reflect when they reach the substrate. In addition, when the Ar flow rate is too low, the particle bombardment on the film's surface is intense, which will also lead to an increase in the surface temperature of the film and the reaction with the residual gas. On the other hand, it will destroy the orderly growth of the film and cause surface defects.

3.2. Mechanical Properties

Hardness and elastic modulus are important indicators affecting the mechanical properties of the coating and prominently influence the tribological properties. Therefore, the nanoindentation hardness and elastic modulus of the deposited coating were measured by nano-indenter, as shown in Figure 2. Compared with pure MoS_2 coating, the doping of Ce and Ti significantly increases the coatings' microhardness and improves the coatings' bearing capacity. The hardness of the coating is almost proportional to the elastic modulus at each Ar flow rate and increases first and then decreases with the increase of the Ar flow rate; the nano-hardness of 50 sccm is the largest, which is 7.85 GPa, and the elastic modulus is 134.99 GPa. According to the data in Table 2, the sputtering efficiency of the doped element at 50 sccm is relatively high, and the solid solution strengthening effect is more obvious. In addition, according to Hall-Petch theory [32], the hardness increase is also related to grain boundary strengthening. When the argon flow rate is 70 and 30 sccm, the hardness of the sample is only 5.94 and 5.49 GPa, respectively. According to Figure 1, the coating defects and large pores are the main factors for low hardness.

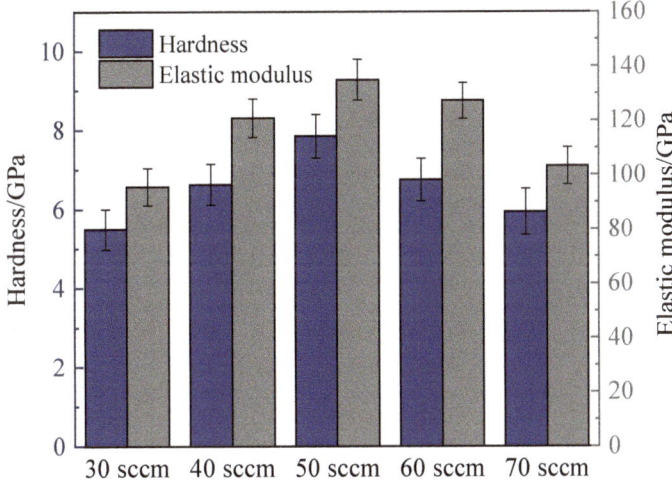

Figure 2. Hardness and elastic modulus of the composite coatings deposited at different Ar flows.

3.3. Phase Analysis

Figure 3 presents the X-ray diffraction patterns of coatings over the 2θ, ranging from 5° to 90°. No pronounced diffraction peak in the spectrum indicates that the Ce-Ti/MoS_2 coating is in an approximate amorphous state. Compared with pure MoS_2 coating [3,33], the characteristic peaks of MoS_2 with (100), (105), and (110) orientations are not obvious, and only the peak of (002) crystal plane growth is left. This growth mode can reduce the number of dangling bonds on the coating surface and effectively organize the oxidation failure of the coating. The peak of Mo_xS_y (0.5 < x:y < 1) was found at a low Ar flow rate (PDF card: No. 51-1004, No. 27-0319), indicating that the S/Mo is less than 2 in the sputtering environment with a large Ar flow rate, and some of the S elements are not attached to the coating. This phenomenon generally occurs in the related research of

MoS$_2$ coating [34], mainly related to the preferential re-sputtering of S elements during the deposition process [35]. In addition, a weak rare earth oxide peak also appeared when the Ar flow rate was 40 sccm.

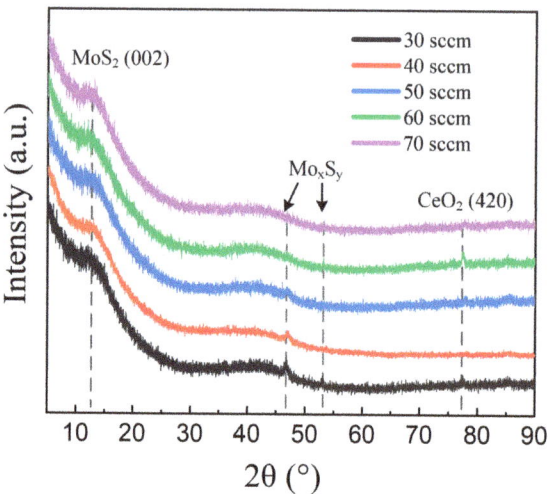

Figure 3. XRD patterns of coating surface at different Ar flows.

The surface of the samples was scanned by XPS, and the spectra of Mo 3d, S 2s, Ce 3d, and Ti 2p orbitals are analyzed, as shown in Figure 4. Except for a peak at the binding energy of 226.5 eV for all samples corresponding to S 2s [36], the Mo 3d and S 2s core lines presented in Figure 4 consists of two (3d5/2–3d3/2) and one (2s3/2–2s1/2) spin-orbit doublets, respectively. The peaks at near 229.0 eV and 232.2 eV are attributed to Mo^{4+} 3d5/2 and Mo^{4+} 3d3/2 of MoS$_2$, respectively [37,38]; the peaks at near 232.9 eV and 236.0 eV are attributed to Mo^{6+} 3d5/2 and Mo^{6+} 3d3/2 of MoO$_3$, respectively. The characteristic peaks of molybdenum oxides are generally due to oxygen adsorption by the dangling bonds on the coating surface. The peaks near 162.4 and 163.6 eV on the S 2p orbital are S 2p3/2 and S 2p1/2 peaks in MoS$_2$ [39,40], corresponding to the Mo 3d orbital results. In addition, Mo, S, and O binding peaks were also detected in the 30 sccm sample, located near the binding energies of 162.6 and 163.7 eV.

In detecting doped metals, the Ce 3d and Ti 2p core lines presented in Figure 4 consists of two (3d5/2–3d3/2) and one (2p3/2–2p1/2) spin-orbit doublets, respectively. The peaks at 886.4 and 904.5 eV are attributed to Ce^{3+} 3d5/2 and Ce^{3+} 3d3/2 of Ce$_2$O$_3$, respectively [41]; the peaks at 882.3 and 900.4 eV are attributed to Ce^{4+} 3d5/2 and Ce^{4+} 3d3/2 of CeO$_2$ respectively [42]. Furthermore, the peaks near 458.78 and 464.32 eV are Ti 2p3/2 and Ti 2p1/2 peaks in TiO$_2$ [43]. It can be seen that the doped metals on the surface of the coating are easy to combine with O to form oxides, including Ce$_2$O$_3$, CeO$_2$, and TiO$_2$. Doped metals can usually prevent further oxidation inside the coating by preferential oxidation. Cerium oxide also inhibits grain growth, coarsening in the material, and assisting lubrication [44,45].

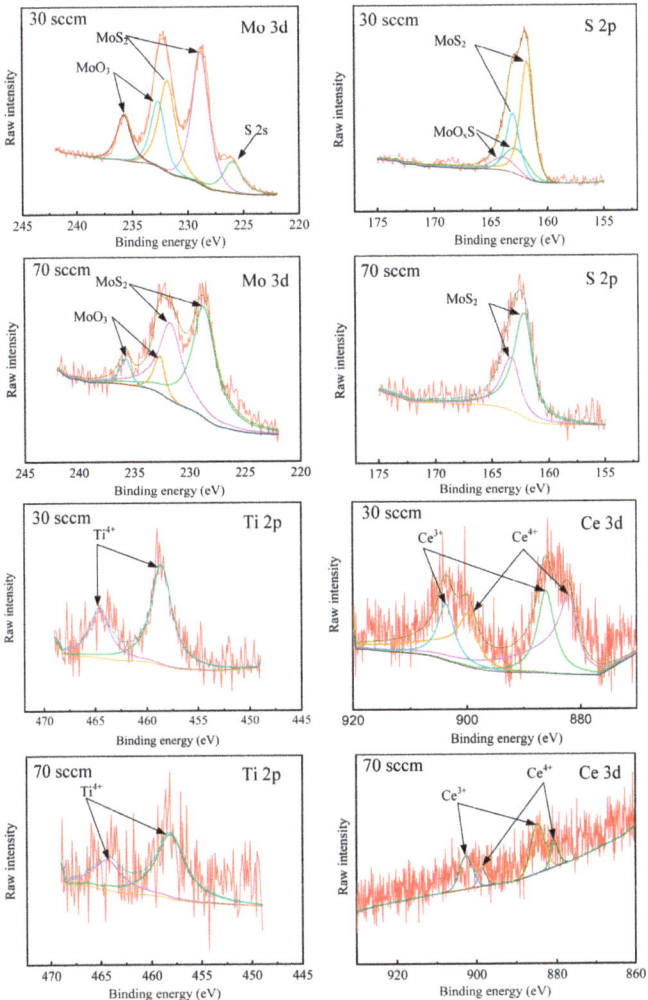

Figure 4. XPS spectra of Ce-Ti/MoS$_2$ coating.

3.4. Tribological Behavior

The friction coefficient curves of the coatings prepared at different Ar flow rates are shown in Figure 5a. The friction coefficient of Ce and Ti-doped MoS$_2$ composite coatings is generally low, and the flow rate can be stabilized at 0.1 and below. Among them, the friction coefficient is the lowest at 40 sccm, and the coefficient fluctuation in the stable friction stage is relatively stable. The addition of Ce and Ti is not only beneficial to the compactness of the coating structure but also retains the lubrication performance of MoS$_2$ itself to a certain extent. The friction coefficient of the highest (70 sccm) and the lowest (30 sccm) flow rate in the experimental group was relatively the highest, and the friction coefficient of 40~60 sccm was kept at a low level, which was only 0.073 when the flow rate was 40 sccm. It is worth noting that the average friction coefficient increases slightly at 50 sccm. Due to the highest hardness of the coating under this process parameter (Figure 2), in addition to solid solution strengthening, the local enrichment of doping elements forms a pinning strengthening effect, which will interfere with the interlayer sliding of MoS$_2$ [34].

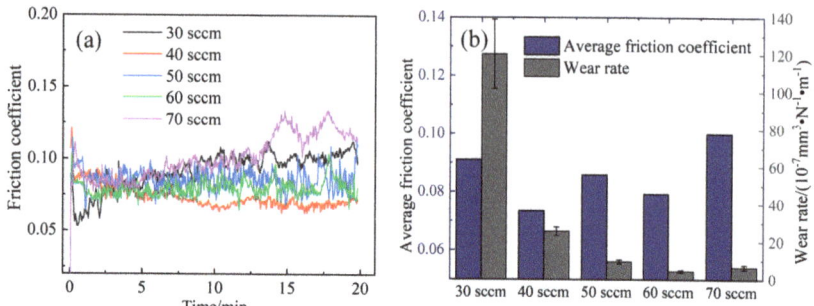

Figure 5. (a) Friction coefficient curves of Ce-Ti/MoS$_2$ coatings; (b) Wear rate of Ce-Ti/MoS$_2$ coatings.

It can be noted from Figure 5b that the wear rate of the coating at 30 sccm is significantly higher than that of other parameter samples. At this time, the coating peels off the substrate. It is speculated from the time when the friction coefficient changes abruptly that the peeling occurs at about the 13th minute of friction. When the Ar flow rate increases to 40~60 sccm, the wear rate reaches the lower value gradually. Especially at 60 sccm, the wear rate is only 4.42×10^{-7} mm^3 N^{-1} m^{-1}. The wear rate increased slightly at 70 sccm, mainly due to the appearance of large particles on the surface (Figure 1). It is considered that the Ar flow rate makes the film growth relatively uniform during the sputtering process and retains more MoS$_2$ coating structures parallel to the (002) crystal plane orientation of the substrate in the flow rate range of 50 to 70.

To further study the wear of the coating, the FESEM morphology and white light interference pattern of the wear track after the coating friction test were tested, as shown in Figure 6. The dark part of the SEM image in Figure 6a is a bare substrate, and the white light image shows that the edge of the wear track contour is perpendicular to the plane, which is approximately rectangular. It shows that the wear depth has penetrated the coating, and the failure mode is the film shedding. Figure 1a confirms that there are too many defects in the coating, and it is easier to form stress concentration at the surface depression under the action of cyclic force, which eventually causes structural damage to the coating. There are two typical wear characteristics when the flow rate is 40 sccm. The main wear area is the part with large Hertz contact stress between the steel ball and the coating. A furrow with a width of 156.9 μm is mainly abrasive wear, and the rest of the wear area is only slight scratches. The coating with a flow rate of 50~70 sccm has similar wear morphology, and the wear track is shallow and smooth. From the SEM image, small pits caused by adhesive wear at the wear track can be observed. When the flow rate is 60 sccm, the width of the wear track is the narrowest, only 119.6 μm.

The solid lubrication coating makes it easy to form a transfer film on the steel ball during the friction process, which has the effect of reducing the shear stress. Therefore, the wear spot of the steel ball is observed, and the EDS mapping is measured, as shown in Figure 7. When the flow rate was 30 sccm, parallel scratches were observed on the surface of the steel ball along the friction direction, and the composition of MoS$_2$ was detected in the scratches. It shows that in the case of coating shedding, the wear debris in the wear scar and the transfer film on the steel ball can still assist lubrication, and the friction coefficient is stabilized within 0.15 (Figure 5). From Figure 7b–e, it can be observed that different degrees of transfer film appear on the steel balls with a flow rate of 40~70 sccm. The size of the transfer film with a flow rate of 60 sccm is the smallest, indicating that the contact area between the coating and the steel ball is small, and it is also confirmed that the wear rate is the lowest under this parameter. It can be seen from the Mapping diagram that the main components of the wear debris and the formed transfer film are Mo and S elements, and the content of doped metal elements is relatively low. In addition, the oxygen element also exists on the wear scar, mainly the oxides of doped metals (Ce, Ti) and MoO$_3$ formed by the oxidation of MoS$_2$.

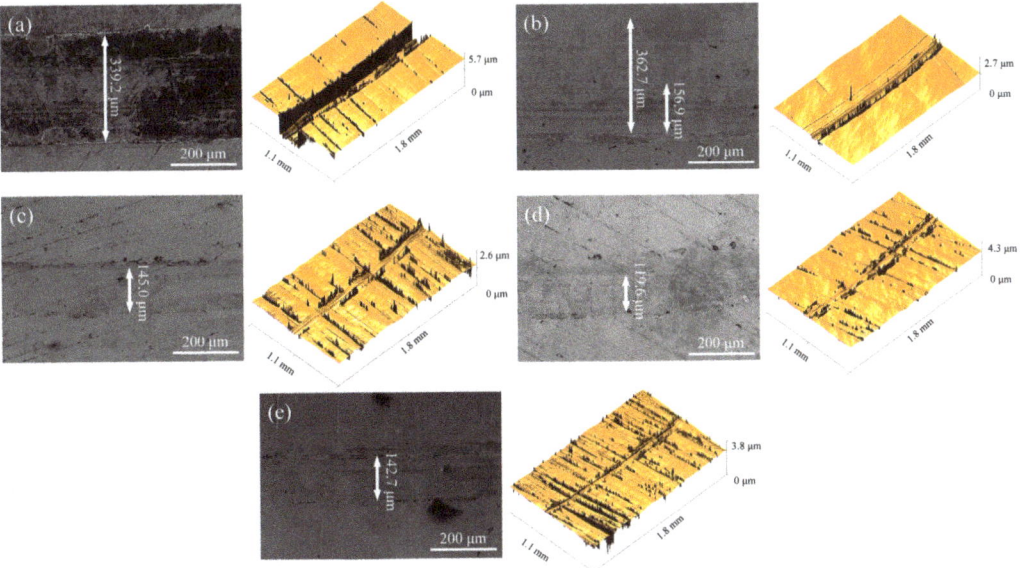

Figure 6. Wear tracks morphologies and white light interferograms at different Ar flows. (**a**) 30 sccm, (**b**) 40 sccm, (**c**) 50 sccm, (**d**) 60 sccm, (**e**) 70 sccm.

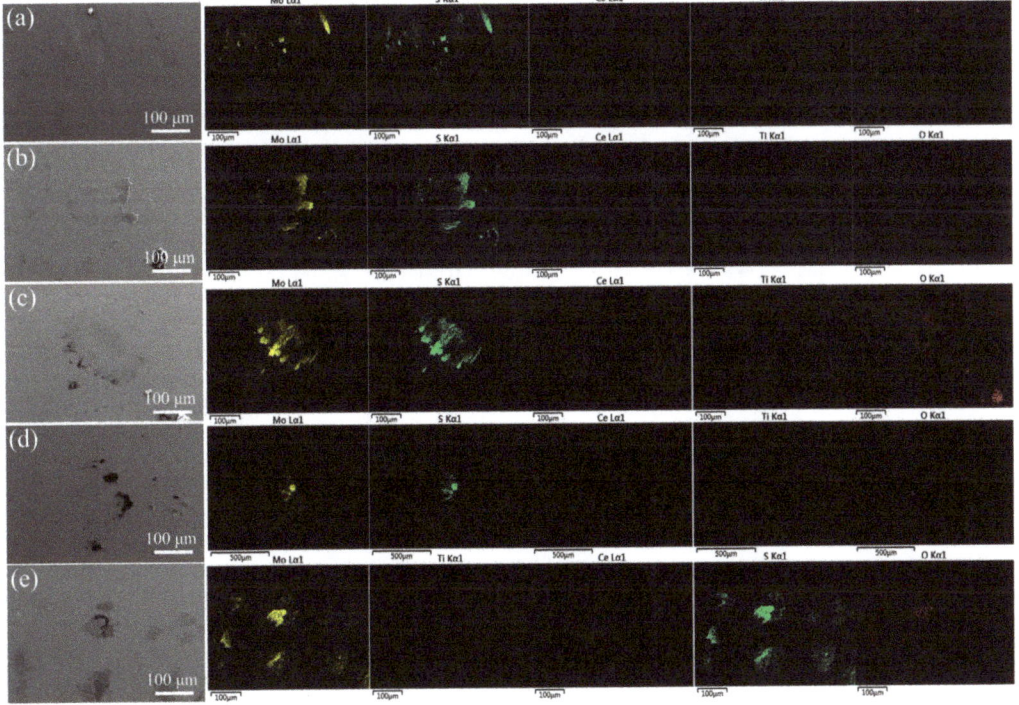

Figure 7. SEM micrographs and EDS mapping images of the wear scars for (**a**) 30 sccm, (**b**) 40 sccm, (**c**) 50 sccm, (**d**) 60 sccm, (**e**) 70 sccm.

To further explore the morphology of the transfer film and wear debris on the wear scar and its role in the friction process, TEM was used to observe the wear debris on the steel ball, as shown in Figure 8. Figure 8a shows the morphology of wear debris. The wear debris scraped on the copper mesh is mainly located at the hole's edge, in the form of sheet single-layer or multilayer stacking. The HRTEM image at high magnification is shown in Figure 8b. Many MoS$_2$ layered crystals are observed, and the interplanar spacing is calculated to be 0.69 nm, which is the (002) crystal orientation of MoS$_2$. The edge of the wear debris is different from the internal crystal arrangement: the MoS$_2$ layered structure at the edge of the wear debris is parallel to the edge line arrangement, the yellow area forms multiple anisotropic nanocrystals, and the internal wear debris has turbulent structure. This structure also appears in the related research of Cu-doped MoS$_2$ coating [34]. In addition, (100) oriented MoS$_2$ nanocrystals with a facet spacing of 0.27 nm were also found to aggregate at the edge of the debris outside the red region. The analysis shows that MoS$_2$ in the wear debris has a layered structure with (002) crystal plane orientation. The mechanical scratching and friction heat generated during the friction process will induce MoS$_2$ crystals to be arranged in parallel within a specific range. This phenomenon can effectively reduce the shear stress of the friction process while preventing the oxygen element from penetrating into the interior.

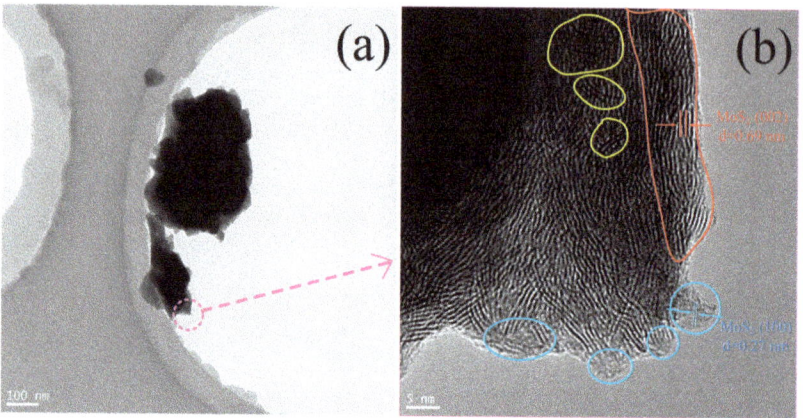

Figure 8. HRTEM patterns at 60 sccm: (**a**) debris topography, (**b**) partial enlarged detail.

4. Conclusions

In summary, unbalanced magnetron sputtering equipment prepared the Ce-Ti/MoS$_2$ coating by controlling the Ar flow rate in the range of 30~70 sccm. The Ar flow rate significantly affected the surface morphology of the coating. The sputtered particles had a suitable sputtering kinetic energy at a flow rate of 40–60 sccm, reducing the surface pores and compacting the structure. The elastic modulus and hardness of the coating increase first and then decrease with the increase of the Ar flow rate. The coating has the highest hardness of 7.85 GPa at 50 sccm due to grain boundary and solid solution strengthening. The coating doped with Ce and Ti is similar to the amorphous state, and only the weak (002) crystal orientation of MoS$_2$ is found. The doping elements in the surface layer exist in the form of metal oxides, which can ensure the integrity of the internal MoS$_2$ structure by preferential oxidation. Due to the surface defects caused by particle bombardment, the low Ar flow sample is prone to coating shedding during friction. The wear rate of the coating is minimum when the flow rate is 60 sccm, and there is no obvious furrow during the friction process, only minor adhesive wear marks. This phenomenon is due to the transfer film on the steel ball under the induction of friction. A layered structure parallel to the substrate reduces the shear force during sliding.

Author Contributions: Conceptualization, Y.X.; Data curation, C.T.; Formal analysis, C.T.; Project administration, H.C. and Y.Y.; Resources, Y.X.; Writing—original draft, C.T. and L.P.; Writing—review & editing, H.C. and Y.X. All authors have read and agreed to the published version of the manuscript.

Funding: This research was funded by the National Key Research and Development program, grant number 2021YFB3400400; Project of Science and Technology Development of Henan Province, grant number 202102210073.

Institutional Review Board Statement: Not applicable.

Informed Consent Statement: Not applicable.

Data Availability Statement: The authors confirm that the data supporting the findings of this study are available within the article.

Conflicts of Interest: The authors declare no conflict of interest.

References

1. Prasad, S.; Zabinski, J. Lubricants: Super slippery solids. *Nature* **1997**, *387*, 761–763. [CrossRef]
2. Vazirisereshk, M.R.; Martini, A.; Strubbe, D.A.; Baykara, M.Z. Solid Lubrication with MoS_2: A Review. *Lubricants* **2019**, *7*, 57. [CrossRef]
3. Xu, Y.; Xie, M.; Li, Y.; Zhang, G.; Zhu, M. The effect of Si content on the structure and tribological performance of MoS_2/Si coatings. *Surf. Coat. Technol.* **2020**, *403*, 126362. [CrossRef]
4. Ye, M.; Zhang, G.; Ba, Y.; Wang, T.; Wang, X.; Liu, Z. Microstructure and tribological properties of MoS_2+Zr composite coatings in high humidity environment. *Appl. Surf. Sci.* **2016**, *367*, 140–146. [CrossRef]
5. Yang, Y.; Zhao, Y.; Mei, H.; Cheng, L.; Zhang, L. 3DN C/SiC-MoS_2 self-lubricating composites with high friction stability and excellent elevated-temperature lubrication. *J. Eur. Ceram. Soc.* **2021**, *41*, 6815–6823. [CrossRef]
6. Liu, X.; Ma, G.J.; Sun, G.; Duan, Y.P.; Liu, S.H. MoSx-Ta composite coatings on steel by d.c magnetron sputtering. *Vacuum* **2013**, *89*, 203–208. [CrossRef]
7. Li, H.; Zhang, G.; Wang, L. The role of tribo-pairs in modifying the tribological behavior of the MoS_2/Ti composite coating. *J. Phys. D Appl. Phys.* **2016**, *49*, 095501. [CrossRef]
8. Jing, W.; Du, S.; Chen, S.; Liu, E.; Du, H.; Cai, H. Tribological Behavior of VN-MoS_2/Ag Composites over a Wide Temperature Range. *Tribol. Trans.* **2021**, *65*, 66–77. [CrossRef]
9. He, J.; Sun, J.; Choi, J.; Wang, C.; Su, D. Synthesis of N-doped carbon quantum dots as lubricant additive to enhance the tribological behavior of MoS_2 nanofluid. *Friction* **2023**, *11*, 441–459. [CrossRef]
10. An, V.; Anisimov, E.; Druzyanova, V.; Burtsev, N.; Shulepov, I.; Khaskelberg, M. Study of tribological behavior of Cu-MoS_2 and Ag-MoS_2 nanocomposite lubricants. *SpringerPlus* **2016**, *5*, 72. [CrossRef]
11. Agmon, L.; Almog, R.; Gaspar, D.; Voscoboynik, G.; Choudhary, M.; Jopp, J.; Klausner, Z.; Ya'Akobovitz, A.; Berkovich, R. Nanoscale contact mechanics of the interactions at monolayer MoS_2 interfaces with Au and Si. *Tribol. Int.* **2022**, *174*, 107734. [CrossRef]
12. Song, W.; An, L.; Lu, Y.; Zhang, X.; Wang, S. Friction behavior of TiN-MoS_2/PTFE composite coatings in dry sliding against SiC. *Ceram. Int.* **2021**, *47*, 24003–24011. [CrossRef]
13. Deng, W.; Tang, L.; Qi, H.; Zhang, C. Investigation on the Tribological Behaviors of As-Sprayed Al_2O_3 Coatings Sealed with MoS_2 Dry Film Lubricant. *J. Therm. Spray Technol.* **2021**, *30*, 1624–1637. [CrossRef]
14. Wang, D.Y.; Chang, C.L.; Chen, Z.Y.; Ho, W.Y. Microstructural and tribological characterization of MoS_2-Ti composite solid lubricating films. *Surf. Coat. Technol.* **1999**, *120–121*, 629–635. [CrossRef]
15. Tunay, R.F.; Poyraz, M. Tribological properties of MoS_2 thin films deposited by RF magnetron sputtering technique. *Int. J. Mater. Prod. Technol.* **2021**, *62*, 284–294. [CrossRef]
16. Fouvry, S.; Paulin, C. An effective friction energy density approach to predict solid lubricant friction endurance: Application to fretting wear. *Wear* **2014**, *319*, 211–226. [CrossRef]
17. Babuska, T.F.; Curry, J.F.; Dugger, M.T.; Jones, M.R.; Delrio, F.W.; Lu, P.; Xin, Y.; Grejtak, T.; Chrostowski, R.; Mangolini, F.; et al. Quality Control Metrics to Assess MoS_2 Sputtered Films for Tribological Applications. *Tribol. Lett.* **2022**, *70*, 103. [CrossRef]
18. Kokalj, D.; Debus, J.; Stangier, D.; Moldenhauer, H.; Nikolov, A.; Wittig, A.; Brummer, A.; Tillmann, W. Controlling the Structural, Mechanical and Frictional Properties of MoS(x) Coatings by High-Power Impulse Magnetron Sputtering. *Coatings* **2020**, *10*, 107734. [CrossRef]
19. Tillmann, W.; Wittig, A.; Stangier, D.; Moldenhauer, H.; Thomann, C.A.; Debus, J.; Aurich, D.; Bruemmer, A. Influence of the bias-voltage, the argon pressure and the heating power on the structure and the tribological properties of HiPIMS sputtered MoSx films. *Surf. Coat. Technol.* **2020**, *385*, 125358. [CrossRef]
20. Ma, Y.; Miao, Q.; Liang, W.; Yu, H.; Yin, M.; Zang, K.; Pang, X.; Wang, X. Microstructure, mechanical and tribological properties of Ta(C, N) coatings deposited on TA15 substrates with different nitrogen partial pressure. *Ind. Lubr. Tribol.* **2023**, *75*, 110–117. [CrossRef]

21. Kehal, A.; Saoula, N.; Abaidia, S.; Nouveau, C. Effect of Ar/N-2 flow ratio on the microstructure and mechanical properties of Ti-Cr-N coatings deposited by DC magnetron sputtering on AISI D2 tool steels. *Surf. Coat. Technol.* **2021**, *421*, 127444. [CrossRef]
22. Meng, M.; Ma, X. Improving the Photoelectric Characteristics of MoS_2 Thin Films by Doping Rare Earth Element Erbium. *Nanoscale Res. Lett.* **2016**, *11*, 513. [CrossRef]
23. Bai, G.; Yuan, S.; Zhao, Y.; Yang, Z.; Choi, S.Y.; Chai, Y.; Yu, S.F.; Lau, S.P.; Hao, J. 2D Layered Materials of Rare-Earth Er-Doped MoS_2 with NIR-to-NIR Down- and Up-Conversion Photoluminescence. *Adv. Mater.* **2016**, *28*, 7472–7477. [CrossRef]
24. Ahmed, S.; Murmu, P.P.P.; Sathish, C.I.; Guan, X.; Geng, R.; Bao, N.; Liu, R.; Kennedy, J.; Ding, J.; Peng, M.; et al. Co- and Nd-Codoping-Induced High Magnetization in Layered MoS_2 Crystals. *Phys. Status Solidi-Rapid Res. Lett.* **2023**, *17*, 2200348. [CrossRef]
25. Yang, J.; Pan, W.; Han, Y.; Zhao, M.; Huang, M.; Wan, C. Mechanical properties, oxygen barrier property, and chemical stability of RE3NbO7 for thermal barrier coating. *J. Am. Ceram. Soc.* **2020**, *103*, 2302–2308. [CrossRef]
26. Tian, H.; Wang, C.; Guo, M.; Gao, J.; Cui, Y.; Wang, F.; Liu, E.; Jin, G. Microstructure and luminescence properties of YSZ-based thermal barrier coatings modified by Eu_2O_3. *Ceram. Int.* **2020**, *46*, 4444–4453. [CrossRef]
27. Cai, H.; Xue, Y.; Li, H.; Ye, J.; Wang, J. Friction and Wear Behavior of Self-lubricating La-Ti/WS_2 Films by Unbalanced Magnetron Sputtering. *Rare Met. Mat. Eng.* **2021**, *50*, 2708–2714.
28. Fu, Y.; He, T.; Yang, W.; Xu, J.; Mu, B.; Pang, X.; Wang, P. Structure, Mechanical and Tribological Properties of MoSN/MoS_2 Multilayer Films. *Coatings* **2019**, *9*, 108. [CrossRef]
29. Shi, X.R.; He, P.H.; Sun, S.Q.; Chen, J.; Beake, B.; Liskiewicz, T.W.; Zhang, X.; Zhou, Z.H. Tailoring the corrosion and tribological performance of Ti-modified MoS_2-based films in simulated seawater. *J. Mater. Res. Technol.* **2022**, *21*, 576–589. [CrossRef]
30. Nemati, F.; Rezaie, M.; Tabesh, H.; Eid, K.; Xu, G.B.; Ganjali, M.R.; Hosseini, M.; Karaman, C.; Erk, N.; Show, P.L.; et al. Cerium functionalized graphene nano-structures and their applications; A review. *Environ. Res.* **2022**, *208*, 112685. [CrossRef] [PubMed]
31. Figueiredo, N.M.; Serra, R.; Manninen, N.K.; Cavaleiro, A. Production of Au clusters by plasma gas condensation and their incorporation in oxide matrixes by sputtering. *Appl. Surf. Sci.* **2018**, *440*, 144–152. [CrossRef]
32. Liu, W.B.; Liu, Y.; Cheng, Y.Y.; Chen, L.R.; Yu, L.; Yi, X.; Duan, H.L. Unified Model for Size-Dependent to Size-Independent Transition in Yield Strength of Crystalline Metallic Materials. *Phys. Rev. Lett.* **2020**, *124*, 235501. [CrossRef] [PubMed]
33. Zhang, R.; Cui, Q.; Weng, L.; Sun, J.; Hu, M.; Fu, Y.; Wang, D.; Jiang, D.; Gao, X. Modification of structure and wear resistance of closed-field unbalanced-magnetron sputtered MoS_2 film by vacuum-heat-treatment. *Surf. Coat. Technol.* **2020**, *401*, 126215. [CrossRef]
34. Sun, W.D.; Wang, J.; Wang, K.W.; Pan, J.J.; Wang, R.; Wen, M.; Zhang, K. Turbulence-like Cu/MoS_2 films: Structure, mechanical and tribological properties. *Surf. Coat. Technol.* **2021**, *422*, 127490. [CrossRef]
35. Wang, T.; Xue, C.; Yu, S.; Chen, W.; Zhang, G. The effect of S/Mo ratio on structure and properties of MoSx-Ti composite coatings deposited by magnetron sputtering. *Mater. Res. Express* **2020**, *7*, 106401. [CrossRef]
36. Qin, X.; Ke, P.; Wang, A.; Kim, K.H. Microstructure, mechanical and tribological behaviors of MoS_2-Ti composite coatings deposited by a hybrid HIPIMS method. *Surf. Coat. Technol.* **2013**, *228*, 275–281. [CrossRef]
37. Anwar, M.; Hogarth, C.A.; Bulpett, R. An XPS study of amorphous MoO_3/SiO films deposited by co-evaporation. *J. Mater. Ence* **1990**, *25*, 1784–1788.
38. Dupin, J.C.; Gonbeau, D.; Martin-Litas, I.; Vinatier, P.; Levasseur, A. Amorphous oxysulfide thin films MOySz (M = W, Mo, Ti) XPS characterization: Structural and electronic pecularities. *Appl. Surf. Sci.* **2001**, *173*, 140.
39. Moulder, J.F.; Stickle, W.F.; Sobol, P.E.; Bomben, K.D. *Handbook of X-ray Photoelectron Spectroscopy*; Physical Electronics Division, Perkin-Elmer Corporation: Eden Prairie, MN, USA, 1992.
40. Shang, K.; Zheng, S.; Ren, S.; Pu, J.; He, D.; Liu, S. Improving the tribological and corrosive properties of MoS_2-based coatings by dual-doping and multilayer construction. *Appl. Surf. Sci. A J. Devoted Prop. Interfaces Relat. Synth. Behav. Mater.* **2018**, *437*, 233–244.
41. Praline, G.; Koel, B.E.; Hance, R.L.; Lee, H.I.; White, J.M. X-ray photoelectron study of the reaction of oxygen with cerium. *J. Electron Spectrosc. Relat. Phenom.* **1980**, *21*, 17–30. [CrossRef]
42. Paparazzo, E.; Ingo, G.M.; Zacchetti, N. X-ray induced reduction effects at CeO_2 surfaces: An X-ray photoelectron spectroscopy study. *J. Vac. Sci. Technol. A Vac. Surf. Film.* **1991**, *9*, 1416–1420. [CrossRef]
43. Ong, J.L.; Lucas, L.C.; Raikar, G.N.; Gregory, J.C. Electrochemical corrosion analyses and characterization of surface-modified titanium. *Appl. Surf. Sci.* **1993**, *72*, 7–13.
44. Wang, L.; Zhu, D.; Wei, Z.; Huang, L.; Song, W.; Chen, Y. The Refinement Effect of Al-Ti-C-RE Master Alloy Prepared by Adding Ce_2O_3 on Pure Al. *Adv. Mater. Res.* **2010**, *139–141*, 227–234.
45. Min, C.; He, Z.; Liu, D.; Jia, W.; Qian, J.; Jin, Y.; Li, S. Ceria/reduced Graphene Oxide Nanocomposite: Synthesis, Characterization, and Its Lubrication Application. *Chemistryselect* **2019**, *4*, 4615–4623. [CrossRef]

Disclaimer/Publisher's Note: The statements, opinions and data contained in all publications are solely those of the individual author(s) and contributor(s) and not of MDPI and/or the editor(s). MDPI and/or the editor(s) disclaim responsibility for any injury to people or property resulting from any ideas, methods, instructions or products referred to in the content.

Article

Tribological Properties of Attapulgite Nanofiber as Lubricant Additive for Electric-Brush Plated Ni Coating

Feng Nan [1,2,3,]* and Dong Wang [4,]*

1. Hubei Provincial Key Laboratory of Chemical Equipment Intensification and Intrinsic Safety, School of Mechanical and Electrical Engineering, Wuhan Institute of Technology, Wuhan 430205, China
2. Hubei Provincial Engineering Technology Research Center of Green Chemical Equipment, School of Mechanical and Electrical Engineering, Wuhan Institute of Technology, Wuhan 430205, China
3. School of Mechanical and Electrical Engineering, Wuhan Institute of Technology, Wuhan 430205, China
4. College of Mechanical Engineering, Zhijiang College of Zhejiang University of Technology, Shaoxing 312030, China
* Correspondence: nanfeng2005@126.com (F.N.); wangdong@zzjc.edu.cn (D.W.)

Abstract: In order to expand the application field of attapulgite in tribology, the tribological properties of attapulgite as a lubricant additive on electric-brush plated Ni coating were investigated using the ball-disc contact mode of a SRV-IV friction and wear tester. The worn surfaces were characterized and analyzed via scanning electron microscope (SEM), energy-dispersive X-ray spectroscopy (EDS), and X-ray photoelectron spectroscopy (XPS). Results indicated that the friction-reducing and antiwear properties of 150 SN lubricating oil on the Ni coating were remarkably improved by an appropriate amount of attapulgite. Tribofilm mainly composed of Ni, NiO, SiO_2, Al_2O_3, graphite, and organic compounds was formed on the worn surface under the action of attapulgite, which was responsible for the reduction of friction and wear.

Keywords: attapulgite; lubricant additive; electric-brush plated Ni coating; tribological properties

Citation: Nan, F.; Wang, D. Tribological Properties of Attapulgite Nanofiber as Lubricant Additive for Electric-Brush Plated Ni Coating. *Lubricants* 2023, 11, 204. https://doi.org/10.3390/lubricants11050204

Received: 27 March 2023
Revised: 29 April 2023
Accepted: 2 May 2023
Published: 5 May 2023

Copyright: © 2023 by the authors. Licensee MDPI, Basel, Switzerland. This article is an open access article distributed under the terms and conditions of the Creative Commons Attribution (CC BY) license (https://creativecommons.org/licenses/by/4.0/).

1. Introduction

One-third to one-half of the energy used on earth is consumed by friction [1]. Additionally, abrasion caused by friction is a main cause of component failures. Therefore, reducing friction and wear is of great significance for human beings. There are many measures to reduce friction and wear, and using lubricating oil is the most common. With the rapid development of nanotechnology, a variety of nanomaterials have been used as lubricating oil additives to decrease friction and wear owing to their small size effect and excellent physicochemical properties [2–8].

It is generally believed that large-scale powders cannot be used as lubricating additives because they tend to agglomerate and cause serious abrasive wear. However, it was found that a kind of micrometric natural mineral powder, serpentine, can be utilized as an additive to remarkably improve the tribological properties of lubricating oil [9,10]. Some research has revealed the tribological mechanism of serpentine by means of microscopic characterizations. Zhang et al. [11] found that a tribolayer mainly composed of diamond like carbon (DLC) was formed under the effect of serpentine additive, which was responsible for the improvement in the lubrication performance of lubricating oil. Zhang et al. [12] attributed the reduction of friction and wear to the formation of an amorphous Si–O film. Yu et al. [13] declared that the tribolayer formed on the rubbing surface by serpentine possesses better mechanical properties than the metallic substrate. In addition, some reports have studied the tribological properties of serpentine lubricating additive under different conditions [14,15]. Yin et al. [16] investigated the friction and wear behaviors of steel/bronze tribopairs lubricated by oil with serpentine additive. It was found that the tribological properties of tin bronze were remarkably improved by the addition of serpentine powder into the oil. In the effect of serpentine, a non-conductive tribofilm consisting of metal oxides, oxide ceramic particles, graphite, and organics was formed on the worn tin bronze surface. Moreover, some studies have reported using other natural mineral materials, including kaolin antigorite, as lubricating additives to improve the tribological properties of oils [17,18].

Attapulgite is another natural mineral that has similar crystal structure to serpentine minerals. In recent years, it was found that attapulgite can also be used as an additive to remarkably improve the friction-reducing and antiwear properties of lubricating oil [19,20]. The friction-reducing and antiwear behavior of attapulgite can be attributed to the formation of tribofilm consisting of oxides, ceramics, silicates, and graphite on the rubbing surface. Such tribofilms have good lubricity, high hardness, and a high hardness/elastic modulus ratio [20]. Additionally, some nanomaterials such as nano-La_2O_3 and nano-Ni have been proved to further improve the friction-reducing and antiwear effect of attapulgite [21,22]. However, the studies are all focused on steel–steel contact. There are few reports on the tribological properties of an attapulgite lubricating additive on other materials. Compared with other mineral materials, attapulgite has the following advantages: (1) The attapulgite materials are natural nanofibers. They can be obtained via simple crystal-bundles separation processing; (2) Attapulgite possesses high reserves in nature; (3) Attapulgite is an environmentally friendly material without toxic and harmful elements. Therefore, attapulgite has promising development prospects in the field of tribology.

Electric-brush plated technology has been applied widely in industry to prepare coatings with abrasion-resistant and corrosion-preventive properties owing to its flexibility, portability, and easy operation. Through a simple and mature process, nanocrystalline Ni coating with high hardness, fine wear, and corrosion resistance can be prepared [23,24]. Additionally, the electric-brush plated Ni coating has been widely applied in the abrasion repairing of mechanical parts such as crankshafts, piston rods, gear shafts, and so on.

In this work, the tribological properties of attapulgite as a lubricant additive on electric-brush plated Ni coating was investigated for the first time. Moreover, the friction-reducing and antiwear mechanism of the attapulgite additive was discussed on the basis of the tribological tests, analysis, and characterization of worn surfaces. The results would provide a reference for the application of attapulgite lubricating additive for the lubrication of electric-brush plated Ni coating. We believe that attapulgite lubricating additive can significantly improve the service life of the electric-brush plated Ni coating under oil lubrication.

2. Materials and Methods

2.1. Materials

The attapulgite powders (ATP) were purchased from Jiuchuan Nanometer material Science and Technology Ltd. (Huai'an, China). The chemical composition of ATP is SiO_2 (58.88 wt%), MgO (12.10 wt%), Al_2O_3 (9.50 wt%), Fe_2O_3 (5.20 wt%), K_2O (1.04 wt%), CaO (0.4 wt%), TiO_2 (0.55 wt%), P_2O_5 (0.18 wt%), MnO (0.05 wt%), Cr_2O_3 (0.04 wt%), and H_2O (12.06 wt%), indicating that the attapulgite is a kind of silicate mainly composed of some metallic oxides such as SiO_2, MgO, Al_2O_3, and Fe_2O_3. The powders were purified by the manufacturer, and we did not perform any further treatment. Oleic acid was obtained from Sinopharm Chemical Reagent Ltd. (Shanghai, China). The 150 SN lubricating oil was purchased from Qingdao Compton Technology Co., Ltd. (Qingdao, China). The chemical composition of 150 SN is paraffin hydrocarbon, naphthenic hydrocarbon, alkyl naphthalene, alkyl benzene, polycyclic aromatic hydrocarbons, sulfide, nitride, and oxide.

2.2. Preparation of Electric-Brush Plated Ni Coating

AISI 1045 steel substrate discs (φ24 mm × 8 mm) with a hardness of HRC 27–30 were polished to a roughness below Ra0.3 μm. Then, Ni coating was prepared on the substrate samples through the process shown in Table 1. The composition of Nickel plating solution is displayed in Table 2. The hardness of the coating was measured using a microhardness tester (HVS-1000). When measuring hardness, the applied load was 100 g and loading time was 15 s. At least six points on the coating were selected and the average value was calculated. The hardness of the prepared Ni coating was about 430 HV. The surface roughness of the coating was measured with a MicroXAM surface mapping profilometer (LEXT OLS4000, Olympus, Tokyo, Japan). The surface roughness of the coating was Ra ≈ 0.5 μm.

Table 1. Process flow and parameters of electro-brush plating.

	Process	Solution	Voltage/V	Plating Time/min	Movement Velocity of Plating pen/m·min^{-1}
1	Electrical cleaning	Electing cleaning solutions	+12	1	6~8
2	Intense activation	No. 2 activation solutions	−12	0.5	8~10
3	Slight activation	No. 3 activation solutions	−18	0.5	6~8
4	Pre-plating	Pre-plating nickel solutions	+12	1	8~10
5	Plating	Nickel plating solutions	+12	5	10~12

Table 2. Major components of Nickel plating solutions.

Component	Amount
$NiSO_4 \cdot 6H_2O$	20 g·L^{-1}
CH_3COONH_4	40 g·L^{-1}
$(NH_4)_3C_6H_5O_7$	45 g·L^{-1}
$NH_3 \cdot H_2O$/(adjust pH to 7.3~7.5)	100~130 mL·L^{-1}

2.3. Friction and Wear Test

0.2 wt%, 0.4 wt%, 0.6 wt%, and 0.8 wt% of ATP were put in a certain amount of 150 SN. In order to obtain good dispersity in the oil, 5.0 wt% (percent of ATP) of oleic acid was also added in the oil to have the ATP modified. The structural formula of oleic acid is $CH_3(CH_2)_7CH=CH(CH_2)_7COOH$, which is an organic surface modifier with long carbon chains. The long carbon chains were grafted onto the surface of attapulgite nanoparticles during mechanical stirring. Therefore, the lipophilicity of attapulgite nanoparticles was significantly enhanced. The attapulgite nanoparticles can be fully dispersed in oil to avoid agglomeration and forming large particles. The mixtures of ATP, oil, and oleic acid were well-mixed through ball mill at 200 r/min for 2 h. Eventually, the 150 SN with an added 0.2 wt%, 0.4 wt%, 0.6 wt%, and 0.8 wt% of ATP were prepared.

An SRV-IV oscillating friction and wear tester was used to evaluate the tribological properties of the oils. During the tribological tests, the upper AISI 52,100 steel balls (φ10 mm) with a hardness of HRC 59–62 slide reciprocally at an amplitude of 1 mm, a frequency of 10 Hz, and a temperature of 50 °C against the stationary discs for 30 min. Firstly, the tribological properties of 150 SN with different added contents of ATP were investigated at a load of 50 N. After determining the optimum content of ATP, the effect of the load on the tribological properties of ATP was studied. The selected loads were 10, 20, 50, and 100 N. Each tribology test was repeated 3 times to minimize data scatter. The friction coefficient curve was recorded by the computer connected to the SRV friction and wear tester and the average value was calculated. The wear volume of each disc was measured with a MicroXAM surface mapping profilometer for at least three times and the mean value was calculated.

2.4. Characterization and Analysis

The phase composition of ATP was characterized by X-ray Diffraction (XRD) (Advance 8, Bruker, Mannheim, Germany). A few attapulgite powders were uniformly dispersed into alcohol via ultrasonic. Then, a little suspension was dropped onto copper mesh. After alcohol volatilization, the microstructure of ATP was observed using TEM.

The morphologies and elemental contents of the Ni coating and worn surfaces on the discs were analyzed via SEM (Nova Nano 650, FEI, Hillsboro, OR, USA) equipped with EDS. The cross-sectional morphology of the coating was analyzed via SEM (Nova Nano 650). The grain size distribution of the ATP powders was measured through a laser particle size analyzer (Malvern, MAL1066490, UK). The chemical state of typical elements on the worn surfaces was characterized using XPS (ESCALAB 250Xi, Thermo Scientific, Waltham, MA, USA). The XPSPEAK 4.1 software was utilized to fit the XPS data. The binding energy resolution was approximately ±0.2 eV.

3. Results

3.1. Characterization of ATP and Electric-Brush Plated Ni Coating

The XRD spectrogram of ATP is shown in Figure 1a. Characteristic diffraction peaks of attapulgite (d(110), d(200), d(040), d(231), etc.) can be obviously seen on the XRD spectrogram. This indicates that the chemical formula of the ATP was $(Mg, Al, Fe)_5Si_8O_{20}(OH_2)_4 \cdot 4H_2O$ (JCPDS No. 21-0957). The ATP contained almost no other impurities. The TEM morphology displayed in Figure 1b indicates that the ATP particles had a nano-fibrous morphology with a diameter of about 50 nm. The length of nanoparticles was nonuniform. Some nanoparticles were about 1 μm. Few nanoparticles were as short as below 100 nm. In addition, the nanoparticles exhibited good dispersion. Only a small amount of nanoparticles gathered together.

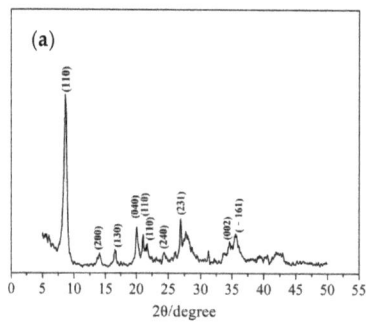

Figure 1. Characterization of ATP: (a) XRD; (b) TEM.

Figure 2a shows the surface morphology of the prepared Ni coating. The Ni coating surface was compact and smooth. Additionally, several typical cauliflower-like structures can be seen on the surface. There were few holes, burrs, cracks, and other macro defects in the coating. The corresponding EDS analysis result of prepared Ni coating shown in Figure 2b indicates that only Ni and C elements can be detected on the coating surface.

Figure 3 shows cross-sectional SEM morphology of the electric-brush plated Ni coating. It was found that the thickness of the prepared Ni coating is 21.4 μm. Additionally, the adhesion between coating and substrate is fine.

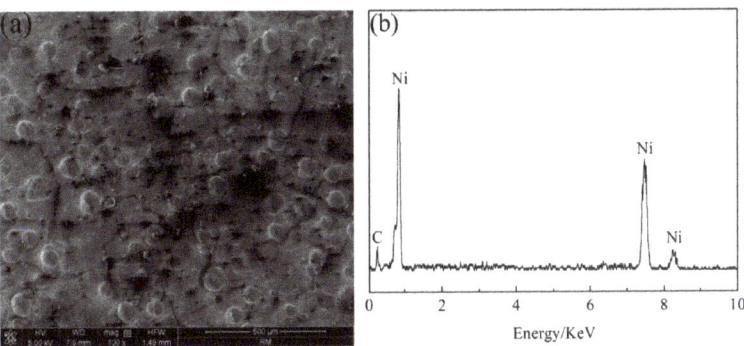

Figure 2. Characterization of the electric-brush plated Ni coating: (**a**) SEM; (**b**) EDS.

Figure 3. Cross-sectional SEM morphology of the electric-brush plated Ni coating. The grain size distribution of the ATP powders is displayed in Figure 4. It was found that the particle size of most powders was from 150 to 900 nm.

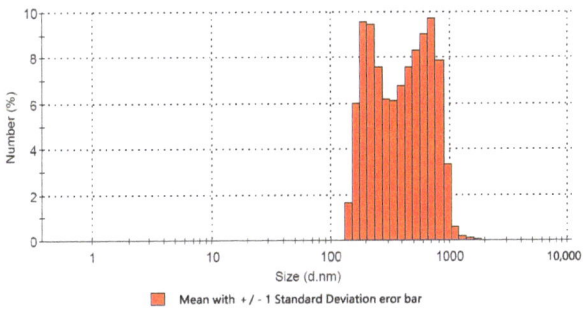

Figure 4. Grain size distribution of the ATP powders.

3.2. Friction and Wear Behavior

Figure 5 shows the tribological properties of 150 SN with different added amounts of ATP at 50 N. Under the lubrication of 150 SN, the friction coefficient and wear volume were 0.214 and 50.12×10^{-3} mm^3, respectively. With the addition of ATP, the friction coefficient and wear volume for 150 SN were both decreased. When the amount of ATP was 0.4 wt%, the friction coefficient and wear volume were 0.169 and 42.58×10^{-3} mm^3, respectively, which were the lowest. Therefore, the optimum amount of ATP is 0.4 wt%. Under the effect of 0.4 wt% ATP, the friction coefficient and wear volume of oil were reduced by 21% and 16%, respectively.

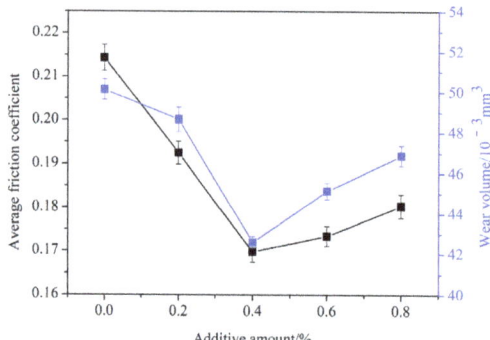

Figure 5. Tribological properties of 150 SN with different added amounts of ATP (50 N, 10 Hz, 30 min).

Furthermore, the effect of load on the tribological properties of 150 SN and 150 SN with added ATP was investigated. Figure 6 shows the friction coefficient curves under the lubrication of 150 SN and 150 SN with added ATP. At the initial stage of the friction test, the friction coefficients of the two kinds of oils both showed a significant increase, which may be attributed to the running-in of the friction pair. Under the lubrication of 150 SN, the friction coefficient fluctuated greatly at 10 and 20 N due to the effect of the cauliflower-like structure. When the load increased to 50 and 100 N, the friction coefficient became more stable than that at 10 and 20 N. Throughout the tribological test, the friction coefficients at 50 and 100 N were basically the same. The friction coefficients both remained around 0.20 after 1200 s. Overall, the friction coefficient at low loads was higher than that at high loads. Additionally, the friction coefficient at 10 N was the highest. As for 150 SN with added ATP, the friction coefficients at all loads were more stable and lower than that at 150 SN. The friction coefficient at 50 and 100 N was more stable than that at 10 and 20 N, which was similar to 150 SN. Additionally, the friction coefficient at all loads can basically achieve stability after 400 s, which is apparently shorter than 150 SN. Similarly, the friction coefficient at low loads was higher than that at high loads. The friction coefficient at 20 N was the highest and at 50 N was the lowest.

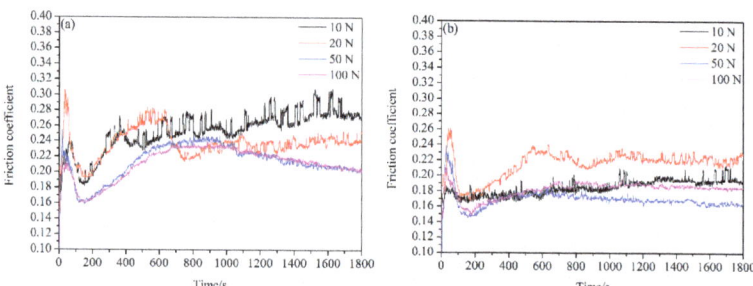

Figure 6. Friction coefficients as a function of sliding time for (**a**) 150 SN; (**b**) ATP at different loads (50 N, 10 Hz, 30 min).

The average friction coefficient and wear volume under the lubrication of 150 SN and 150 SN with added ATP at different loads is shown in Figure 7. Under the lubrication of 150 SN, the friction coefficient decreased gradually along with the increased load, while for 150 SN with added ATP, the friction coefficient at 20 N was the highest and at 50 N was the lowest. In addition, the friction coefficient of 150 SN with added ATP was lower than that of 150 SN at all selected loads. From Figure 7b, it can be seen that the wear volume increased gradually along with the increased load for both the two oils. Additionally, the wear volume of 150 SN with added ATP was lower than that of 150 SN at all selected loads. In summary, the friction-reducing and antiwear properties of 150 SN can be remarkably improved by adding moderate ATP at all selected loads. The friction-reducing and antiwear effect of ATP is most obvious at the applied load of 10 N. At 10 N, the average friction coefficient for 150 SN was decreased from 0.26 to 0.18, which is a 30.8% reduction. The wear volume for 150 SN was decreased from 21.8×10^{-3} mm^3 to 10.2×10^{-3} mm^3, which is a 53.2% reduction.

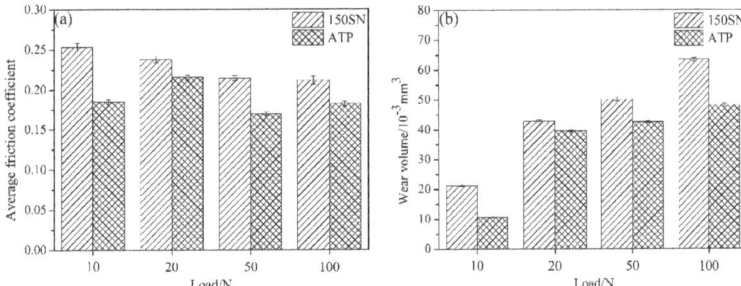

Figure 7. Effect of load on (**a**) average friction coefficient; (**b**) wear volume for 150 SN and ATP (10 Hz, 30 min).

3.3. Worn Surface Analysis

The morphologies of the worn surfaces lubricated with 150 SN and 150 SN with added ATP at 50 N are displayed in Figure 8. Under the lubrication of 150 SN, it is obvious that the cauliflower-like structure of the Ni coating disappeared. There existed a large area of material peeling and some pits on the worn surface of Ni coating, while for 150 SN with added ATP, the worn surface became much smoother and flatter. Only shallow furrows and a few pits could be seen on the worn surface of Ni coating.

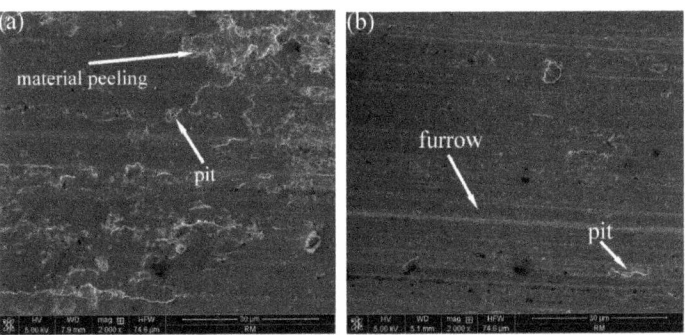

Figure 8. SEM images of the worn surfaces lubricated with (**a**) 150 SN; (**b**) ATP (50 N, 10 Hz, 30 min).

Figure 9 and Table 3 show the element composition and content on the worn surfaces displayed in Figure 6. C and Ni elements can be detected on the worn surface lubricated with 150 SN. There is no Fe element on the worn surface, indicating that the Ni coating is not completely worn off. In addition, the Ni content on the worn surface is lower than that on the substrate. As for the worn surface lubricated with 150 SN with added ATP, besides C and Ni elements, the new elements of O, Si, and Al can be found on it. Moreover, the Ni content of the worn surface lubricated with 150 SN with added ATP is basically the same as that with 150 SN. The above results demonstrate that tribofilm consisting of Ni, C, O, Si, and Al has formed on a worn surface under the lubrication of the 150 SN with added ATP.

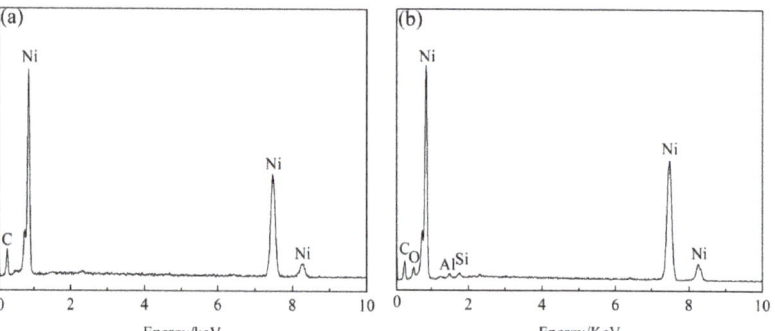

Figure 9. EDS analysis results of the worn surfaces lubricated with (**a**) 150 SN; (**b**) ATP (50 N, 10 Hz, 30 min).

Table 3. Elemental content of Ni coating and worn surfaces on it (50 N, 10 Hz, 30 min).

	Atomic%				
	C	Ni	O	Si	Al
Substrate	57.80	42.20	0	0	0
Lubricated with 150 SN	64.14	35.86	0	0	0
Lubricated with 150 SN with added ATP	55.47	36.21	6.92	0.63	0.78

XPS analysis on the worn surface lubricated with 150 SN with added ATP was performed to further confirm the composition of the formed tribofim. The results are presented in Figure 10. The analysis of the $Ni2p_{3/2}$ peak indicates that Ni (853 eV) and NiO (854.1 eV) exist in the tribofilm [25–27]. The spectrum of C1s can be fitted into three subpeaks at 283.6, 284.8, and 285.5 eV, corresponding to graphite, C-C (pollution carbon), and organic compounds [14,27]. The subpeaks of O1s at around 530, 531.2, 532.3, and 533.4 eV are associated with NiO, Al_2O_3, SiO_2, and organic compounds, respectively [14,25,27]. In addition, the spectrum of Si2p can be identified SiO_2 (103.2 eV) [14,27]. The spectra of the Al2p peak suggests that Al_2O_3 (74.5 eV) can be found in the tribofilm [27]. Hence, it can be confirmed from XPS analysis results that the tribofilm formed on the worn surface of Ni coating is mainly composed of Ni, NiO, Al_2O_3, SiO_2, graphite, and organic compounds.

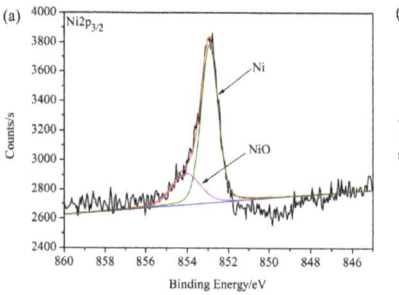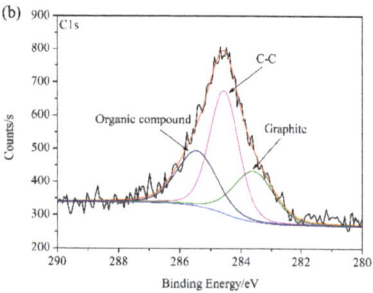

Figure 10. *Cont.*

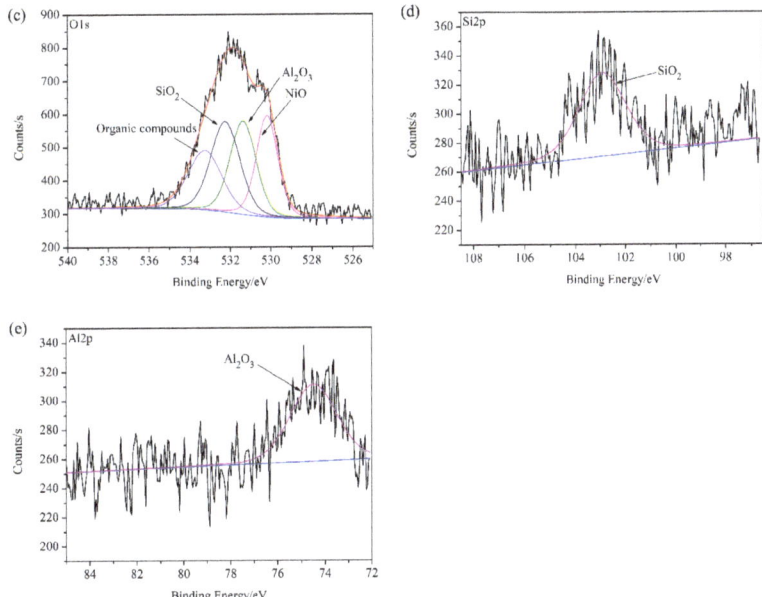

Figure 10. XPS patterns of worn surface lubricated with ATP: (**a**) Ni2p$_{3/2}$; (**b**) C1s; (**c**) O1s; (**d**) Si2p; (**e**) Al2p (50 N, 10 Hz, 30 min).

4. Discussion

It can be indicated from the above experimental results that ATP can obviously improve the friction-reducing and antiwear properties of 150 SN for electric-brush plated Ni coating. Under the lubrication of 150 SN, the cauliflower-like structure of Ni coating is constantly worn away at first, due to the poor lubricity of oil. Along with the Ni coating continuing to be worn out, the cauliflower-like structure disappears, and a large area of material peeling and some pits appear on the worn surface. As for 150 SN with added ATP, a protective tribofilm is formed on the worn surface under the action of ATP. This behavior is directly related to the crystal structure of attapulgite and friction heat and high pressure. The chemical formula of attapulgite is (Mg, Al, Fe)$_5$Si$_8$O$_{20}$(OH$_2$)$_4$·4H$_2$O. It possesses a layered chain structure formed by a continuous Si-O tetrahedral layer and a discontinuous Mg(Al)-OH/O octahedral layer, as shown in Figure 11 [20,28]. Active oxygen atoms and unsaturated bonds (Si–O, Si–OH, Mg–O, Mg–OH, Al–O, Al–OH) exist in the tetrahedral layers and octahedral layers. Therefore, attapulgite possess several good physicochemical properties, including high surface activity and adsorption ability. During the friction procedure, attapulgite particles suspended in oil can be easily absorbed and deposited onto the Ni coating surface. Then, following tribochemical reactions among Ni coatings, attapulgite particles and oil occur under the effect of high contact stress and shearing force: (1) Microstructural instability of attapulgite occurred, leading to the release of plentiful active oxygen atoms, which take reaction with the Ni coating to form NiO [22,26]; (2) Some unsaturated bonds decomposed and recombined to form SiO$_2$ and Al$_2$O$_3$ [22]; (3) The cracking of oil could form graphite and organic compounds. Finally, tribofilm mainly composed of Ni, NiO, Al$_2$O$_3$, SiO$_2$, graphite, and organic compounds was formed on the rubbing surface. The tribofilm can fill and repair the wear region to reduce the surface roughness of the rubbing surface, thus reducing friction and wear. In addition, as the thickness of the tribofilm continuously increases, the metal–Ni coating friction pair changed into a tribofilm–tribofilm friction pair. The tribofilm is mainly composed of ceramic phases including NiO, Al$_2$O$_3$, and SiO$_2$, which possess some excellent physicochemical properties, such as high strength, favorable antioxidation, and anticorrosion properties. Moreover, the ductility of such metallic oxides is low. Therefore, under the same tribology test conditions, the contact area of the tribofilm–tribofilm friction pair is smaller than that of the metal–Ni coating friction pair [29]. Consequently, the friction and wear can be remarkably reduced. However, the content and microstructure of each substance in the tribofilm cannot be determined; thus, the formation mechanism of tribofilm needs further research and exploration.

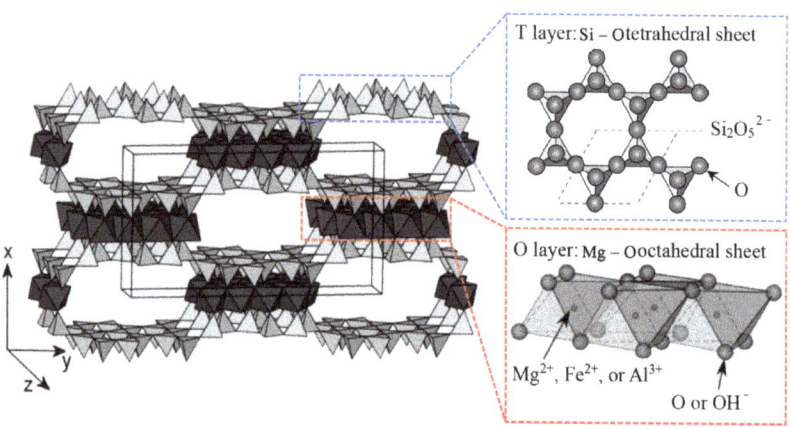

Figure 11. Crystal structure model of attapulgite.

It was found that the optimum content of attapulgite additive in oil is 0.4 wt% when the load is 50 N.

It can be seen from Figure 6 that the friction coefficient under the lubrication of 150 SN fluctuated greatly at 10 and 20 N. This is attributed to the high roughness of the coating surface caused by the cauliflower-like structure. When the load increased to 50 and 100 N, the hardness of the coating was not high enough to resist the shear and extrusion of heavy loads. The cauliflower-like structure was quickly worn away to bring about significant reduction of coating surface roughness. Thus, the friction coefficient became more stable than light loads. Under the lubrication of 150 SN with added ATP, tribofilm can form on the worn surface to reduce coating surface roughness. Therefore, the friction coefficient was more stable and lower. Compared with the Ni coating, the ability of tribofilm to bear heavy loads is stronger. Thus, the friction coefficient and wear loss of 150 SN can be remarkably reduced even at the heavy load of 100 N. The tribological properties of attapulgite additive under different loads are dissimilar, which may be attributed to the competition between the formation and abrasion of tribofilm. With the increase in load, the energy provided by friction increases. However, at the same time, wear is also intensified [30,31]. It was found that the friction-reducing and antiwear effect of ATP was most obvious at the applied load of 10 N. This result demonstrates that the formation of tribofilm occupies a dominant position at the applied load of 10 N. The contact mode of the friction pair is ball on disc. Therefore, the contact stress of the friction pair was high. Thus, the energy provided by friction was high. If the contact mode changes to another form that cannot provide such high energy, it is doubtful that attapulgite still possesses friction-reducing and antiwear effects for the electric-brush plated Ni coating. This requires further research.

5. Conclusions

(1) The attapulgite nanofibers can be used as additives to improve the tribological properties of 150 SN for electric-brush plated Ni coating. The oil with an added 0.4 wt% attapulgite was found to be the most efficient in reducing friction and wear under the ball-disc model at a load of 50 N, a frequency of 10 Hz, and a temperature of 50 °C.

(2) The attapulgite additive can improve the friction-reducing and antiwear properties of oil at all selected loads—10, 20, 50, and 100 N. Under the lubrication of 150 SN, the friction coefficient decreased gradually along with the increased load, while for 150 SN with added ATP, the friction coefficient at 20 N is the highest and at 50 N is the lowest.

(3) Under the action of attapulgite, tribofilm mainly composed of Ni, NiO, Al_2O_3, SiO_2, graphite, and organic compounds formed on the worn surface of Ni coating, which contributed to the decrease in friction and wear.

Author Contributions: Conceptualization, F.N.; methodology, F.N.; software, D.W.; validation, F.N.; formal analysis, D.W.; investigation, F.N.; resources, F.N.; data curation, D.W.; writing—original draft preparation, F.N.; writing—review and editing, D.W.; visualization, D.W.; supervision, D.W.; project administration, F.N.; funding acquisition, F.N. All authors have read and agreed to the published version of the manuscript.

Funding: This research was funded by the National Natural Science Foundation of China (51705511) and the Scientific Research Foundation of the Wuhan Institute of Technology (K202013).

Data Availability Statement: The data presented in this study are available on request from the corresponding author. The data are not publicly available due to privacy or ethical restrictions.

Conflicts of Interest: The authors declare no conflict of interest.

References

1. Wen, S.Z.; Huang, P. *Principles of Tribology*, 3rd ed.; Tsinghua University Press: Beijing, China, 2008. (In Chinese)
2. Jia, X.; Huang, J.; Li, Y.; Yang, J.; Song, H. Monodisperse Cu nanoparticles@MoS2 nanosheets as a lubricant additive for improved tribological properties. *Appl. Surf. Sci.* **2019**, *494*, 430–439. [CrossRef]
3. Jia, Z.; Wang, Z.; Liu, C.; Zhao, L.; Ni, J.; Li, Y.; Shao, X.; Wang, C. The synthesis and tribological properties of Ag/polydopamine nanocomposites as additives in polyalphaolefin. *Tribol. Int.* **2017**, *114*, 282–289. [CrossRef]
4. Huang, J.; Tan, J.; Fang, H.; Gong, F.; Wang, J. Tribological and wear performances of graphene-oil nanofluid under industrial high-speed rotation. *Tribol. Int.* **2019**, *135*, 112–120. [CrossRef]
5. Zhai, W.; Lu, W.; Liu, X.; Zhou, L. Nanodiamond as an effective additive in oil to dramatically reduce friction and wear for fretting steel/copper interfaces. *Tribol. Int.* **2019**, *129*, 75–81. [CrossRef]
6. Asnida, M.; Hisham, S.; Awang, N.W.; Amirruddin, A.K.; Noor, M.M.; Kadirgama, K.; Ramasamy, D.; Najafi, G.; Tarlochan, F. Copper (II) oxide nanoparticles as additive in engine oil to increase the durability of piston-liner contact. *Fuel* **2018**, *212*, 656–667. [CrossRef]
7. Kotia, A.; Ghosh, G.K.; Srivastava, I.; Deval, P.; Ghosh, S.K. Mechanism for improvement of friction/wear by using Al_2O_3 and SiO_2/Gear oil nanolubricants. *J. Alloys Compd.* **2019**, *782*, 592–599. [CrossRef]
8. Wang, Y.; Wan, Z.; Lu, L.; Zhang, Z.; Tang, Y. Friction and wear mechanisms of castor oil with addition of hexagonal boron nitride nanoparticles. *Tribol. Int.* **2018**, *124*, 10–22. [CrossRef]
9. Jin, Y.S.; Li, S.H.; Zhang, Z.Y.; Yang, H.; Wang, F. In situ mechanochemical reconditioning of worn ferrous surfaces. *Tribol. Int.* **2004**, *37*, 561–567.
10. Yu, Y.; Gu, J.; Kang, F.; Kong, X.; Mo, W. Surface restoration induced by lubricant additive of natural minerals. *Appl. Surf. Sci.* **2007**, *253*, 7549–7553. [CrossRef]
11. Zhang, J.; Tian, B.; Wang, C. Long-term surface restoration effect introduced by advanced silicate based lubricant additive. *Tribol. Int.* **2013**, *57*, 31–37. [CrossRef]
12. Zhang, B.; Xu, B.; Xu, Y.; Ba, Z.; Wang, Z. An amorphous Si–O film tribo-induced by natural hydrosilicate powders on ferrous surface. *Appl. Surf. Sci.* **2013**, *285*, 759–765. [CrossRef]
13. Yu, H.L.; Xu, Y.; Shi, P.J.; Wang, H.M.; Zhao, Y.; Xu, B.S.; Bai, Z.M. Tribological behaviors of surface-coated serpentine ultrafine powders as lubricant additive. *Tribol. Int.* **2010**, *43*, 667–675. [CrossRef]
14. Zhang, B.; Xu, Y.; Gao, F.; Shi, P.; Xu, B.; Wu, Y. Sliding friction and wear behaviors of surface-coated natural serpentine mineral powders as lubricant additive. *Appl. Surf. Sci.* **2011**, *257*, 2540–2549. [CrossRef]
15. Zhang, Y.; Li, Z.; Yan, J.; Ren, T.; Zhao, Y. Tribological behaviours of surface modified serpentine powder as lubricant additive. *Ind. Lubric. Tribol.* **2016**, *68*, 1–8. [CrossRef]
16. Yin, Y.L.; Yu, H.L.; Wang, H.M.; Song, Z.Y.; Zhang, Z.; Ji, X.C.; Cui, T.H.; Wei, M.; Zhang, W. Friction and wear behaviors of steel/bronze tribopairs lubricated by oil with serpentine natural mineral additive. *Wear* **2020**, *456*, 203387. [CrossRef]
17. Rao, X.; Sheng, C.; Guo, Z.; Zhang, X.; Yin, H.; Xu, C.; Yuan, C. Anti-friction and self-repairing abilities of ultrafine serpentine, attapulgite and kaolin in oil for the cylinder liner-piston ring tribo-systems. *Lubr. Sci.* **2022**, *3*, 34. [CrossRef]
18. Bai, Z.; Li, G.; Zhao, F.; Yu, H. Tribological Performance and Application of Antigorite as Lubrication Materials. *Lubricants* **2020**, *8*, 93. [CrossRef]

19. Nan, F.; Xu, Y.; Xu, B.; Gao, F.; Wu, Y.; Tang, X. Effect of natural attapulgite powders as lubrication additive on the friction and wear performance of a steel tribo-pair. *Appl. Surf. Sci.* **2014**, *307*, 86–91. [CrossRef]
20. Yu, H.L.; Wang, H.M.; Yin, Y.L.; Song, Z.Y.; Zhou, X.Y.; Ji, X.C.; Wei, M.; Shi, P.J.; Bai, Z.M.; Zhang, W. Tribological behaviors of natural attapulgite nanofibers as an additive for mineral oil investigated by orthogonal test method. *Tribol. Int.* **2021**, *153*, 106562. [CrossRef]
21. Nan, F.; Zhou, K.; Liu, S.; Pu, J.; Fang, Y.; Ding, W. Tribological properties of attapulgite/La_2O_3 nanocomposite as lubricant additive for a steel/steel contact. *RSC Adv.* **2018**, *8*, 16947–16956. [CrossRef]
22. Nan, F.; Xu, Y.; Xu, B.; Gao, F.; Wu, Y.; Li, Z. Tribological performance of attapulgite nano-fiber/spherical nano-Ni as lubricant additive. *Tribol. Lett.* **2014**, *56*, 531–541. [CrossRef]
23. Subramanian, B.; Mohan, S.; Jayakrishnan, S.; Jayachandran, M. Structural and electrochemical characterization of Ni nanostructure films on steels with brush plating and sputter deposition. *Curr. Appl. Phys.* **2007**, *7*, 305–313. [CrossRef]
24. Shang, T.; Zhang, H.Z.; Jiang, H.; Dai, L.; Zhang, G.L.; Liu, X.L. Fabrication of nanocrystalline nickel coatings by brush plating. *Surf. Eng.* **2013**, *5*, 342–345. [CrossRef]
25. Wang, L.; Liu, W.; Wang, X. The preparation and tribological investigation of Ni–P amorphous alloy nanoparticles. *Tribol. Lett.* **2010**, *2*, 381–387. [CrossRef]
26. Qiu, S.; Zhou, Z.; Dong, J.; Chen, G. Preparation of Ni nanoparticles and evaluation of their tribological performance as potential additives in oils. *J. Tribol.* **2001**, *123*, 441–443. [CrossRef]
27. Wagner, C.D.; Riggs, W.M.; Davis, L.E.; Moulder, J.F.; Muilenberg, G.E. *Handbook of X-ray Photoelectron Spectroscopy*; Perkin-Elmer Corporation: Eden Prairie, MN, USA, 1979.
28. Branley, W.F. The structural scheme of attapulgite. *Am. Mineral.* **1940**, *25*, 405–410.
29. Bhushan, B. *GeSR. Introduction to Tribology*; China Machine Press: Beijing, China, 2007. (In Chinese)
30. Nan, F.; Xu, Y.; Xu, B.; Gao, F.; Wu, Y.; Li, Z. Effect of Cu Nanoparticles on the Tribological Performance of Attapulgite Base Grease. *Tribol. Trans.* **2015**, *58*, 1031–1038. [CrossRef]
31. Lee, J.; Cho, S.; Hwang, Y.; Lee, C.; Kim, S.H. Enhancement of lubrication properties of nano-oil by controlling the amount of fullerene nanoparticle additives. *Tribol. Lett.* **2007**, *28*, 203–208. [CrossRef]

Disclaimer/Publisher's Note: The statements, opinions and data contained in all publications are solely those of the individual author(s) and contributor(s) and not of MDPI and/or the editor(s). MDPI and/or the editor(s) disclaim responsibility for any injury to people or property resulting from any ideas, methods, instructions or products referred to in the content.

Article

An Experimental Investigation of the Tribological Performance and Dispersibility of 2D Nanoparticles as Oil Additives

Kishan Nath Sidh, Dharmender Jangra and Harish Hirani *

Department of Mechanical Engineering, Indian Institute of Technology, Delhi 110016, India; mez228327@mech.iitd.ac.in (K.N.S.)
* Correspondence: hirani@mech.iitd.ac.in

Abstract: The present study aims to investigate the tribological performance of 2D nanoparticles such as graphene (G), molybdenum disulfide (MoS_2), hexagonal boron nitride (hBN), and reduced graphene oxide (rGO) as gear lubricant additives. A new method of additive doping in gear lubricants was proposed and examined in terms of the degradation of lubricants. The additives were energized by ultrasonication, thermal agitation, and mechanical shearing to enhance the dispersibility and stability, which were confirmed using visual and rheological analysis. Further, the tribological performance of the nano-additives was studied by doping them in fresh lubricants, chemically degraded lubricants, and chemically degraded lubricants with surfactants. The results indicate that surface roughness and the method of mixing play a crucial role in reducing wear. The nano-additives exhibit an inverse relationship with the roughness, and their agglomeration results in a decline in performance. To mitigate agglomeration, oleic acid surfactant was employed, which diminished the effects of nano-additives and degraded the lubricant. The attenuated total reflectance-Fourier transform infrared spectroscopy (ATR-FTIR) analysis revealed that the oleic acid and deteriorating reagent work synergistically, leading to enhanced wear volume and reduced friction. The nano-additives were characterized using field emission scanning electron microscopy (FESEM) and transmission electron microscopy (TEM). Overall, the study presents a comprehensive plan for new method of additive mixing, stability, dispersibility and tribological performance of the selected 2D nanoparticles.

Keywords: nano-additives; gear lubricant; TEM; FESEM; wear; surfactant

Citation: Sidh, K.N.; Jangra, D.; Hirani, H. An Experimental Investigation of the Tribological Performance and Dispersibility of 2D Nanoparticles as Oil Additives. *Lubricants* 2023, 11, 179. https://doi.org/10.3390/lubricants11040179

Received: 22 March 2023
Revised: 13 April 2023
Accepted: 14 April 2023
Published: 17 April 2023

Copyright: © 2023 by the authors. Licensee MDPI, Basel, Switzerland. This article is an open access article distributed under the terms and conditions of the Creative Commons Attribution (CC BY) license (https://creativecommons.org/licenses/by/4.0/).

1. Introduction

The growing global demand for energy has emphasized the need for more energy-efficient components of moving machinery. Wear and friction are major contributors to machinery inefficiency and material loss [1,2]. Several researchers [3–11] have proposed using various solid/liquid lubricants and low shear material coatings, chosen based on the application and its responsiveness, to address this issue. Spikes et al. [12] provided a detailed classification and listing of liquid lubricants formulated with molecular additives to confer specific properties based on their intended use.

In recent years, researchers [13–16] have focused on the use of solid lubricants as compounds in liquid lubricants to study their composite effects, which can be synergistic or antagonistic [12]. There has been a growing interest among researchers in the synthesis of 2D nanoparticles [17,18] and the investigation of their tribological performance [19] under various operating conditions. This is a result of the exceptional ultra-low shear strength exhibited by 2D nanoparticles, which is a result of their material isotropy and controlled layer orientation [17].

The tribological efficacy of 2D nanomaterials in boundary lubrication is proportional to the number of atomic layers and the interlayer shear force [17,20], and the tribological behaviour is determined using techniques such as atomic force microscopy (AFM) and friction force microscopy (FFM) [15,17,18]. The number of layers has a direct relationship with friction force [20]. However, boundary lubrication has limitations, particularly with the critical thickness of the coated layer, which can be worn away over time [18]. The reactivity of the nanomaterial with water and corrosive environments can lead to the breakdown of the deposited layer [21].

When dispersed in base oils, synthesized nanomaterials exhibit a high tendency to agglomerate [18,22,23], necessitating physical and chemical treatments [23,24] to enhance their stability. Temperature, time, and dispersion rate have a significant impact on the stability of dispersed nanoparticles. In addition, oxidation and acidification of the lubricant have a significant impact on the nanoparticles' stability [25]. Physical treatments, such as ultrasonication and homogenization, and chemical treatments, such as surface modification and the use of surfactants, have been used to increase the nanoparticles' stability [17]. It has been discovered that ultrasonication and homogenization effectively reduce the aggregation of nanoparticles by dispersing the clusters into smaller particles [24]. Surface modification entails the coating of nanoparticles with functional groups to enhance their dispersion and stability in lubricants [23]. It has been discovered that the use of surfactants [26–29] improves the dispersion and stability of nanoparticles in lubricants by reducing interparticle forces and averting agglomeration.

Numerous studies [8,11,18,30–47] have investigated the tribological efficacy of nano-additives in base oil. According to these studies, the incorporation of nano-additives improves the load carrying capacity, frictional properties, and wear resistance of lubricants significantly. The working mechanisms of nano-additive modified lubricants are currently based on four main theories [40,47]: (a) the ball effect (nanoparticles function as a rolling ball between the contacting surfaces); (b) tribochemical reactions form a thin tribofilm separating contacting surfaces; (c) the adsorption film theory; and (d) the mending effect. The mending effect theory proposes that nano-additives can restore and renew worn machine surfaces during operation [30].

It is possible to improve the properties of widely available lubricants by dispersing granules of solid lubricant that are either layered, flake-like, or spherical [14,32–34]. Synergistic improvements in tribological performance [43–45] can be achieved by dispersing granules of molybdenum disulfide, graphite, boron nitride, graphite, fullerene, Cu, CuO, functionalized graphene, silicon oxide, carbon nanotubes, etc., in lubricants [17]. Several studies [14,17,19] have investigated the impacts of oils containing dispersed solid lubricants on wear and friction. The optimum concentration, morphology, size, oil solubility, and particle hardness were found to have a significant impact on the tribological efficacy of these additives [17,19]. Few researchers categorize nano-additives from 0-D to 3-D [47], based on the number of dimensions they have. Finding the optimal concentration and dimension of nano-additives is crucial for optimal performance. However, the use of nano-additives is constrained by operational variables such as temperature, load, and contact type [18]. Therefore, when using nano-additives for tribological uses, it is important to consider the aforementioned factors.

The effects of load, shear, particle size, and humidity on lubricant behaviour have been studied [18,48–53], and optimization methods [49] have been used to select appropriate lubricant formulations. Lubricants' responses to load and stress are crucially important [48]. Because of the deformation and tension that high loads put on contacting surfaces, friction and wear can be made worse. A lubricant's viscosity, thickness, and film strength all contribute to how well it reduces friction and wear under heavy weights; high shear rates can cause the lubricant to experience substantial mechanical and thermal stresses, lowering its effectiveness [50]. These problems can be mitigated, and the lubricant's performance can be improved by picking the right additives and basic oils. Furthermore, humidity is a significant element that can affect lubricant performance. When the humidity is too high, acids and other corrosive species form in the lubricant, which can cause serious harm to the mating surfaces. Scientists have created cutting-edge formulations with improved water separation and rust protection properties [51] to lessen the lubricant's vulnerability to the effects of moisture. Additionally, the impacts of polytetrafluoroethylene (PTFE) particles as grease additives were studied by Kumar et al. [30]. Results showed that tiny, spherical particles performed better than other shapes when tested. In addition, Kumar et al. [31] looked into how talc nanoparticles in oil would affect tribological performance. The results showed that anti-wear properties were improved by 43% at an optimum concentration of 2 wt%, while extreme pressure properties were improved by 17%. The significance of particle size, shape, and concentration in the selection of nano-additives for tribological uses is demonstrated by these findings. The efficiency of the grease and the longevity of the equipment both benefit from careful consideration of these factors. With an eye toward oil-based, water-based, and metal-derivative additives, Xiao and Liu [14] performed a thorough review of 2D nanomaterial lubricant additives. Graphene, with a thickness of only 0.335 nm and a hexagonal structure, was reported as the thinnest nanomaterial in the review, which focused mainly on the role of 2D materials in lowering friction and wear. The authors also described how functionalizing graphene with GO, iron oxide, and zirconium oxide improved its performance by creating a hybrid with shearing and rolling characteristics. Some nanoparticles, like WS_2 and MX_2, tend to aggregate in base oil because of their high density, but MoS_2 showed excellent dispersibility and relatively low load-bearing capacity, as noted by the authors. To enhance tribological performance in different lubrication applications, these results highlight the significance of choosing the appropriate 2D nanomaterial and its functionalization. Research by Huang et al. [32] on the effects of doping paraffin oil with graphene nanosheets found that the inclusion of these nanosheets increased the oil's resistance to friction, wear, and load. At a concentration of 0.05% graphene, How et al. [33] found a 33% and 34% decrease in friction and wear in synthetic oil.

In addition, Xie et al. [34], who investigated the tribological performance of SiO_2/graphene mixtures, reported that a mixture of 0.1 wt% nano-SiO_2 and 0.4 wt% graphene in water decreased the friction coefficient by about 48% and the wear volume by about 79%. These studies lay the groundwork for the synergistic use of a combination of nano-additives in lubrication applications, in which the tribological performance of the lubricant can be significantly improved through the proper selection of additive concentration and combination. The effect of two-dimensional (2D) materials on interfacial friction was the subject of a recent systematic study by Guo et al. [15] and Marian et al. [43]. Nanomaterial integration, the authors note, can ease surface changes that lessen surface roughness. Furthermore, suitable base lubricants can encourage surface self-healing, which further reduces friction and wear. These results indicate that surface modification incorporating 2D materials can enhance interfacial friction and wear properties.

Recent research has largely centred on finding optimal nano-additives to improve lubricant performance. However, there is a lack of published work on restoring the effectiveness of lubricants that have deteriorated during storage or while in use. Limited studies have been conducted on degraded grease [54] and liquid lubricants [55]. Given this, the current research set out to examine how four different 2D nanoparticles—G, MoS_2, hBN, and rGO—improve the tribological performance of chemically deteriorated lubricants. Chemically degraded gear lubricants of API GL4 grade were used in the experiments. The impact of surface roughness and surfactant addition on the performance of the nano-additives investigated, and a new methodology for effectively dispersing the 2D nano-additives in the oil was developed.

2. Materials and Methods
2.1. Materials
2.1.1. Lubricant

The study employs the commercially available API GL4 EP90 (VI = 90) grade gear lubricant, Maharashtra, India. At a shear rate of $100~s^{-1}$, the lubricant has a dynamic viscosity of 0.21 Pa·s.

Artificial Lubricant Degradation

The lubricant is degraded artificially by mixing aqueous HCl in the manner described in reference [55]. The lubricant's life can be estimated using the following relationship:

$$\text{Time} = 10^{18.02} \times (\% \text{ aqueous HCl })^{4.75} \quad (1)$$

In the current study, aqueous HCl doped at 0.0025% v/v degrades the GL4 EP90 oil to an equivalent degraded life of ~7520 h.

2.1.2. Test Samples

The tests were performed on a 12 mm × 12 mm × 12 mm cuboid block made of EN24 steel. The hardness of the block is maintained at 40 ± 1 HRC. While looking for better nano-additives for liquid lubricants, it is necessary to use superfinish tribo-surfaces because specific film thickness, which determines whether lubrication regimes are boundary, mixed, or hydrodynamic, is determined by the surface roughnesses of the tribo-surfaces. Therefore, all of the test surfaces were superfinished using grit numbers 600, 1000, and 2000 abrasive sandpapers. The average surface roughness of all surfaces decreased from "Ra = 0.1817 ± 0.05 μm, and Rq = 0.3615 ± 0.119 μm" to "Ra = 0.032 ± 0.005 μm, and Rq = 0.051 ± 0.0009 μm". A disc of 40 mm diameter with an average hardness of 60 HRC and an average roughness of "Rq = 0.046 ± 0.007 μm" was used as the counter surface.

2.1.3. Nano-Additive Particles

Potential lubricants must provide superlubricity and maximum wear resistance to increase service life and ensure equipment dependability. Four distinct nanomaterials (graphene (G), reduced graphene oxide (rGO), hexagonal boron nitride (hBN), and molybdenum disulfide (MoS_2)) with high bond strength between atoms in the same layer and low shear strength between layers were chosen as potential candidates for lubricating tribo-surfaces. The purpose of this research is to develop a new generation of lubricants with improved tribological (superlubricity and zero-wear) properties to improve the energy efficiency and reliability of existing mechanical systems. Table 1 contains the specifics.

Table 1. Details of selected nanoparticles (as per the OEM_ Vedayukt India Pvt. Ltd., Jamshedpur, India).

S. No.	Particle Nomenclature	Average Size of Particles (nm)
1	Graphene (G)	
2	MoS_2	50–60 nm with the purity of 99.99%
3	hBN	
4	rGO	

2.2. Methodology

Figure 1 depicts the current study's overall framework. In the study, the framework consists of four steps. Initially, the 2D nanoparticles were dried at 100 °C. Subsequently, the 2D nanoparticles were mixed into the lubricant using the method proposed in the literature [56]. Additionally, a new method for dispersing the 2D nanoparticles in the lubricant is proposed. Both methods are described in greater detail in subsequent sections.

After preparing the colloidal solution, it was used for tribological testing in a lubricity tester and rheological analysis. The dried 2D nanoparticles and lubricity tester samples were processed for morphological and elemental characterisation.

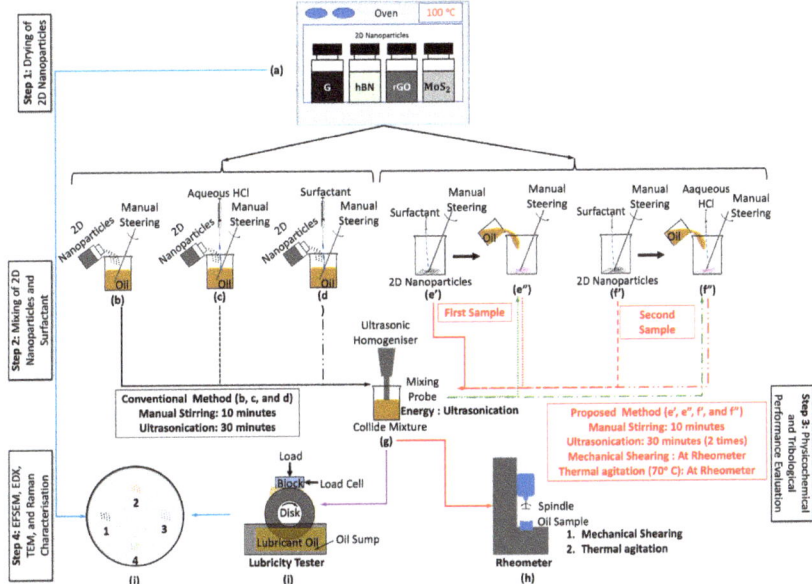

Figure 1. Schematic of comprehensive framework. (**a**) Drying of the 2D nanoparticles using the oven at 100 °C, (**b–d**) samples prepared by employing the conventional method, (**e′, e″, f′,** and **f″**) samples prepared by employing the proposed method, (**g**) ultrasonic homogenizer, (**h**) rheometer (MCR-102, Anton Paar India made), (**i**) lubricity tester (Ducom India made), and (**j**) characterization of the dried powder and post-test samples (samples 1–4).

2.2.1. Preparation of the Nanoparticle Contained Improved Lubricants
Process of Mixing of the Nanoparticles in Liquid Lubricant—Conventional Method

In this method, a volume of 40 mL of lubricant is placed in a glass beaker, and a predetermined weight percentage of nanoparticles is added to the lubricant. The mixture is then manually stirred for a duration of 10 min to ensure thorough mixing of the nanoparticles within the lubricant. In total, three samples were prepared using this method:

- Figure 1b depicts the samples prepared using the conventional method, where the nanoparticles are mixed with fresh oil, and the solution is manually stirred.
- Figure 1c, the nanoparticles are mixed with fresh oil and manually stirred, and the solution is chemically degraded using aqueous HCl.
- Figure 1d, the nanoparticles are mixed with fresh oil, and a surfactant, oleic acid, is added after manually stirring the solution.

After mixing, the mixture is sonicated for 30 min using an ultrasonic sonicator with a probe diameter of 10 mm and a pulse on/off duration of 3 s. In order to homogenise the mixture and improve the dispersion of nanoparticles within the lubricant, this stage is crucial.

Process of Mixing of the Nanoparticles in Liquid Lubricant—Proposed Method

When using 2D nanomaterials, dispersibility in oil is a major issue. Because of the high surface-to-volume ratio of 2D materials and compelling van der Waals forces, nanomaterials appear to have a high tendency to aggregate. The interaction of nanoparticles with the liquid lubricant, as well as the interaction of nanoparticles with each other, can be managed to control nanoparticle dispersion. Energy (ultrasonication), thermal agitation, and mechanical shearing can all be used to reduce particle agglomeration. In fact, heterostructures (combining one nano-additive with other kinds of nano-additives: hBN and MoS_2) will aid in the deagglomeration process because the surface of one type of nano-additive can be coated with another, reducing the cohesive force among similar nano-additive particles. Nanoparticles may deagglomerate during the regular sonication process, but over time, because of a stronger binding force, they often re-agglomerate. The concept of building heterostructures out of two or more nanomaterials to improve their durability was proposed as a solution to this issue. Under mechanical shearing, uncoiled nanoparticles exfoliate to produce nano-sheets of bimaterials or trimaterials with weaker cohesion forces. The nanoparticles are less likely to rebound under these conditions, and the dispersibility of nanomaterials will be effective.

To test this hypothesis, a small sample was prepared using 3 mL of lubricant and a corresponding concentration of nanoparticles. The sample was sonicated conventionally and then subjected to mechanical shearing for 10 min using a rheometer with a shear rate of $1000~s^{-1}$ and a temperature of 70 °C. The result of consistent viscosity over time at $1000~s^{-1}$ shear rate hinted at the efficacy of the proposed method. However, a different setup was required to scale up because the rheometer can only hold 2–3 mL of oil sample.

To test the same hypothesis, a larger sample was made with 5 mL of lubricant and nanoparticles corresponding to 40 mL of the sample, as shown in Figure 1(e' and f'). Three samples (Case 1—Fresh lubricant, Case 2—0.4 wt% of rGO, MoS_2, hBN of each mixed with fresh lubricant and case 3—0.4 wt% graphene, 0.4 wt% rGo and 0.2 wt% of MoS_2 mixed with fresh lubricant) were prepared to maintain heterogeneity. The mixture was sonicated for 30 min before being mixed with the remaining 35 mL of oil (Figure 1(e")). The sample was sonicated once more to achieve a temperature of around 70 °C. These samples were tested in a lubricity tester at a high mechanical shear rate ($10,000~s^{-1}$) while maintaining a very thin film thickness, and we compared these friction results. It's worth noting that friction force decreases over time (as shown in Figure 2). The physical mechanisms of generating nanosheets and forming protective 2D films of nanomaterials on steel tribo-interfaces are responsible for the possibility of such positive friction results.

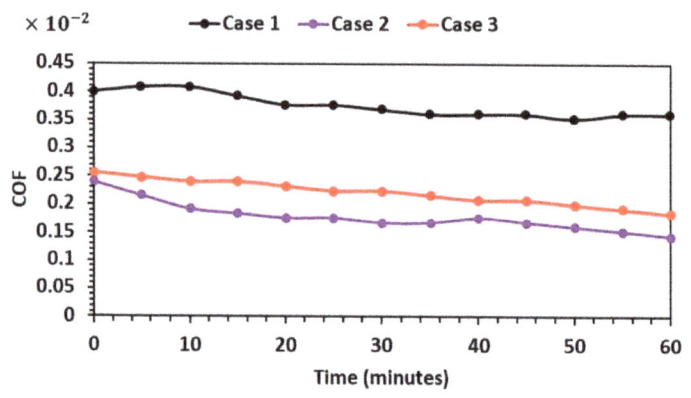

Figure 2. Variation of COF with time.

To summarize, the proposed method of generating heterostructures (as seen in Section 3) of the selected nanomaterials via mechanical exfoliation following sonication shows promise for improving the stability of 2D nano lubricants.

2.2.2. Experimental Design and Methodology

For the purpose of estimating the efficacy of nano-additives, a series of experiments (three levels of each nano-additive) were created using the L9 orthogonal array [57]. The chosen factors, their values, and the experimental strategy are displayed in Table 2.

Table 2. Experimental plan and levels of different variables.

Variable	Level		
	1	2	3
Graphene (G)	0	0.2 wt%	0.4 wt%
rGO	0	0.2 wt%	0.4 wt%
MoS_2	0	0.2 wt%	0.4 wt%
hBN	0	0.2 wt%	0.4 wt%
Experiment Design			
Experiment No./Sample ID	Graphene (in wt%)	rGO (in wt%)	MoS_2 (in wt%)	hBN (in wt%)
---	---	---	---	---
L1	0	0	0	0
L2	0	0.2	0.2	0.2
L3	0	0.4	0.4	0.4
L4	0.2	0	0.2	0.4
L5	0.2	0.2	0.4	0
L6	0.2	0.4	0	0.2
L7	0.4	0	0.4	0.2
L8	0.4	0.2	0	0.4
L9	0.4	0.4	0.2	0

The lubricating ability of a nano-lubricant is well known to be extremely sensitive to the concentration of nanoparticles. While low concentrations of particles may not help to improve tribological performance, high concentrations of nanoparticles increase the likelihood of particle agglomeration. According to the literature that is currently accessible, tribological performance has been enhanced using graphene in the range of 0.1% to 4% w/w [58], rGO in the range of 0.03% to 0.5% w/w [59], MoS_2 in the range of 0.02% to 2% w/w [15,60], and hBN in the range of 0.1 to 0.5% w/w [61]. In the present study, nanoparticles have been maintained in the range of 0 to 0.4% w/w.

When using the lubricity testing apparatus (See Figure 3), the block (test specimen) was pressed against the spinning disc to gauge the tribological performance of the synthesized lubricants. Prior to and following the test, the test specimens' weights were measured using a balance with a minimum count of 0.00001 g. Low rotational speed (100 rpm) of the disc and high normal load (125.86 N) on the block were kept, maintaining harsh operating conditions (boundary to mixed lubrication regimes). The detailed parameters selected for the study are listed in Table 3.

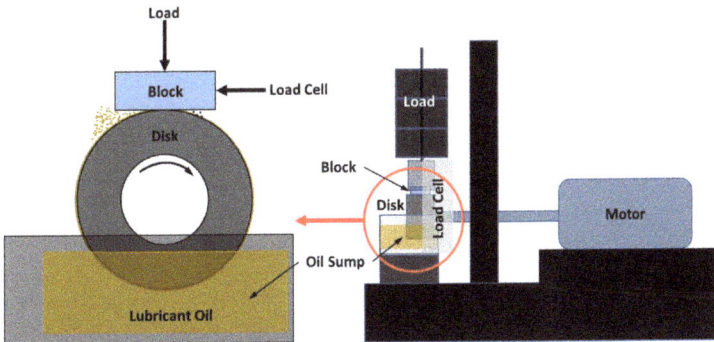

Figure 3. Schematic of the lubricity tester.

Table 3. Parameters and their levels for wear and friction analysis.

Parameters	Levels
Test sample	EN24 with 40 ± 1 HRC
Roughness (μm)	Ra: 0.032 ± 0.005
	Rq: 0.051 ± 0.009
Load (N)	125.86
Time (s)	3600
Speed (rpm)	100
Temperature (°C)	40

2.3. Experimental Setup

The tribological performance of the nanoparticles was evaluated using a lubricity tester manufactured by Ducom, India. Figure 3 depicts the test setup, which engages a cubical EN24 steel block on a hardened steel disc (counter surface). The induction motor drives this disc, which is partially immersed in lubricant. The lubricant inside the tank was kept at 40 degrees Celsius by an inbuilt heater and thermal cut-off switch.

2.4. 2D Nanoparticles Morphological Analysis (FESEM, EDX, TEM and Raman Analysis)

This paper examines the tribological performance of two-dimensional (2D) nanomaterials as lubricant additives. The underlying physical mechanisms, the creation of nanosheets, the development of heterostructures from the chosen nanomaterials, the development of a protective 2D film on steel tribo-interfaces, and the stable dispersion will depend mainly on intrinsic factors, such as the thickness, surface characteristics, and structural defects of the purchased 2D materials. Field emission scanning electron microscopy (EFSEM), energy dispersive X-ray spectroscopy (EDX), transmission electron microscopy (TEM), and Raman analysis were used to shed light on the morphology and elemental makeup of the particles.

2.4.1. FESEM, EDX and TEM Analysis of 2D Nanoparticles

The morphology and topography of the nanosheets can be seen in FESEM pictures, and their elemental makeup and any impurities that may already be present can be determined by EDX analysis. The MoS_2 and hBN were discovered to be aggregated and to have an irregular and spherical form, respectively, as shown in Figure 4 (a–d, FESEM's images). High magnification images of MoS_2 and hBN revealed a small, irregular sheet agglomerated shape. The rGO and graphene, on the other hand, were discovered in the shape of flakes, with layers agglomerated on top of one another. The strips of rGO and graphene were discovered in a crumpled and wrinkly state. It is also possible that rGO and graphene's crumpled and wrinkled appearance is due to the particles' extreme thinness, which makes them bend and crumple readily.

Furthermore, EDX spectra showed that the particles were primarily composed of their base elements, except for graphene, which had a manganese impurity of 0.21%. (Mn). The 0.21% impurity level, though low, is not likely to have a major effect on their performance. Strong van der Waals pressures caused the particles to aggregate into small sheets, as was observed.

In order to validate the nature, crystallinity, and size distribution of the nanosheets, as well as to find any impurities within the 2D sheets that might have an impact on their performance, the particles were also subjected to TEM analysis. The results of the TEM examination in Figure 5a–d showed that the particles were layered, with a thickness of about 1 nm and a width of a few nanometres (as shown in the figure). As a result, it was verified that the particles were 2D.

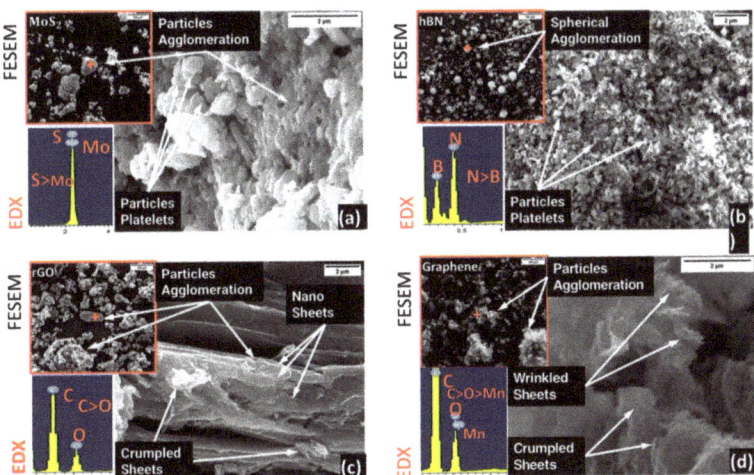

Figure 4. FESEM images and EDX of the (**a**) MoS$_2$, (**b**) hBN, (**c**) rGO, and (**d**) graphene.

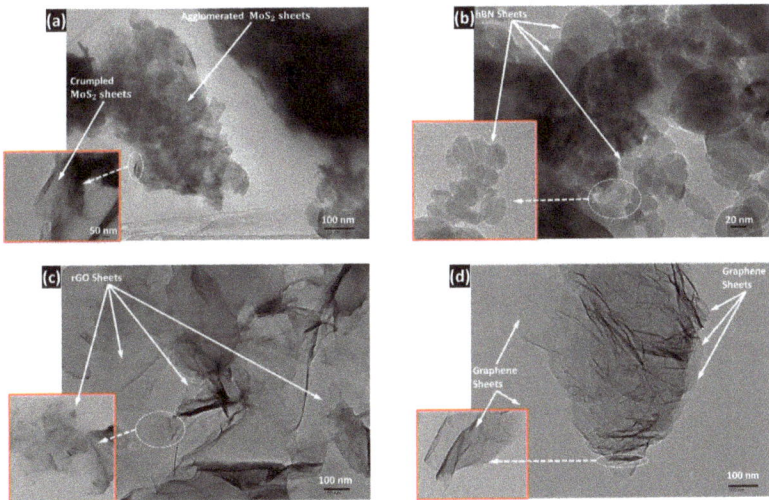

Figure 5. TEM images of the (**a**) MoS$_2$, (**b**) hBN, (**c**) rGO, and (**d**) graphene.

2.4.2. Raman Analysis of 2D Nanoparticles

The Raman spectra of four 2D nanoparticles (shown in Figure 6), including graphene, reduced graphene oxide (rGO), molybdenum disulfide (MoS$_2$), and hexagonal boron nitride (hBN), were obtained using a microscopic Raman spectrometer with a 532 nm laser wavelength. For graphene and rGO, the D, G, and 2D bands were observed, with the D band attributed to defects and disordered carbon and the G band representing the vibration of ordered sp^2–C in a 2D hexagonal lattice. The G, D, and 2D bands are affected by the number of layers and the graphene material's quality. The G band is typically found at around 1586 cm^{-1} and corresponds to the stretching vibration of graphene's sp^2 carbon–carbon bonds. The 2D band, which is sensitive to the number of graphene layers, is located at around 2894 cm^{-1}. The ratio of 2D to G band intensities (233/2259) can be used to estimate the number of graphene layers, with higher ratios indicating fewer layers.

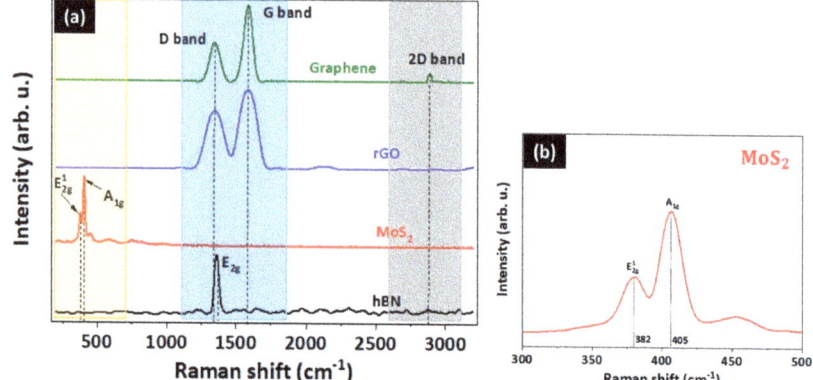

Figure 6. (a) Raman spectra of the graphene, rGO, MoS$_2$, and hBN, and (b) enlarged spectra of the MoS$_2$.

The ID/IG (Graphene—'2259/1124' and rGO—'2359/1735') value was used to assess the degree of disorder and defect, and rGO's lower ID/IG value compared to graphene indicated a lower degree of disordering and defects in the GO structure. The analysis of MoS$_2$ revealed two distinct peaks at 382 cm^{-1} and 405 cm^{-1}, corresponding to the E_{2g}^1 and A_{1g} mode of the material, respectively. The number of layers present in the sample was estimated by calculating the Raman shift difference between the E_{2g}^1 and A_{1g} vibration modes, which was found to be 23 cm^{-1} (shown in Figure 6b). Utilizing an empirical relationship proposed by Lee et al. [62], which relates the Raman shift corresponding to E_{2g}^1 and A_{1g} band and their difference to the thickness of the material, the thickness of the MoS$_2$ layer was determined to be between 3 and 4 layers for a 23 cm^{-1} peak difference. The Raman spectrum of hBN showed a single peak at 1367 cm^{-1} corresponding to the E_{2g} mode, which was attributed to the in-plane vibrational motion of the boron and nitrogen atoms. No other peaks in the spectrum demonstrate the purity of hBN nanoparticles.

3. Results and Discussions

In accordance with the designed experiments outlined in Section 2.2.2, tests were conducted on a lubricity tester. Prior to these tests, the impact of sample surface roughness was evaluated through two experiments utilizing blocks with differing roughness values, Rq 0.362 and 0.053, while utilizing parameters specified in Table 3. Wear was observed and quantified through mass loss and material density, resulting in wear volumes of 28.182×10^{-12} and 7.686×10^{-12} m^3, respectively, as depicted in Figure 7. The reduction of surface roughness Rq from 0.362 µm to 0.053 µm was found to lead to a ~73% decrease in wear volume. Because high surface roughness causes the system to operate under the boundary lubrication regime (specific film thickness 0.49, Appendix A), more asperity contact occurs, resulting in increased wear. The schematic representation of the contact is shown in Figure 8, where asperities are present in the contact, leading to increased wear. Nanoparticles will relatively be more useful on rough surfaces, where the 2D nanoparticles fill the valleys, reducing the overall roughness and wear of the contacting surfaces.

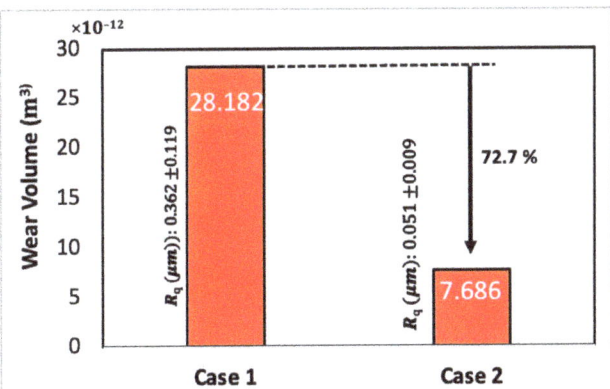

Figure 7. Effect of surface roughness on the wear volume.

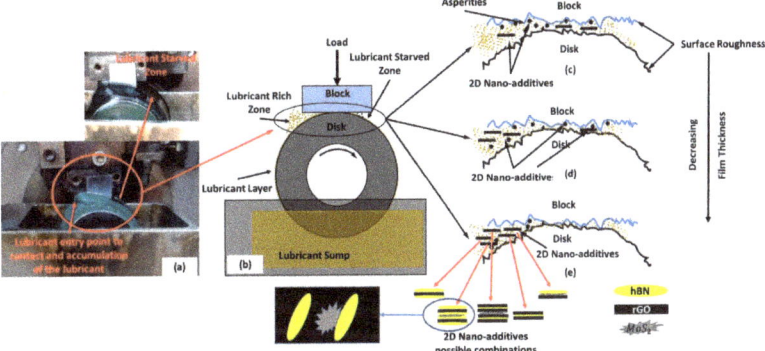

Figure 8. The images and schematic show the running condition (**a**,**b**) as lubricant get blocked at the entry point and (**c**–**e**) the effect of surface roughness and particles deposition on each other that enhance its stability.

Furthermore, as the film thickness decreases, the shear rate approaches $10,000\ s^{-1}$. Under such high shear rates, hetero-nanomaterials may rub against each other, transfer layers, and form heterostructures (Figure 8e) of nanoparticles that aid in friction reduction.

3.1. Specific Film Thickness Calculation

The viscosity of a lubricant is an important factor. Temperature and shear rate have a significant impact on viscosity, which is important for determining film thickening. The rheometer was used to determine the viscosity of three selected lubricant samples, L1, L3, and L9, as listed in Table 2. As shown in Figure 9a, viscosity measurements were obtained by modulating the shear rate. Changes in viscosity with varying shear rates revealed that viscosity curves are a result of both Newtonian and non-Newtonian behaviour in certain locations. To better understand the behaviour of these three nanofluids, we ran further tests on the viscosity of the nanofluids in three different regions—Region 1 shear rate $100–500\ s^{-1}$ (Figure 9b), Region 2 shear rate $1000–1500\ s^{-1}$ (Figure 9c), and Region 3 shear rate $1500–2000\ s^{-1}$ (Figure 9d).

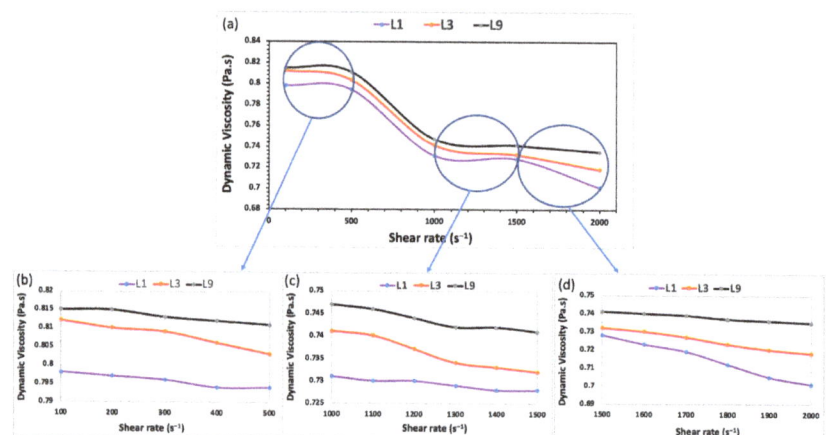

Figure 9. Dynamic viscosity variation to shear rate (**a**) shear rate 100–2000 s^{-1} (**b**) shear rate 100–500 s^{-1} (**c**) shear rate 1000–1500 s^{-1} (**d**) shear rate 1500–2000 s^{-1}.

Our findings demonstrate that all the samples display non-Newtonian behaviour, despite the fact that the degree of non-Newtonian behaviour varies for various regions in contrast to the region where the shear rate is 500–1000 s^{-1}. Investigating the viscoelastic behaviour of such non-Newtonian samples is worthwhile, as is contrasted with the behaviour of pure (base) oil. The characteristics of the base oil are unknown to the authors because the OEM did not disclose data relating to the base oil of the commercial gear oil [63] used in the current study. As a result, we were unable to test the viscoelastic behaviour of synthesized lubricants. Furthermore, as shown in Figure 10, it was challenging to obtain accurate data when we attempted to expand experiments to shear rates higher than 2000 s^{-1} due to lubricant splashing out.

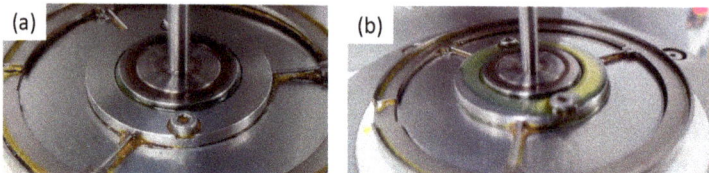

Figure 10. Rheometer setup (**a**) shear rate 1000 s^{-1} (**b**) shear rate 2500 s^{-1}.

By incorporating nanoparticles into oil, the viscosity of the lubricant is raised (improving anti-wear effectiveness) while also increasing the oil's shear thinning tendency. Lubricant effective viscosity at operating temperature and shear rate must be specified for the experimental task. The lubricant's dynamic viscosity was measured at 1000 s^{-1} at two temperatures (40 and 70), as listed in Table 4. Walther's relation [64,65] was used to calculate the lubricant's dynamic viscosity at 60 °C. It should be noted that the lubricant's viscosity varies with shear rate. The shear rate at the contact region was around 0.2 million s^{-1}, implying that the viscosity would decrease further under these conditions.

The procedure of calculating a particular film thickness detailed in Appendix A.

The calculated specific film thickness is in the range of 2–3 and the computed contact pressure is less than 150 MPa, confirming the mixed lubrication condition.

Table 4. Specific film thickness for different samples.

	Specific Film Thickness			
Experiment No./Sample ID	Viscosity (Pa·s) (@40 °C)	Viscosity (Pa·s) (@70 °C)	Viscosity (Pa·s) (@60 °C, Using Equation (A8))	Specific Film Thickness
L1	0.731	0.157	0.247	2.529
L3	0.741	0.171	0.264	2.649
L9	0.747	0.174	0.268	2.671

3.2. Tribological Evaluation

The tribological performances of nanoparticles, for both combining nanoparticles in fresh oil (without aqueous HCl) and mixing nanoparticles in chemically degraded (with aqueous HCl mixed) lubricants, were evaluated using the lubricity tester (as shown in Figure 3). The wear mass was calculated by weighing the test specimen before and after the test with an accuracy of 0.00001 g. Figure 11 depicts the obtained results for wear volume and coefficient of friction. The sample ID corresponds to the designed experiments specified in Table 2. The wear volume increased with the addition of the aqueous HCl, as shown in Figure 11a. From Figure 11b, it can be observed that the coefficient of friction (COF) for the chemically degraded oil shows a lower value as compared to fresh oil. The observed anomaly is attributed to the development of an oxide layer [12] on the test specimen, which facilitates the attachment of friction modifiers and subsequently leads to a reduction in friction. To investigate this, we selected four samples (L1 and L3, with fresh and degraded lubricant) to determine why the performance of L3 with fresh lubricant is superior and does not adhere as well with degraded lubricant. Surfaces of test blocks were investigated by conducting FESEM and EDX analyses on all four samples. More detailed information about the experimental setup, methods, and results can be found in Section 3.3.

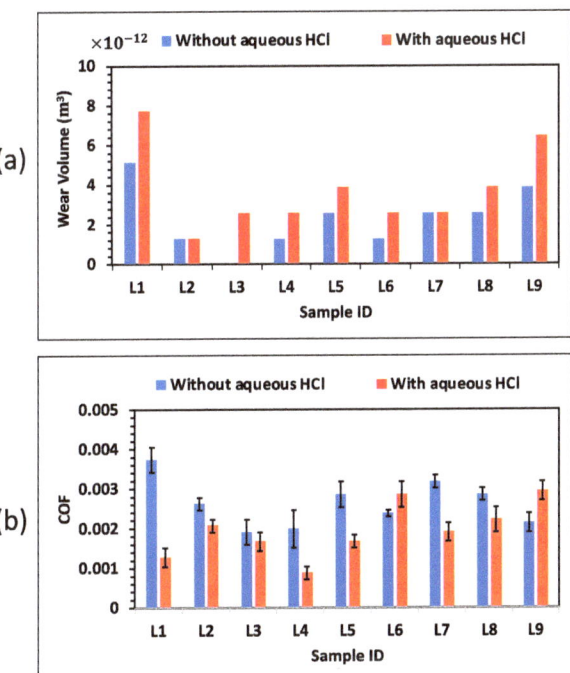

Figure 11. (**a**) Wear volume, and (**b**) coefficient of friction for fresh oil (without aqueous HCl) and chemically degraded lubricant (with aqueous HCl mixed).

Sample L3 exhibited the greatest 'wear resistance' of the nano-additives mixed in conventional gear oil, as the measured wear is nearly zero. Moreover, sample L9 exhibits the worst performance, as the wear volume for nano-additives mixed in chemically degraded oil was 6.405×10^{-12} m^3. To investigate these results and comprehend the function of nanoparticles, it was decided to enhance the efficacy of nano-additives with surfactants like oleic acid. For a comprehensive understanding, both samples (L3 with the highest performance and L9 with the lowest performance) were chosen for testing. The oleic acid is expected to drop the agglomeration. However, as shown in Figure 12, the addition of oleic acid decreases the COF while increasing the volume of wear. The data presented in Figure 12a indicates a substantial reduction in wear, with a maximum decrease of 75% when compared to the wear rate of the initial base lubricant. Moreover, the addition of nano-additives to degraded lubricants can minimize wear by approximately 83%. In Figure 12b, the findings demonstrate that the nano-additives effectively decrease the average coefficient of friction in the range of 23.4% to 42.53% compared to the base lubricant.

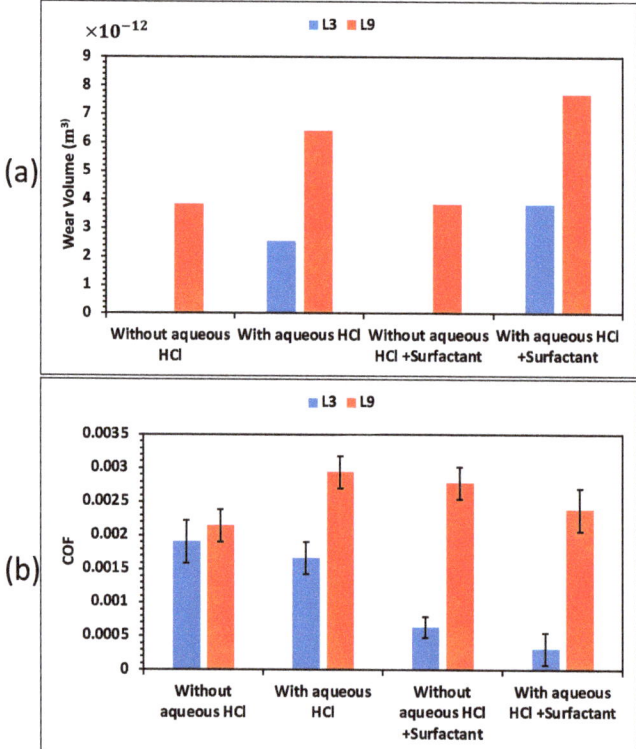

Figure 12. (a) Wear volume, and (b) COF for the best performance and worst performance cases for both set of experiments.

According to the literature, oleic acid enhances the dispersibility and stability of nano-additive particles through polar bonding with them. However, the increased wear suggests that oleic acid (OA) may react with either the test surface or the chemically aged oil. To confirm this, four samples were selected for attenuated total reflectance-Fourier transform infrared spectroscopy (ATR-FTIR) analysis: 'pure oil', 'best performing combination sample ID L3 mixed with aqueous HCl (L3@HCl)', 'L3 mixed with OA (L3@OA)', and 'L3@HCl + OA'. Figure 13 shows the ATR-FTIR spectra of the samples.

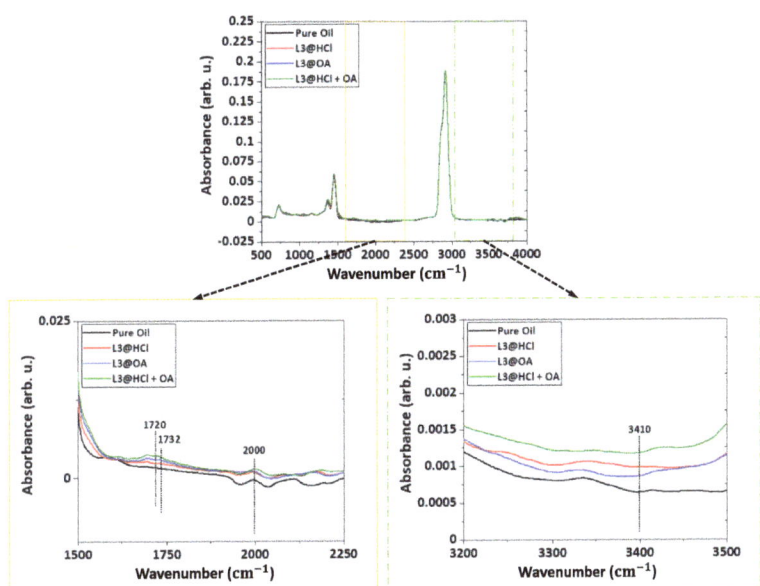

Figure 13. ATR-FTIR spectra of the prepared lubricant.

It can be observed from Figure 13 that the addition of aqueous HCl and OA leads to soot formation in the gear oil and acidification of the gear oil, as evidenced by an increase in the peak at about 2000 cm^{-1} and 1720–1732 cm^{-1}. Increase in the peak at 3410 cm^{-1} indicates the formation of moisture. Increased moisture results from the disintegration of OA into carboxylic compounds, free oxygen, and hydrogen. It is also observed that aqueous HCl and OA act synergistically to degrade the lubricant, leading to an increase in wear volume and a decrease in friction. The formed oxide layer was ideal for the attachment as friction modifiers [12], resulting in a decrease in overall friction.

Based on the observations, it is possible to conclude that the formation of oxide layers on the surface decreases the coefficient of friction (COF) but increases wear in the form of oxides when the lubricant has degraded. The study also suggests that the nano-additives' ineffectiveness may be due to their aggregation, which inhibits their ability to improve wear performance. In terms of wear, the study indicates that oleic acid did not exhibit positive results. To enhance the efficacy of nano-additives, improved dispersion techniques must be researched.

By calculating the mean wear mass of the blocks, the wear performance of the four selected nanomaterials was examined at three different lubricant concentration levels (0, 0.2, and 0.4 wt%). The "smaller is better" criterion [55] has been applied to the measured wear mass to assess the specific effects of each 2D nano additive. The results have been plotted in Figure 14. This figure indicates that, among the selected nano-additives, the hBN nanomaterial consistently exhibits the lowest wear values. The hBN nanomaterial showed the lowest wear values while using fresh gear lubrication for both concentration levels (0.2 and 0.4 wt%); however, when using degraded lubricant, it showed the lowest wear values at the second level of lubricant concentration (0.2 wt%). Notably, the minimum mean wear for hBN, which is the lowest of all nano-additives, is still greater than the best performing heterogeneous nano-lubricant combinations (L3 with fresh oil and L2 with degraded oil). This indicates that heterogeneity in 2D nonadditive behaviour plays a significant role in limiting agglomeration and positively influencing wear and friction performance. In conclusion, these findings indicate that the hBN nanomaterial is the most effective of the selected nano-additives tested in this research at reducing wear, although it is less effective than the heterogeneous mixture.

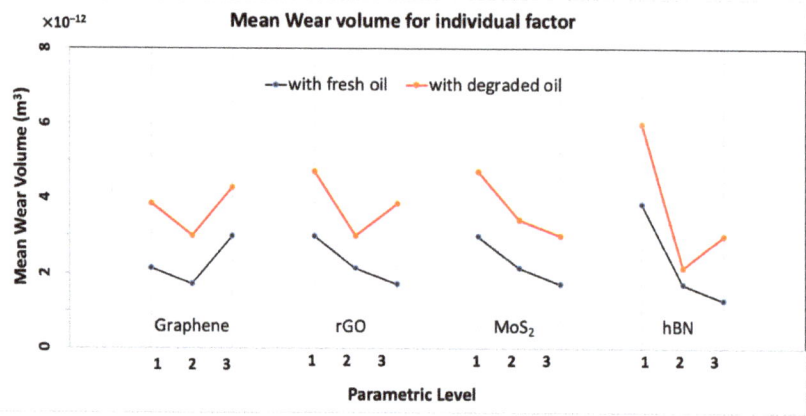

Figure 14. Comparison among nano-additives.

3.3. FESEM and EDX Analysis of the Worn out Surfaces

Two samples from each set of experiments were taken based on the aforementioned findings and processed for FESEM imaging and EDX elemental analysis. L1 and L3 are the chosen sample IDs for comparing the two cases.

According to Figure 15a,c, a thin oxide layer in the worn track is formed for the sample ID "L1" aqueous HCl mixed lubricant without any nano-additive. The oxide layer decreases the overall friction value but increases the wear rate. The EDX spectra shows a tiny amount of oxide in the form of elevated oxygen levels between 3.20 and 7.62 wt%, as well as an extra chloride component at 0.19 wt%. Additionally, it may be deduced from FESEM images that pure gear oil exhibits more plastic deformation than oil that has undergone chemical deterioration.

As seen in Figure 15b,d, sample ID "L3" exhibits zero wear when fresh oil is combined with 2D nano-additives but increases wear when aqueous HCl is added to degrade it. It can be observed that, for fresh lubricant, a smooth topography was found (FESEM images), but a more metallic passive layer was observed under degraded lubricant usage, which gets scrapped and reformed by successive sliding between the surface and relaxation.

Hence, it can be said that the HCl reacts with the test specimen and form a passive oxide layer that leads to increased wear. Further, as reported in reference [55], it also deteriorates the lubricant by increased oxide formation.

We carried out more experiments on a sample with ID "L1" to investigate the phenomena of disc wear. Prior to processing the disc for FESEM imaging and EDX elemental analysis, the disc was weighed before and after the experiment to assess the wear. The geometry of the disc poses certain challenges for FESEM and EDX. To overcome this, two cutting techniques—wire cut electro discharge machining (EDM) and abrasive cutting—were utilised to segment the disc profile, as illustrated in Figure 16.

The findings from Figure 17a,b demonstrate that only a small oxide layer forms in the worn track when using gear oil (sample ID "L1") doped with aqueous HCl. The EDX spectra show that there is very little oxide, with oxygen levels of 2.4 wt% when the disc was segmented with wire EDM and 0.9 wt% when the disc was segmented with an abrasive cutter. The chloride component was discovered to be approximately 0.2 wt%. Despite the minimal oxide formation, the disc exhibited no measurable wear.

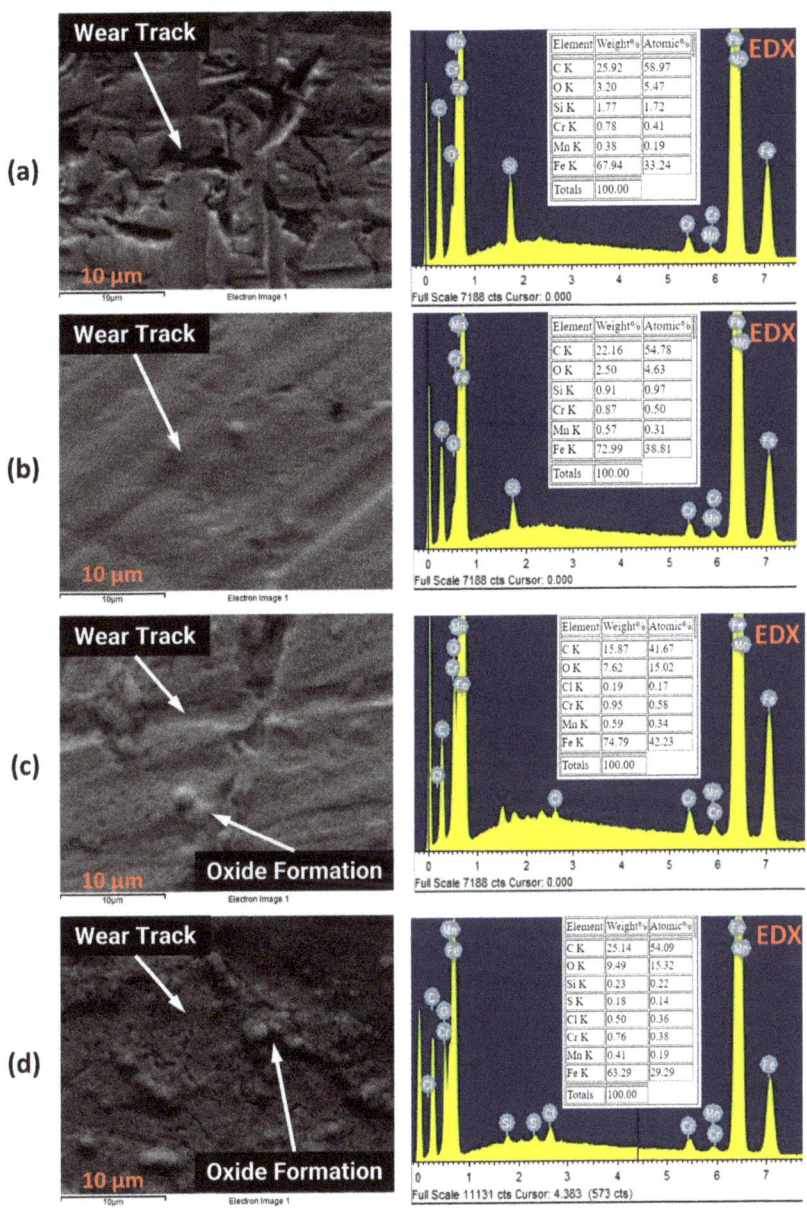

Figure 15. FESEM and EDX of the tested blocks using two samples of fresh oil (ID (**a**) L1 and (**b**) L3) and two samples of chemically degraded oil (ID (**c**) L1 and (**d**) L3).

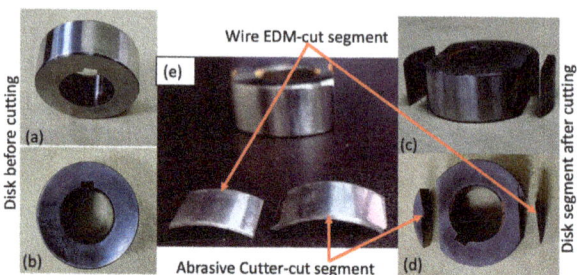

Figure 16. Segmentation of disk profile for FESEM and EDX analysis; (**a**,**b**) disk before cutting; (**c**,**d**) disk after segmentation; (**e**) profilometry of segmented disk.

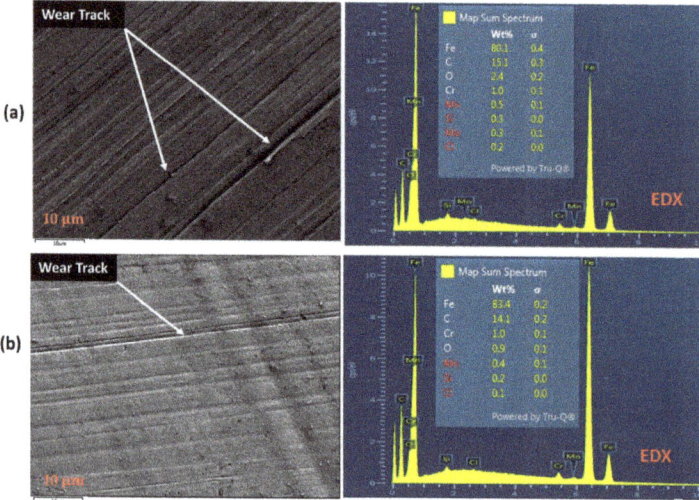

Figure 17. FESEM imaging and EDX of the disk profile lubricated with chemically degraded oil sample ID L1; (**a**) wire EDM cut segment; (**b**) abrasive cut segment.

In conclusion, our findings support the hypothesis that the formation of oxide layers on the surface can reduce the coefficient of friction (COF) but can also increase wear in the form of oxides when lubricated with degraded oil. The additional experiments to track the damage to the disk profile provided further insights into this phenomenon, despite the challenges posed by the disk's geometry.

3.4. Comparison of the Conventional Methodology of Mixing the 2D Nano Additives and the New Proposed Methodology

To evaluate the efficacy of the new mixing methodology that has been proposed, sedimentation/agglomeration phenomena related to nanoparticles were investigated. Three methods are used to assess sedimentation: visual examination, optical imaging, and evaluation of the viscosity/rheological properties.

3.4.1. Visual Inspection

In order to conduct a visual examination, the nanoparticles were mixed in a fixed ratio of 0.4 wt% and allowed to settle for 24 h. As seen in Figure 18a, all nanoparticles exhibit some degree of sedimentation when mixed using the conventional technique. The sedimentation range for all the particulates was between 2.5 scale point height (rGO) to 5 scale point height (hBN and graphene) and scale point height for MoS_2 was 3.

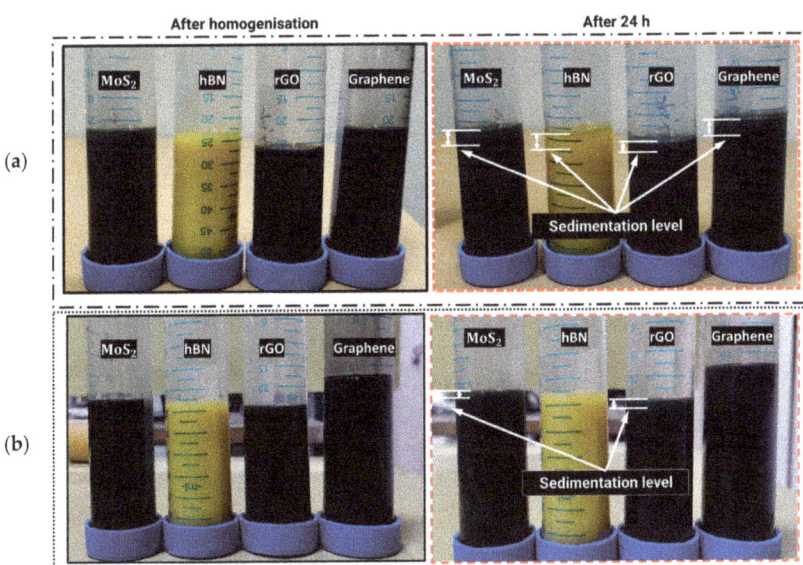

Figure 18. Effect of mixing methodology on the particle sedimentation; (**a**) old method of mixing, and (**b**) new proposed mixing method.

The nanoparticles were mixed utilising the proposed technique and all prepared samples are illustrated in Figure 18b. Only for MoS$_2$ and rGO was deposition of 1.5 scale point height discovered. Therefore, it is pretty apparent from this that the new mixing methodology provides better stability than the earlier method.

3.4.2. Inspection through Optical Imaging

In this instance, a digital microscope with an 800× magnification was used to assess the stability of the nano additives. The solution is made up by mixing the 0.4 wt% of graphene, 0.4 wt% of rGO and 0.2 wt% MoS$_2$, and prepared by both methods. Pictures of the solutions were captured for three hours.

As seen in Figure 19, the old technique of mixing the particles results in the particles clumping together in various locations. When compared to the oil technique, the new proposed method's particles are quite stable and do not exhibit any significant agglomeration.

3.4.3. Viscosity Measurement

To investigate the homogeneity and dispersion stability of the mixed 2D nanoparticles over time, the dynamic viscosities of the particles were studied. For 300 s, the dynamic viscosity was recorded at a shear rate of 1000 s^{-1}. The two oils were synthesized by blending gear oil with the nanoparticles (0.4 wt% of graphene, 0.4 wt% of rGO, 0.2 wt% MoS$_2$) and aqueous HCl using the conventional method (OM) and the proposed method (NPM). The third oil was pure oil (PO). Each sample underwent three tests, as depicted in Figure 20. The viscosity of PO originally decreased with time, but after 70 s it remains almost constant. The viscosity of NPM oil exhibits the same pattern. When it comes to the OM, viscosity changes erratically over time, before becoming nearly constant after 100 s. It is clear from Figure 20 that the NPM oil has a greater viscosity value, which may help to reduce the wear value. These findings clearly show that the proposed mixing approach exhibits a high degree of homogeneous blending and consistent stability over time.

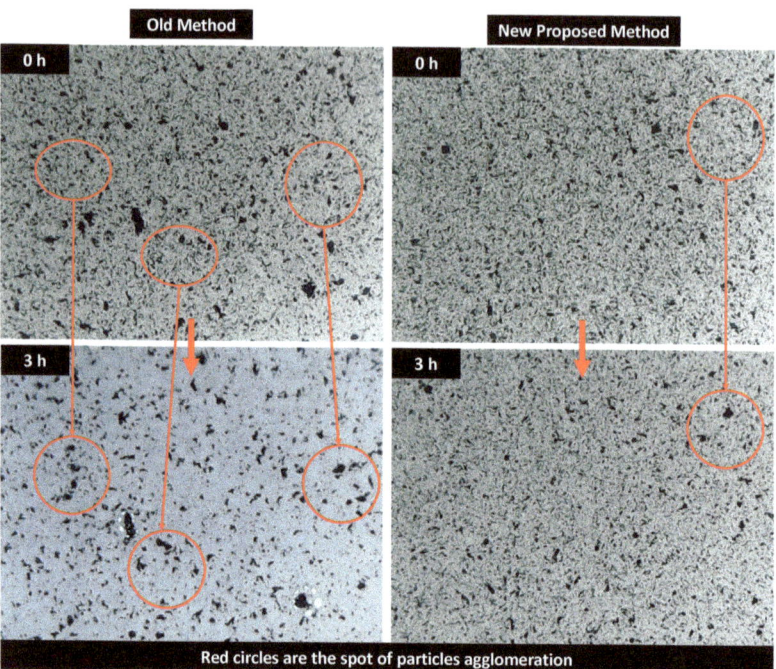

Figure 19. Images of the homogenised particles initially and after 3 h for both the old method and the newly proposed method.

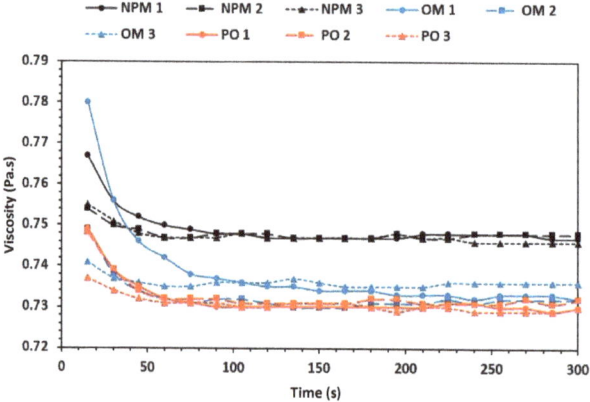

Figure 20. Dynamic viscosity at shear rate 1000 s^{-1} and for a duration of 300 s.

We also examined whether the new method has an impact on performance of sample ID L3 (as described in Table 2) with aqueous HCl which was mixed in lubricant. The results of the test showed that the wear of the sample was reduced by 50% compared to previous tests, as illustrated in Figure 21a. Additionally, the friction of the sample was reduced by 33.33%, as shown in Figure 21b.

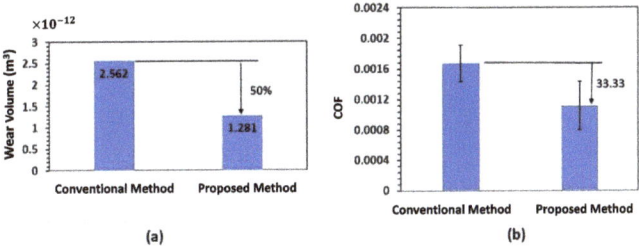

Figure 21. (a) Wear and (b) Coefficient of friction comparison between conventional and proposed method.

4. Conclusions

The current research focused on the effect of 2D nanoparticles on the tribological performance of liquid lubricants and proposes a new method of mixing 2D nanoparticles into the lubricant. Based on the findings, the following conclusions can be drawn:

(a) Surface roughness has a direct relationship with wear volume, with increased roughness resulting in more boundary lubrication and increased asperity contact. Lowering surface roughness by 85% can decrease wear volume by 72.7%.

(b) Graphene-based nanolubricants remain ineffective in improving the performance of chemically degraded lubricants.

(c) The proposed dispersion method for mixing the 2D nanoparticles was confirmed to reduce agglomeration and enhance the lubricant consistency. The results confirm a wear reduction of 50% and a friction reduction of 33.33%, compared those results obtained from sample synthesised by the conventional method.

(d) The average coefficient of friction reduction achieved by the nano-additives, compared to the base fluid, ranges from 23.4% to 42.53%.

(e) The implementation of two-dimensional (2D) nano-additives with exceptionally thin longitudinal dimensions has demonstrated a significant reduction in wear, of up to 75% compared to the fresh base lubricant, and up to approximately 83% when utilized with deteriorated lubricant.

Based on the study's findings, it can be concluded that the efficacy of 2D nanoparticles is affected by both the mixing method, aging of lubricant, and the lubrication regime. To tap the potential of 2D nanolubricants, more research is needed, particularly in the field of dispersion of 2D materials in a liquid phase, utilizing advanced surface analysis techniques and molecular dynamics simulations practices. Both microscopic experiments and nanoscopic tests are required to correctly assess the different interactions at the interfaces of the nano-additives, liquid, and solid surfaces. To reutilize the developed knowledge, there is a need to build a model library using machine learning techniques to extract the relevant features of the constructed model to apply the optimized model to selection of nanoparticles percentage, sonication, and high strain rate at higher temperature to improve the performance of nanosheets.

Author Contributions: Conceptualization, H.H.; methodology, H.H.; resources, H.H.; supervision, H.H.; writing-editing of the manuscript H.H. and D.J.; data curation, D.J. and K.N.S.; formal analysis, D.J. and K.N.S.; writing—original draft preparation, D.J. and K.N.S. All authors have read and agreed to the published version of the manuscript.

Funding: This research received no external funding.

Data Availability Statement: Data can be shared on request.

Conflicts of Interest: The authors declare no conflict of interest.

Appendix A

Appendix A.1. Specific Film Thickness

The specific film thickness was computed by using the following relation [66]:

$$\lambda = \frac{h_{min}}{\sqrt{R_{q1}^2 + R_{q2}^2}} \tag{A1}$$

The oil film thickness h_{min} is calculated using the Dowson–Higginson formula [66].

$$h_{min} = 0.985 \times (U_*)^{0.7} \times (G_*)^{0.6} \times (W_*)^{-0.11} \tag{A2}$$

where U_*, G_*, W_* are dimensionless parameters. These can be computed using the following relations:

$$G_* = \alpha \times E' \tag{A3}$$

$$U_* = \mu_0 \times \frac{(u_1 + u_2)}{E' \times R_x} \tag{A4}$$

$$W_* = \frac{W}{E' \times R_x \times L} \tag{A5}$$

$$a = \frac{8 \times W \times R_x}{E' \times \pi \times L} \tag{A6}$$

$$P_0 = \frac{2 \times W}{\pi \times a \times L} \tag{A7}$$

where:

α = Pressure viscosity coefficient
L = Face width of the gear
W = Load normal to contact
μ_0 = Dynamic viscosity
R_q = Peak-to-valley surface roughness
R_x = Effective radius
P_0 = Hertzian pressure
a = Contact half width
E' = Equivalent modulus of elasticity

The measured temperature (using an infrared thermometer) at the contact area in the current research was found to be around 60 °C. Walther's relation [65] is used to calculate the dynamic viscosity of the lubricant at this temperature. To use this equation for the present study, the lubricant density was assumed to be constant across all samples, at around 900 kg/m^3. The equation is as follows:

$$\text{loglog}(cSt + 0.6) = A - B\log(T) \tag{A8}$$

where cSt is the kinematic viscosity, T is the temperature in Kelvin, and A and B are constants.

Parameters used for study:

$\alpha = 1.2 \times 10^{-8}$ m^2/N
$L = 0.012$ m
$W = 125$ N
$\mu_0 = 0.24$ to 0.26 Pa·s
$R_x = 0.02$ m
$E' = 226$ GPa

So, $h_{min} = \sim 0.18$ μm
Specific film thickness (for roughness of block $R_q = 0.362$ μm and roughness of ring 0.046 μm) = 0.49 (boundary lubrication).

References

1. Holmberg, K.; Andersson, P.; Erdemir, A. Global energy consumption due to friction in passenger cars. *Tribol. Int.* **2012**, *47*, 221–234. [CrossRef]
2. Holmberg, K.; Andersson, P.; Nylund, N.-O.; Mäkelä, K.; Erdemir, A. Global energy consumption due to friction in trucks and buses. *Tribol. Int.* **2014**, *78*, 94–114. [CrossRef]
3. Grutzmacher, P.G.; Suarez, S.; Tolosa, A.; Gachot, C.; Song, G.; Wang, B.; Presser, V.; Mücklich, F.; Anasori, B.; Rosenkranz, A. Superior wear-resistance of Ti3C2T × multilayer coatings. *ACS Nano* **2021**, *15*, 8216–8224. [CrossRef]
4. Singer, I.L. How Third-Body Processes Affect Friction and Wear. *MRS Bull.* **1998**, *23*, 37–40. [CrossRef]
5. Holmberg, K.; Ronkainen, H.; Laukkanen, A.; Wallin, K. Friction and wear of coated surfaces—Scales, modelling and simulation of tribomechanisms. *Surface Coat. Technol.* **2007**, *202*, 1034–1049. [CrossRef]
6. Khadem, M.; Penkov, O.V.; Yang, H.-K.; Kim, D.-E. Tribology of multilayer coatings for wear reduction: A review. *Friction* **2017**, *5*, 248–262. [CrossRef]
7. Subramanian, C.; Strafford, A.K. Review of multicomponent and multilayer coatings for tribological applications. *Wear* **1993**, *165*, 85–95. [CrossRef]
8. Bartz, W.J. Tribology, lubricants and lubrication engineering—A review. *Wear* **1978**, *49*, 1–18. [CrossRef]
9. Boyde, S. Green lubricants. Environmental benefits and impacts of lubrication. *Green Chem.* **2002**, *4*, 293–307. [CrossRef]
10. Syahir, A.Z.; Zulkifli, N.W.M.; Masjuki, H.H.; Kalam, M.A.; Alabdulkarem, A.; Gulzar, M.; Khuong, L.S.; Harith, M.H. A review on bio-based lubricants and their applications. *J. Clean. Prod.* **2017**, *168*, 997–1016. [CrossRef]
11. Cai, M.; Yu, Q.; Liu, W.; Zhou, F. Ionic liquid lubricants: When chemistry meets tribology. *Chem. Soc. Rev.* **2020**, *49*, 7753–7818. [CrossRef]
12. Spikes, H.A. Additive-additive and additive-surface interactions in lubrication. *Lubr. Sci.* **1989**, *2*, 3–23. [CrossRef]
13. Yan, Z.; Liu, J.; Wang, C.; Lu, X.; Hao, J. Synergistic effect and long-term lubricating mechanism of WS2/oil combinations determined via oil molecule structures in vacuum. *Tribol. Int.* **2023**, *179*, 107997. [CrossRef]
14. Xiao, H.; Liu, S. 2D nanomaterials as lubricant additive: A review. *Mater. Des.* **2017**, *135*, 319–332. [CrossRef]
15. Guo, Y.; Zhou, X.; Lee, K.; Yoon, H.C.; Xu, Q.; Wang, D. Recent development in friction of 2D materials: From mechanisms to applications. *Nanotechnology* **2021**, *32*, 312002. [CrossRef] [PubMed]
16. Liu, L.; Zhou, M.; Li, X.; Jin, L.; Su, G.; Mo, Y.; Li, L.; Zhu, H.; Tian, Y. Research Progress in Application of 2D Materials in Liquid-Phase Lubrication System. *Materials* **2018**, *11*, 1314. [CrossRef] [PubMed]
17. Zaharin, H.; Ghazali, M.; Thachnatharen, N.; Ezzah, F.; Walvekar, R.; Khalid, M. Progress in 2D materials based Nanolubricants: A review. *Flatchem* **2023**, *38*, 100485. [CrossRef]
18. Uzoma, P.C.; Hu, H.; Khadem, M.; Penkov, O.V. Tribology of 2D Nanomaterials: A Review. *Coatings* **2020**, *10*, 897. [CrossRef]
19. Gulzar, M.; Masjuki, H.H.; Kalam, M.A.; Varman, M.; Zulkifli, N.W.M.; Mufti, R.A.; Zahid, R. Tribological performance of nanoparticles as lubricating oil additives. *J. Nanoparticle Res.* **2016**, *18*, 223. [CrossRef]
20. Pantano, M.F.; Kuljanishvili, I. Advances in mechanical characterization of 1D and 2D nanomaterials: Progress and prospects. *Nano Express* **2020**, *1*, 022001. [CrossRef]
21. Pritchard, C.; Midgley, J. The effect of humidity on the friction and life of unbonded molybdenum disulphide films. *Wear* **1969**, *13*, 39–50. [CrossRef]
22. Jazaa, Y.; Lan, T.; Padalkar, S.; Sundararajan, S. The Effect of Agglomeration Reduction on the Tribological Behavior of WS2 and MoS2 Nanoparticle Additives in the Boundary Lubrication Regime. *Lubricants* **2018**, *6*, 106. [CrossRef]
23. Hwang, Y.; Lee, J.-K.; Lee, J.-K.; Jeong, Y.-M.; Cheong, S.-I.; Ahn, Y.-C.; Kim, S.H. Production and dispersion stability of nanoparticles in nanofluids. *Powder Technol.* **2008**, *186*, 145–153. [CrossRef]
24. Feng, J.; Mao, J.; Wen, X.; Tu, M. Ultrasonic-assisted in situ synthesis and characterization of superparamagnetic Fe_3O_4 nanoparticles. *J. Alloy. Compd.* **2011**, *509*, 9093–9097. [CrossRef]
25. Bakunin, V.; Suslov, A.; Kuzmina, G.; Parenago, O.; Topchiev, A. Synthesis and Application of Inorganic Nanoparticles as Lubricant Components—A Review. *J. Nanoparticle Res.* **2004**, *6*, 273–284. [CrossRef]
26. Shiao, S.; Chhabra, V.; Patist, A.; Free, M.; Huibers, P.; Gregory, A.; Patel, S.; Shah, D. Chain length compatibility effects in mixed surfactant systems for technological applications. *Adv. Colloid Interface Sci.* **1998**, *74*, 1–29. [CrossRef]
27. Negin, C.; Ali, S.; Xie, Q. Most common surfactants employed in chemical enhanced oil recovery. *Petroleum* **2017**, *3*, 197–211. [CrossRef]
28. Gbadamosi, A.O.; Junin, R.; Manan, M.A.; Agi, A.; Yusuff, A.S. An overview of chemical enhanced oil recovery: Recent advances and prospects. *Int. Nano Lett.* **2019**, *9*, 171–202. [CrossRef]
29. Raffa, P.; Broekhuis, A.A.; Picchioni, F. Polymeric surfactants for enhanced oil recovery: A review. *J. Pet. Sci. Eng.* **2016**, *145*, 723–733. [CrossRef]
30. Kumar, N.; Saini, V.; Bijwe, J. Performance properties of lithium greases with PTFE particles as additive: Controlling parameter-size or shape? *Tribol. Int.* **2020**, *148*, 106302. [CrossRef]
31. Kumar, N.; Saini, V.; Bijwe, J. Exploration of Talc nanoparticles to enhance the performance of Lithium grease. *Tribol. Int.* **2021**, *162*, 107107. [CrossRef]
32. Huang, H.; Tu, J.; Gan, L.; Li, C. An investigation on tribological properties of graphite nanosheets as oil additive. *Wear* **2006**, *261*, 140–144. [CrossRef]

33. How, H.G.; Jason, Y.J.J.; Teoh, Y.H.; Chuah, H.G. Investigation of Tribological Properties of Graphene Nanoplatelets in Synthetic Oil. *J. Adv. Res. Fluid Mech. Therm. Sci.* **2022**, *96*, 115–126. [CrossRef]
34. Xie, H.; Dang, S.; Jiang, B.; Xiang, L.; Zhou, S.; Sheng, H.; Yang, T.; Pan, F. Tribological performances of SiO2/graphene combinations as water-based lubricant additives for magnesium alloy rolling. *Appl. Surface Sci.* **2019**, *475*, 847–856. [CrossRef]
35. Singh, A.; Chauhan, P.; Mamatha, T.G. A review on tribological performance of lubricants with nanoparticles additives. *Mater. Today: Proc.* **2020**, *25*, 586–591. [CrossRef]
36. Patel, J.; Pereira, G.; Irvine, D.; Kiani, A. Friction and wear properties of base oil enhanced by different forms of reduced graphene. *AIP Adv.* **2019**, *9*, 045011. [CrossRef]
37. Ali, I.; Basheer, A.A.; Kucherova, A.; Memetov, N.; Pasko, T.; Ovchinnikov, K.; Pershin, V.; Kuznetsov, D.; Galunin, E.; Grachev, V.; et al. Advances in carbon nanomaterials as lubricants modifiers. *J. Mol. Liq.* **2019**, *279*, 251–266. [CrossRef]
38. Patel, J.; Kiani, A. Effects of Reduced Graphene Oxide (rGO) at Different Concentrations on Tribological Properties of Liquid Base Lubricants. *Lubricants* **2019**, *7*, 11. [CrossRef]
39. Thampi, A.D.; Prasanth, M.; Anandu, A.; Sneha, E.; Sasidharan, B.; Rani, S. The effect of nanoparticle additives on the tribological properties of various lubricating oils—Review. *Mater. Today Proc.* **2021**, *47*, 4919–4924. [CrossRef]
40. He, T.; Chen, N.; Fang, J.; Cai, G.; Wang, J.; Chen, B.; Liang, Q. Micro/nano carbon spheres as liquid lubricant additive: Achievements and prospects. *J. Mol. Liq.* **2022**, *357*, 119090. [CrossRef]
41. Rahman, M.; Islam, M.; Roy, R.; Younis, H.; AlNahyan, M.; Younes, H. Carbon Nanomaterial-Based Lubricants: Review of Recent Developments. *Lubricants* **2022**, *10*, 281. [CrossRef]
42. Morshed, A.; Wu, H.; Jiang, Z. A Comprehensive Review of Water-Based Nanolubricants. *Lubricants* **2021**, *9*, 89. [CrossRef]
43. Marian, M.; Berman, D.; Rota, A.; Jackson, R.L.; Rosenkranz, A. Layered 2D Nanomaterials to Tailor Friction and Wear in Machine Elements—A Review. *Adv. Mater. Interfaces* **2021**, *9*, 2101622. [CrossRef]
44. Marian, M.; Berman, D.; Nečas, D.; Emami, N.; Ruggiero, A.; Rosenkranz, A. Roadmap for 2D materials in biotribological/biomedical applications—A review. *Adv. Colloid Interface Sci.* **2022**, *307*, 102747. [CrossRef] [PubMed]
45. Rosenkranz, A.; Liu, Y.; Yang, L.; Chen, L. 2D nano-materials beyond graphene: From synthesis to tribological studies. *Appl. Nanosci.* **2020**, *10*, 3353–3388. [CrossRef]
46. Ashraf, A.; Shafi, W.K.; Ul Haq, M.I.; Raina, A. Dispersion stability of nano additives in lubricating oils—An overview of mechanisms, theories and methodologies. *Tribol. Mater. Surf. Interfaces* **2022**, *16*, 34–56. [CrossRef]
47. Shahnazar, S.; Bagheri, S.; Hamid, S.B.A. Enhancing lubricant properties by nanoparticle additives. *Int. J. Hydrog. Energy* **2016**, *41*, 3153–3170. [CrossRef]
48. Muzakkir, S.M.; Hirani, H.; Thakre, G.D. Lubricant for Heavily Loaded Slow-Speed Journal Bearing. *Tribol. Trans.* **2013**, *56*, 1060–1068. [CrossRef]
49. Shinde, A.B.; Pawar, P.M. Multi-objective optimization of surface textured journal bearing by Taguchi based Grey relational analysis. *Tribol. Int.* **2017**, *114*, 349–357. [CrossRef]
50. Hirani, H.; Athre, K.; Biswas, S. Lubricant Shear Thinning Analysis of Engine Journal Bearings. *Tribol. Trans.* **2001**, *44*, 125–131. [CrossRef]
51. Das, R.; Bej, S.; Hirani, H.; Banerjee, P. Trace-level humidity sensing from commercial organic solvents and food products by an AIE/ESIPT-triggered piezochromic luminogen and ppb-level "OFF–ON–OFF" sensing of Cu^{2+}: A combined experimental and theoretical outcome. *Acs Omega* **2021**, *6*, 14104–14121. [CrossRef] [PubMed]
52. Sengupta, S.; Murmu, M.; Mandal, S.; Hirani, H.; Banerjee, P. Competitive corrosion inhibition performance of alkyl/acyl substituted 2-(2-hydroxybenzylideneamino)phenol protecting mild steel used in adverse acidic medium: A dual approach analysis using FMOs/molecular dynamics simulation corroborated experimental findings. *Colloids Surf. A: Physicochem. Eng. Asp.* **2021**, *617*, 126314. [CrossRef]
53. Bej, S.; Das, R.; Hirani, H.; Ghosh, S.; Banerjee, P. "Naked-eye" detection of CN^- from aqueous phase and other extracellular matrices: An experimental and theoretical approach mimicking the logic gate concept. *New J. Chem.* **2019**, *43*, 18098–18109. [CrossRef]
54. Rezasoltani, A.; Khonsari, M. Experimental investigation of the chemical degradation of lubricating grease from an energy point of view. *Tribol. Int.* **2019**, *137*, 289–302. [CrossRef]
55. Hirani, H.; Jangra, D.; Sidh, K.N. Experimental Investigation on the Wear Performance of Nano-Additives on Degraded Gear Lubricant. *Lubricants* **2023**, *11*, 51. [CrossRef]
56. Gupta, M.K.; Bijwe, J. A complex interdependence of dispersant in nano-suspensions with varying amount of graphite particles on its stability and tribological performance. *Tribol. Int.* **2020**, *142*, 105968. [CrossRef]
57. Box, G.E.; Hunter, J.S.; Hunter, W.G. Statistics for experimenters. In *Wiley Series in Probability and Statistics*; Wiley: Hoboken, NJ, USA, 2005.
58. Guo, P.; Chen, L.; Wang, J.; Geng, Z.; Lu, Z.; Zhang, G. Enhanced Tribological Performance of Aminated Nano-Silica Modified Graphene Oxide as Water-Based Lubricant Additive. *ACS Appl. Nano Mater.* **2018**, *1*, 6444–6453. [CrossRef]
59. Bao, Y.Y.; Sun, J.L.; Kong, L.H. Tribological properties and lubricating mechanism of SiO_2 nanoparticles in water-based fluid. In *IOP Conference Series: Materials Science and Engineering*; IOP Publishing: Bristol, UK, 2017; Volume 182, p. 12025. [CrossRef]
60. Fan, X.; Li, X.; Zhao, Z.; Yue, Z.; Feng, P.; Ma, X.; Li, H.; Ye, X.; Zhu, M. Heterostructured rGO/MoS_2 nanocomposites toward enhancing lubrication function of industrial gear oils. *Carbon* **2022**, *191*, 84–97. [CrossRef]

61. Reyes, L.; Loganathan, A.; Boesl, B.; Agarwal, A. Effect of 2D Boron Nitride Nanoplate Additive on Tribological Properties of Natural Oils. *Tribol. Lett.* **2016**, *64*, 41. [CrossRef]
62. Lee, C.; Yan, H.; Brus, L.E.; Heinz, T.F.; Hone, J.; Ryu, S. Anomalous Lattice Vibrations of Single- and Few-Layer MoS_2. *ACS Nano* **2010**, *4*, 2695–2700. [CrossRef] [PubMed]
63. HPCL India. Available online: https://www.hplubricants.in/sites/default/files/hp-gear-drive-ep-90.pdf (accessed on 10 April 2023).
64. Booser, E.R. *Applied Tribology: Bearing Design and Lubrication*; John Wiley & Sons: Hoboken, NJ, USA, 2008.
65. Ash, D.C.; Joyce, M.J.; Barnes, C.; Booth, C.J.; Jefferies, A.C. Viscosity measurement of industrial oils using the droplet quartz crystal microbalance. *Meas. Sci. Technol.* **2003**, *14*, 1955–1962. [CrossRef]
66. Amarnath, M.; Lee, S.-K. Assessment of surface contact fatigue failure in a spur geared system based on the tribological and vibration parameter analysis. *Measurement* **2015**, *76*, 32–44. [CrossRef]

Disclaimer/Publisher's Note: The statements, opinions and data contained in all publications are solely those of the individual author(s) and contributor(s) and not of MDPI and/or the editor(s). MDPI and/or the editor(s) disclaim responsibility for any injury to people or property resulting from any ideas, methods, instructions or products referred to in the content.

Article

Non-Similar Solutions of Dissipative Buoyancy Flow and Heat Transfer Induced by Water-Based Graphene Oxide Nanofluid through a Yawed Cylinder

Umair Khan [1,2], Aurang Zaib [3], Sakhinah Abu Bakar [1], Siti Khuzaimah Soid [4], Anuar Ishak [1,*], Samia Elattar [5] and Ahmed M. Abed [6,7]

1. Department of Mathematical Sciences, Faculty of Science and Technology, Universiti Kebangsaan Malaysia, Bangi 43600, Malaysia
2. Department of Mathematics and Social Sciences, Sukkur IBA University, Sukkur 65200, Pakistan
3. Department of Mathematical Sciences, Federal Urdu University of Arts, Science & Technology, Karachi 75300, Pakistan
4. Faculty of Computer and Mathematical Sciences, Universiti Teknologi MARA, Shah Alam 40450, Malaysia
5. Department of Industrial & Systems Engineering, College of Engineering, Princess Nourah bint Abdulrahman University, Riyadh 11671, Saudi Arabia
6. Department of Industrial Engineering, College of Engineering, Prince Sattam Bin Abdulaziz University, Alkharj 16273, Saudi Arabia
7. Industrial Engineering Department, Faculty of Engineering, Zagazig University, Zagazig 44519, Egypt
* Correspondence: anuar_mi@ukm.edu.my

Citation: Khan, U.; Zaib, A.; Abu Bakar, S.; Soid, S.K.; Ishak, A.; Elattar, S.; Abed, A.M. Non-Similar Solutions of Dissipative Buoyancy Flow and Heat Transfer Induced by Water-Based Graphene Oxide Nanofluid through a Yawed Cylinder. *Lubricants* 2023, 11, 60. https://doi.org/10.3390/lubricants11020060

Received: 24 November 2022
Revised: 26 January 2023
Accepted: 30 January 2023
Published: 2 February 2023

Copyright: © 2023 by the authors. Licensee MDPI, Basel, Switzerland. This article is an open access article distributed under the terms and conditions of the Creative Commons Attribution (CC BY) license (https://creativecommons.org/licenses/by/4.0/).

Abstract: The fluid flow through blunt bodies that are yawed and un-yawed frequently happens in many engineering applications. The practical significance of deep-water applications such as propagation control, splitting the boundary layer over submerged blocks, and preventing recirculation bubbles is explained by the fluid flow across a yawed cylinder. The current work examined the mixed convective flow and convective heat transfer by incorporating water-based graphene oxide nanofluid around a yawed cylinder with viscous dissipation and irregular heat source/sink. To investigate the heat diffusion across the system of buoyancy effects, the mathematical formulation of the problem was modeled in terms of coupled, nonlinear partial differential equations. The boundary value problem of the fourth-order (bvp4c) solver was operated to find the non-similarity solution. The outcomes indicated that the velocity in both directions enlarged owing to the higher impacts of yaw angle for the phenomenon of assisting flow but decreased for the instance of opposing flow, while the temperature of nanofluid increased because of heightened estimations of yaw angle for both assisting and opposing flows. In addition, with larger impacts of nanoparticle volume fraction, the shear stresses were enhanced by about 0.76% and 0.93% for the case of assisting flow, while for the case of opposing flow, they improved by almost 0.65% and 1.38%, respectively.

Keywords: nanofluid; yawed cylinder; mixed convection; convective heat condition

1. Introduction

Several significant research papers have been published over the past nearly six decades on the subject of mixed convection (MCN). Numerous scholars from around the world have produced amazing findings on mixed convection or buoyancy flow along a range of geometries. Surprisingly, very little research has investigated the mixed convective flow across a yawed cylinder (YC). The physical structure of heat exchangers must take into account the assessment of flow through yawed cylinders. In reality, this research aids in comprehending how the yaw angle parameter of a vertical cylinder affects the transport properties of the heat exchanger. The literature review reveals that few research articles on yawed cylinders have been published. The first author to explore boundary layer analysis along a yawed cylinder is Sears [1], who examined a variety of cylinder types in this circumstance. The separation point was then demonstrated to be independent of the yaw angle by Chiu and Lienhard [2] in their study of the flow past a yawed cylinder. Later, Bucker and Lueptow [3] conducted an experimental analysis of the YC under turbulent flow conditions of the boundary layer. However, the research carried out by several researchers has made the many facets of the YC in the forced convection regime more accessible. For example, Roy [4] and Subhashini et al. [5] inspected the compressible flow through a cylinder with non-uniform enthalpy without and with suction, respectively, while Roy and Saikrishnan [6] used a non-uniform suction/injection slot past a yawed cylinder by incorporating water boundary-layer. Saikrishnan [7] and Revathi et al. [8] analyzed the steady and time-dependent problems through a mass transfer slot across a yawed cylinder. Recently, Patil et al. [9] scrutinized the free and forced convective flow through a yawed cylinder and presented a non-similar solution. They noticed that the yaw angle is a significant factor of raising the motion of the prescribed fluid in both the chordwise (CW) and spanwise (SW) directions for the instance of vertical heated sheet.

A nanofluid is a continuous phase of a solid–liquid mixture containing a nanometer-sized nanoparticle scattered in regular base fluids. The precise measurement of the thermal and physical parameters of the nanofluids, such as their specific heat, viscosity, and thermal conductivity is needed to understand their heat transfer behavior. Researchers frequently use well-known expressions or correlations to anticipate physical and thermal features of nanofluids to determine their convective heat transfer behaviors. In their studies, each researcher employed a distinct model of the thermophysical features. The study of convective heat transport comprising nanofluid is an area of interest in engineering and science. Several common fluids, such as ethylene glycol, water, mineral oils, toluene, etc., have relatively poor thermal conductivity in heat transfer operations. The nanofluid is a modern sort of fluid that contains nanometer-sized particles or fibres dispersed in the regular fluid. It was first introduced by Choi [10]. More about the significance of nanofluid can be observed in the book by Avramenko et al. [11].

Nanofluids unquestionably have benefits in their ability to be more stable, appropriate in terms of viscosity, and have better spreading, wetting, and dispersion capabilities on solid surfaces. Nanofluids are employed in a variety of technical fields, including microfluidics, microelectronics, transportation, solid-state lighting, and biomedicine. Additionally, the dispersion of metal nanoparticles is being produced for various additional uses, such as cancer treatment applications in medicine. The continuous impacts of buoyancy past a vertical/orthogonal flat plate encased in a porous media containing nanofluids were investigated by Ahmed and Pop [12]. Nazar et al. [13] investigated mixed convective flow that contained nanofluids through a circular cylinder immersed in a porous medium. Rashad et al. [14] investigated the mixed convection flow across a circular horizontal cylinder in a vertically upward stream encased in porous media containing a nanofluid. The significance of heat transfer using the single and the two-phase models of nanofluids was inspected by Turkyilmazoglu [15]. He presented an analytic solution and also showed that Ag/water is a good conductor of heat transfer. Sulochana and Naramgiri [16] considered the stagnation point flow of water-based copper nanoparticles along an exponential and horizontal movable cylinder with heat absorption/generation. Rekha et al. [17] examined the impacts of thermal radiation on heat transfer flow including nanoparticles through various geometries with thermophoretic particle deposition effects. The stagnation point flow induced by a uniform rotation of the disk and magnetohydrodynamic and radial stretching with hybrid nanoparticles over a spiraling disk via an asymptotic approach was discovered by Sarfaraz et al. [18]. Mabood et al. [19] utilized a water-based hybrid nanofluid to investigate the fully developed steady forced convection flow past a stretchable surface with melting heat transfer and irregular radiation effects. The 3D stagnation point flows with a buoyancy effect through a hybrid nanofluid past a vertical plate with suction and slips impact was examined by Wahid et al. [20]. Zangooee et al. [21] utilized the impact of slip surface and stagnation-point flow through a vertical plate with hybrid nanoparticles (Cu and Al_2O_3). Recently, Malekshah et al. [22] investigated experimentally as well as numerically the impact of free convective flow and heat transfer within a cavity induced by a nanofluid. They observed that the thermal performance of the cavity is significantly impacted by the use of nanofluid and the thermal configuration of fins.

Convective heat transfer plays a significant role in processes involving high temperatures. Examples include nuclear power plants, gas turbines, thermal energy storage, etc. The convective boundary conditions (CBCs) are applicable to various technical and industrial processes, such as transpiration cooling, material drying, etc. Numerous scholars have investigated and published findings on this subject for viscous fluid because of the practical significance of CBCs. The Blasius flow and Sakiadis flow by utilizing CBCs in a viscous fluid were studied by Bataller [23]. Aziz [24] examined the steady flow and heat transfer by incorporating the CBCs and observed that a similarity solution is possible when the hot fluid is proportional to $x^{-1/2}$. Makinde [25] expanded the abovementioned work by including the buoyancy force with heat and mass transfer mechanisms. Wahid et al. [26] were able to obtain dual solutions for buoyancy effects on the magnetic time-dependent radiative flow of hybrid nanoparticles with the impact of CBCs. The features of fluid and heat transfer characteristics flow in the presence of Sisko fluid through a non-isothermal stretchable sheet with the convective condition were examined by Malik et al. [27]. Ramesh et al. [28] considered the 2D flow of dusty fluid through a convective boundary condition subject to the stretchable sheet. They observed that the temperature of the normal fluid as well as the dust fluid increased due to convective conditions. Rashid et al. [27] inspected the steady non-linear radiative tangent hyperbolic flow in hybrid nanoparticles past a stretchable sheet with CBCs and Lorentz forces. They discovered that as the magnetic parameter increases, the temperature and velocity accelerate and decelerate, respectively. Recently, Prasad et al. [28] utilized convective and surface conditions to investigate the magnetohydrodynamic flow past a Riga radially stretchable plate with a chemical reaction.

According to the aforementioned literature analysis, the investigation of buoyancy or mixed convection flow induced by water-based graphene oxide nanofluid involving irregular heat source/sink and convective boundary conditions through a yawed cylinder has not yet been conducted. Therefore, the goal of this study was to analyze the steady convection boundary-layer flow that is occurring across a yawed cylinder. The following are the novelties in the current analysis:

➢ Convective flow is driven by buoyancy force across a yawed cylinder.
➢ Impact of irregular heat source/sink, viscous dissipation and convective boundary condition is also one of the main objectives of the yawed problem.
➢ The influences of yawed angle on the velocity and temperature profiles.
➢ The contribution of water-based graphene oxide nanofluid enhances the requisite thermal characteristics or properties.
➢ The quantitative analysis of the shear stress and heat transfer for the influence of various distinct influential parameters.

Furthermore, the leading partial differential equations and boundary restriction are simplified using the non-similar variables and the local non-similarity solution technique. Then, these equations numerically solved using a boundary value problem solver. In addition, it is believed that the current study will be helpful in the practical significance of subsea applications such as boundary layer separation the above submerged blocks, transference control, and suppressing recirculating bubbles.

2. Materials and Methods

The physical model of the problem is shown schematically in Figure 1. Meanwhile, the influence of buoyancy or mixed convection flow conveying water-based graphene oxide nanofluid is assumed to flow past a vertically yawed cylinder (YC) with radius R_a, so that a yaw angle is taken between 0° and 60°. In addition, the irregular or non-uniform heat source/sink, convective boundary conditions and viscous dissipation impact is also incorporated in the given investigation. Since, the modeling is based on effects of mixed convection, the cylinder must be positioned inclined or vertically manner to experience the effects of buoyancy force or mixed convection flow. The horizontal and vertical cylinders are specified by $\theta_a = 90^0$, and $\theta_a = 0^0$, respectively. In addition, the values of the yaw angle parameter above 60° would be closer to a stagnation-point flow analysis, which is not the goal of the existing buoyancy or mixed convection flow analysis. The boundary layer formation is assumed along the $y_a\bar{\,}$direction and the respective $z_a\bar{\,}$axis direction is selected along the surface of the YC. The mainstream velocity is mathematically expressed as $u_E(x_a) = 2u_\infty \cos(R_a^{-1} x_a)$ with $w_\infty = u_\infty(\sin\theta_a)^{-1}$, in which u_∞ and w_∞ are constant velocities of the free-stream in $x_a\bar{\,}$ and $z_a\bar{\,}$directions, respectively. Given that the cylinder's yawed shape is considered, the far field is expected to flow both the chord-wise (along the $x_a\bar{\,}$coordinates) and span-wise (along the $z_a\bar{\,}$coordinates) directions. The free-stream velocity $w_E = w_\infty \cos\theta_a$ is considered in the direction of $x_a\bar{\,}$axes. The velocities of fluid u_a, v_a, and w_a, taking into account the following $x_a\bar{\,}, y_a\bar{\,}$, and $z_a\bar{\,}$directions, respectively. All flows taken into consideration are fully developed and have an infinite spanwise extent. Thus, we are looking here for the outcomes with velocity and temperature fields independent of the coordinate z_a. In contrast to the fluid $T_f > T_\infty$ around it, the yawed cylinder surface should be hot. The following presumptions are used to further simplify the analysis of the problem.

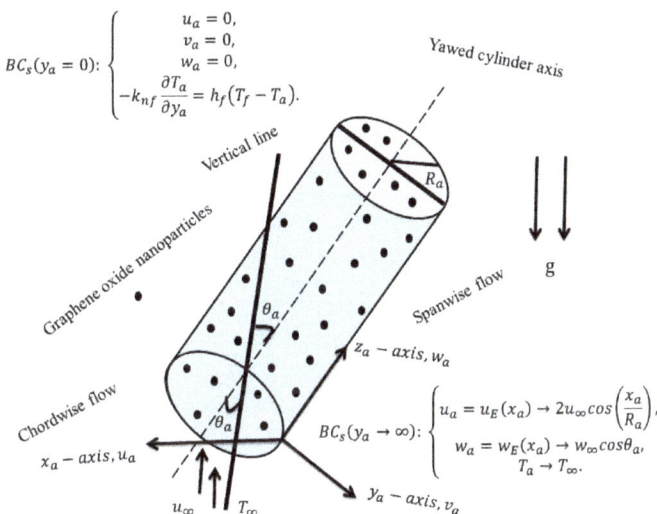

Figure 1. Physical model of the yawed cylinder embedded nanofluid.

- Steady flow
- Stagnation-point flow
- Convective boundary conditions
- Incompressible fluid
- Buoyancy or Mixed convection flow (Assisting and Opposing flows)
- Irregular heat sink/source
- Viscous dissipation
- Nanofluid
- Yawed Cylinder
- Boundary-layer approximations
- Boussinesq approximation

Using the above stated assumptions, the governing PDEs that differentiate the trend of the boundary-layer flow problem are given by [8,9]

$$\frac{\partial u_a}{\partial x_a} + \frac{\partial v_a}{\partial y_a} = 0, \quad (1)$$

$$u_a \frac{\partial u_a}{\partial x_a} + v_a \frac{\partial u_a}{\partial y_a} = v_{nf} \frac{\partial^2 u_a}{\partial y_a^2} + u_E \frac{du_E}{dx_a} + g\beta_f \frac{(\rho\beta)_{nf}/(\rho\beta)_f}{\rho_{nf}/\rho_f}(T_a - T_\infty)\sin\theta_a, \quad (2)$$

$$u_a \frac{\partial w_a}{\partial x_a} + v_a \frac{\partial w_a}{\partial y_a} = v_{nf} \frac{\partial^2 w_a}{\partial y_a^2} + g\beta_f \frac{(\rho\beta)_{nf}/(\rho\beta)_f}{\rho_{nf}/\rho_f}(T_a - T_\infty)\cos\theta_a, \quad (3)$$

$$u_a \frac{\partial T_a}{\partial x_a} + v_a \frac{\partial T_a}{\partial y_a} = \frac{k_{nf}}{(\rho c_p)_{nf}} \frac{\partial^2 T_a}{\partial y_a^2} + \frac{\mu_{nf}}{(\rho c_p)_{nf}} \left[\left(\frac{\partial u_a}{\partial y_a}\right)^2 + \left(\frac{\partial w_a}{\partial y_a}\right)^2 \right] + \frac{Q^{bbb}}{(\rho c_p)_{nf}}, \quad (4)$$

with the boundary conditions (BCs)

$$u_a = 0, \ v_a = 0, \ w_a = 0, \ -k_{nf}\frac{\partial T_a}{\partial y_a} = h_f\left(T_f - T_a\right) \text{ at } y_a = 0,$$
$$u_a \to u_E(x_a), \ w_a \to w_E = (\cos\theta_a)w_\infty, \ T_a \to T_\infty \text{ as } y_a \to \infty. \quad (5)$$

The mathematical symbols used in the aforesaid equations, such as $v_{nf}, \beta_{nf}, k_{nf}, (\rho c_p)_{nf}, \rho_{nf}$ and h_f signify the kinematic viscosity, coefficient of thermal expansion, thermal conductivity, heat capacity, density, and convective heat transfer of nanofluids. The final terms in Equations (2) and (3) show the effect of buoyancy force, which has a positive sign, indicating the buoyancy-assisting flow (BAF) while the buoyancy-opposing flow (BOF) is shown by a negative sign. Moreover, the last two terms of the right hand of Equation (4) indicate the viscous dissipation and irregular heat source/sink, whilst the heat source/sink is defined as [29]

$$Q^{bbb} = \frac{k_{nf}u_E}{x_a v_{nf}}\left[A_a\left(T_f - T_\infty\right)e^{-\eta_a} + B_a(T_a - T_\infty)\right] \quad (6)$$

where the temperature-dependent heat sink/source and the exponentially decaying space coefficients are denoted by B_a and A_a, respectively. The phenomena of a heat source relate to the positive values of A_a and B_a, while the phenomenon of a heat sink correlates to the negative values of A_a and B_a. Additionally, we attempt a unique method to alter how researchers have recently approached the study of heat transfer in fluids. Generally, the nanofluids are used through the mixed composition of regular fluid and nanoparticles. It is believed that the present method will be an effective technique to modify the process of heat transfer in fluids. As a result, Table 1 lists the thermophysical properties of the regular fluid and graphene oxide nanoparticles. In the meantime, Table 2 offers the experimental characteristics of thermophysical properties.

Table 1. Thermophysical properties of nanofluids and regular fluid.

Properties	Nanofluid
Density	$\rho_{nf} = \{(1-\varphi)\rho_f + \varphi\rho_{snp}\}$
Viscosity	$\mu_{nf} = \frac{\mu_f}{(1-\varphi)^{2.5}}$
Thermal expansion	$(\rho\beta)_{nf} = \left[(1-\varphi)(\rho\beta)_f + \varphi(\rho\beta)_{snp}\right]$
Thermal conductivity	$\frac{k_{nf}}{k_f} = \frac{k_{snp}+2k_f-2\varphi(k_f-k_{snp})}{k_{snp}+2k_f+\varphi(k_f-k_{snp})}$
Heat capacity	$(\rho c_p)_{nf} = \left[(1-\varphi)(\rho c_p)_f + \varphi(\rho c_p)_{snp}\right]$

Table 2. Thermophysical data of regular fluid and GO nanoparticle [30].

Characteristic Properties	H$_2$O	GO
ρ	997.1	1800
c_p	4179	717
k	0.613	5000
β	21	2.84×10^{-4}
Pr	6.2	-

Consequently, Equations (1) through (6) are transformed into a non-dimensional form using the following non-similarity variables [9]:

$$\xi_a = \frac{1}{3}\int_0^{x_a}\left(\frac{u_E}{u_\infty}\right)d\left(\frac{x_a}{R_a}\right), \eta_a = y_a\left(\frac{u_E}{u_\infty}\right)\left(\frac{u_\infty}{2\nu_f\xi_a R_a}\right)^{1/2}, w_a = w_E G(\xi_a, \eta_a), \tag{7}$$

$$\psi(x_a, y_a) = u_\infty\left(\frac{2\nu_f\xi_a R_a}{u_\infty}\right)^{1/2} F(\xi_a, \eta_a), S = (T_a - T_\infty)/(T_f - T_\infty).$$

Equation (1) is satisfied while the other equations are transformed to

$$\frac{\mu_{nf}/\mu_f}{\rho_{nf}/\rho_f}F_{\eta_a\eta_a\eta_a} + \frac{1}{3}(2\xi_a F_{\xi_a} + F)F_{\eta_a\eta_a} + \frac{\beta(\xi_a)}{3}(1 - F_{\eta_a}^2) - \frac{2}{3}\xi_a F_{\eta_a}F_{\xi_a\eta_a} + \frac{(\rho\beta)_{nf}/(\rho\beta)_f}{\rho_{nf}/\rho_f}Ri_a p_a(\xi_a) S \sin\theta_a = 0, \tag{8}$$

$$\frac{\mu_{nf}/\mu_f}{\rho_{nf}/\rho_f}G_{\eta_a\eta_a} + \frac{1}{3}(2\xi_a F_{\xi_a} + F)G_{\eta_a} - \frac{2}{3}\xi_a F_{\eta_a}G_{\xi_a} + \frac{(\rho\beta)_{nf}/(\rho\beta)_f}{\rho_{nf}/\rho_f}Ri_a q_a(\xi_a) S \sin\theta_a = 0, \tag{9}$$

$$\frac{k_{nf}}{k_f}S_{\eta_a\eta_a} + Pr_a\frac{(\rho c_p)_{nf}/(\rho c_p)_f}{3}FS_{\eta_a} + 2\xi_a Pr_a\frac{(\rho c_p)_{nf}/(\rho c_p)_f}{3}(F_{\xi_a}S_{\eta_a} - F_{\eta_a}S_{\xi_a}) + \frac{\mu_{nf}}{\mu_f}Pr_a Ec_a\left(4\sin^2(\xi_a)\sin^2(\theta_a)F_{\eta_a\eta_a}^2 + \cos^2(\theta_a)G_{\eta_a}^2\right) + \frac{(k_{nf}/k_f)(\rho_{nf}/\rho_{nf})}{(\mu_{nf}/\mu_f)}Ja(\xi_a)(A_a e^{-\eta_a} + B_a) = 0, \tag{10}$$

with the transformed BCs

$$\begin{cases} \text{At } \eta_a = 0 &: F(\xi_a, 0) + 2\xi_a\frac{\partial F(\xi_a,0)}{\partial \xi_a} = 0, \quad F_{\eta_a}(\xi_a, 0) = 0, \\ G(\xi_a, 0) = 0, \quad \frac{k_{nf}}{k_f}S_{\eta_a}(\xi_a, 0) = -Bi_a L_a(\xi_a)(1 - S(\xi_a, 0)), \\ \text{As } \eta_a \to \infty &: F_{\eta_a}(\xi_a, \infty) \to 1, \quad G(\xi_a, \infty) \to 1, \quad S(\xi_a, \infty) \to 0, \end{cases} \tag{11}$$

where η_a and ξ_a are the transformation coordinates, $F(\xi_a, \eta_a)$ and ψ indicate the stream functions, Ri_a the mixed convective, Pr_a the Prandtl number, Bi_a the Biot number, Ec_a the Eckert number, G and F specify the NDL (non-dimensional) velocities in the requisite directions of span-wise (SW) and chord-wise (CW), respectively, and S implies as the NDL temperature distribution. Moreover, the mathematical terminology of Ri_a, $\beta_a(\xi_a)$, $p(\xi_a)$, Pr_a, $Ja(\xi_a)$, Ec_a, and $q_a(\xi_a)$ are defined as

$$Ri_a = \frac{Gr}{Re^2} = \frac{R_a g \beta_f(T_f - T_\infty)}{u_\infty^2}, p_a(\xi_a) = \frac{\xi_a}{4\cos^3 x_b}, \beta_a(\xi_a) = \frac{2\xi_a}{u_E}\frac{du_E}{d\xi_a}, q_a(\xi_a) = \frac{\xi_a}{2\cos^2 x_b},$$

$$Ja(\xi_a) = \frac{\xi_a}{\cos x_b}, Pr_a = \frac{\nu_f}{\alpha_f}, Ec_a = \frac{w_\infty^2}{(c_p)_f(T_f - T_\infty)}, Bi_a = \frac{h_f R_a}{\sqrt{2}Re^{1/2}k_f}, L_a(\xi_a) = \frac{\sqrt{\xi_a}}{\cos x_b},$$

$$Gr = \frac{g\beta_f(T_f - T_\infty)R_a^3}{\nu_f^2}, Re = \frac{u_\infty R_a}{\nu_f}, x_b = \frac{x_a}{R_a}.$$

Likewise, the components of velocity v_a, u_a, and w_a in simplified forms are as follows:

$$v_a = -\frac{u_\infty}{3}\left(\frac{2\nu_f \xi_a}{u_\infty R_a}\right)^{1/2}\left\{2F_{\xi_a}\cos x_b + \frac{F}{\xi_a}(\cos x_b) - 3\eta_a F_{\eta_a}\tan x_b - \frac{\eta_a}{\xi_a}\cos x_b F_{\eta_a}\right\}, u_a = u_E F_{\eta_a}(\xi_a, \eta_a), \text{ and } w_a = w_E G(\xi_a, \eta_a).$$

The aforesaid expressions of ξ_a, $\beta_a(\xi_a)$, $q_a(\xi_a)$, $J_a(\xi_a)$, $p_a(\xi_a)$ and $L_a(\xi_a)$ in terms of x_b are given as:

$$\xi_a = \frac{2}{3}\sin x_b, \beta_a(x_b) = -2\tan^2 x_b, q_a(x_b) = \frac{\sin x_b}{3\cos^2 x_b}, J_a(x_b) = \frac{2}{3}\tan x_b, \quad L_a(x_b) = \sqrt{\frac{2}{3}\frac{\sqrt{\sin x_b}}{\cos x_b}}, p_a(x_b) = \frac{\sin x_b}{6\cos^3 x_b}. \quad (12)$$

Furthermore, the relation between x_b and ξ_a is signifying as

$$\xi_a \frac{\partial}{\partial \xi_a} = C(x_b)\frac{\partial}{\partial x_b}, \text{ where } C(x_b) = \tan x_b. \quad (13)$$

Consequently, using Equations (12) and (13), Equations (8)–(10) through the boundary conditions are expressed as

$$\frac{\mu_{nf}/\mu_f}{\rho_{nf}/\rho_f}F_{\eta_a\eta_a\eta_a} + \frac{1}{3}FF_{\eta_a\eta_a} + \frac{\beta_a(x_b)}{3}(1-F_{\eta_a}^2) + \frac{(\rho\beta)_{nf}/(\rho\beta)_f}{\rho_{nf}/\rho_f}Ri_a p_a(x_b)S\sin\theta_a = \frac{2C(x_b)}{3}(F_{\eta_a}F_{x_b\eta_a} - F_{\eta_a\eta_a}F_{x_b}), \quad (14)$$

$$\frac{\mu_{nf}/\mu_f}{\rho_{nf}/\rho_f}G_{\eta_a\eta_a} + \frac{1}{3}FG_{\eta_a} + \frac{(\rho\beta)_{nf}/(\rho\beta)_f}{\rho_{nf}/\rho_f}Ri_a q_a(x_b)S\sin\theta_a = \frac{2C(x_b)}{3}(F_{\eta_a}G_{x_b} - F_{x_b}G_{\eta_a}), \quad (15)$$

$$\frac{k_{nf}}{k_f}S_{\eta_a\eta_a} + \Pr a\frac{(\rho c_p)_{nf}/(\rho c_p)_f}{3}FS_{\eta_a} + \frac{\mu_{nf}}{\mu_f}\Pr a Ec_a\left(4\sin^2(x_b)\sin^2(\theta_a)F_{\eta_a\eta_a}^2 + \cos^2(\theta_a)G_{\eta_a}^2\right) + \frac{(k_{nf}/k_f)(\rho_{nf}/\rho_{nf})}{(\mu_{nf}/\mu_f)}J_a(x_b)(A_a e^{-\eta_a} + B_a) = 2\Pr a\frac{(\rho c_p)_{nf}/(\rho c_p)_f}{3}C(x_b)(F_{\eta_a}S_{x_b} - F_{x_b}S_{\eta_a}), \quad (16)$$

with BCs (11) and (13) become

$$\begin{cases} \text{At} \quad \eta_a = 0 \; : \; F(x_b,0) + 2C(x_b)\frac{\partial F(x_b,0)}{\partial x_b} = 0, \; F_{\eta_a}(x_b,0) = 0, \\ G(x_b,0) = 0, \; \frac{k_{nf}}{k_f}S_{\eta_a}(x_b,0) = -Bi_a L_a(x_b)(1 - S(x_b,0)), \\ \text{As} \quad \eta_a \to \infty \; : \; F_{\eta_a}(x_b,\infty) \to 1, \quad G(x_b,\infty) \to 1, \quad S(x_b,\infty) \to 0, \end{cases} \quad (17)$$

One can see that the leading Equations (14)–(16) with BCs (17) remain partial differential equations upon modification, with the $\frac{\partial}{\partial x_b}$ factor on the corresponding right-hand side (RHS) acting as the principal barrier to a solution. Additionally, the type of NS (non-similarity) velocity model, which in turn results in the NS thermal and momentum boundary layers.

3. Local Similarity Technique (Equation-1 Model)

It is helpful to examine Equations (14)–(17) from the perspective of local similarity before beginning the method to the local NS solution. According to this method, it is assumed that the RHS of Equations (14)–(17) are sufficiently small; therefore, they can be approximated by zero. It gives

$$\frac{\mu_{nf}/\mu_f}{\rho_{nf}/\rho_f}F_{\eta_a\eta_a\eta_a} + \frac{1}{3}FF_{\eta_a\eta_a} + \frac{\beta_a(x_b)}{3}(1-F_{\eta_a}^2) + \frac{(\rho\beta)_{nf}/(\rho\beta)_f}{\rho_{nf}/\rho_f}Ri_a p_a(x_b)S\sin\theta_a = 0, \quad (18)$$

$$\frac{\mu_{nf}/\mu_f}{\rho_{nf}/\rho_f}G_{\eta_a\eta_a} + \frac{1}{3}FG_{\eta_a} + \frac{(\rho\beta)_{nf}/(\rho\beta)_f}{\rho_{nf}/\rho_f}Ri_a q_a(x_b)S\sin\theta_a = 0, \quad (19)$$

$$\frac{k_{nf}}{k_f} S_{\eta_a\eta_a} + \text{Pr}_a \frac{(\rho c_p)_{nf}/(\rho c_p)_f}{3} FS_{\eta_a} + \frac{\mu_{nf}}{\mu_f} \text{Pr}_a Ec_a \left(4\sin^2(x_b)\sin^2(\theta_a) F_{\eta_a\eta_a}^2 + \cos^2(\theta_a) G_{\eta_a}^2\right) +$$
$$\frac{(k_{nf}/k_f)(\rho_{nf}/\rho_{nf})}{(\mu_{nf}/\mu_f)} J_a(x_b)(A_a e^{-\eta_a} + B_a) = 0, \qquad (20)$$

subject to BCs

$$\begin{cases} \text{At} \quad \eta_a = 0 \quad : \quad F(x_b,0) = 0, \; F_{\eta_a}(x_b,0) = 0, \; G(x_b,0) = 0, \\ \qquad \frac{k_{nf}}{k_f} S_{\eta_a}(x_b,0) = -Bi_a L_a(x_b)(1 - S(x_b,0)), \\ \text{As} \quad \eta_a \to \infty \quad : \quad F_{\eta_a}(x_b,\infty) \to 1, \quad G(x_b,\infty) \to 1, \quad S(x_b,\infty) \to 0, \end{cases} \qquad (21)$$

At any direction of flow at the place, the quantity x_b can be seen as an inevitable parameter. Because of this, even though the equations $F_{\eta_a\eta_a\eta_a}$, $G_{\eta_a\eta_a}$ and $S_{\eta_a\eta_a}$ are the requisite posited PDEs, they can be handled as ODEs and are solved using tried-and-true or any similarity techniques that work for boundary layers with similarity when the quantities involved, β_a, p_a, J_a, L_a and q_a, can be discovered for a determined constant factor at any specified or fixed position x_a (or x_b). The flow phenomenon in the SW and CW dependency of the ND temperature and ND velocity profiles can be expressed as a progression of x_b quantity. In order for Equations (14)–(17) and Equations (18)–(21) to be simplified to verify without requiring that x_a be small, it must be assumed that the amounts shown in the right-hand side brackets are insignificant. The Equation-1 model has a weakness because it is unclear whether or not this premise is true.

4. Local NS Solution Technique (Equation-2 Model)

First, it is helpful to define the x_b derivatives of F, G and S to remove their explicit occurrence in relation to the outcome of the LNS of Equations (14)–(17). Let

$$K_a(x_b,\eta_a) = \frac{\partial F}{\partial x_b}, \; h_a(x_b,\eta_a) = \frac{\partial G}{\partial x_b} \text{ and } X_a(x_b,\eta_a) = \frac{\partial S}{\partial x_b}. \qquad (22)$$

Equations (14)–(17), which result from substituting these factors, become as

$$\frac{\mu_{nf}/\mu_f}{\rho_{nf}/\rho_f} F''' + \frac{1}{3} FF'' + \frac{\beta_a(x_b)}{3}(1 - F'^2) +$$
$$\frac{(\rho\beta)_{nf}/(\rho\beta)_f}{\rho_{nf}/\rho_f} Ri_a p_a(x_b) S \sin\theta_a = \frac{2C(x_b)}{3}(F' K_a' - F'' K_a), \qquad (23)$$

$$\frac{\mu_{nf}/\mu_f}{\rho_{nf}/\rho_f} G'' + \frac{1}{3} FG' + \frac{(\rho\beta)_{nf}/(\rho\beta)_f}{\rho_{nf}/\rho_f} Ri_a q_a(x_b) S \sin\theta_a = \frac{2C(x_b)}{3}(F' h_a - K_a G'), \qquad (24)$$

$$\frac{k_{nf}}{k_f} S'' + \text{Pr}_a \frac{(\rho c_p)_{nf}/(\rho c_p)_f}{3} FS' + \frac{\mu_{nf}}{\mu_f} \text{Pr}_a Ec_a \left(4\sin^2(x_b)\sin^2(\theta_a) F''^2 + \cos^2(\theta_a) G'^2\right) +$$
$$\frac{(k_{nf}/k_f)(\rho_{nf}/\rho_{nf})}{(\mu_{nf}/\mu_f)} J_a(x_b)(A_a e^{-\eta_a} + B_a) = 2\text{Pr}_a \frac{(\rho c_p)_{nf}/(\rho c_p)_f}{3} C(x_b)(F' X_a - K_a S'). \qquad (25)$$

Additionally, BCs (17) in reference to (22) are modified as:

$$\begin{cases} \text{At} \quad \eta_a = 0 \quad : \quad F(x_b,0) = -2C(x_b) K_a(x_b,0), \; F'(x_b,0) = 0, \\ G(x_b,0) = 0, \; \frac{k_{nf}}{k_f} S'(x_b,0) = -Bi_a L_a(x_b)(1 - S(x_b,0)), \\ \text{As} \quad \eta_a \to \infty \quad : \quad F'(x_b,\infty) \to 1, \quad G(x_b,\infty) \to 1, \quad S(x_b,\infty) \to 0 \end{cases} \qquad (26)$$

Here, primes signify the partial change with respect to the variable η_a. Then, by differentiating Equations (23)–(26) in terms of a non-similarity variable x_b, we obtain

$$\frac{\mu_{nf}/\mu_f}{\rho_{nf}/\rho_f} K_a''' + \frac{1}{3} F K_a'' + \frac{F'' K_a}{3}\left(1 + 2\frac{dC}{dx_b}\right) + \frac{1}{3}\frac{d\beta_a}{dx_b}(1 - F'^2) - \frac{2}{3} F' K_a'\left(\beta_a + \frac{dC}{dx_b}\right)$$
$$\frac{(\rho\beta)_{nf}/(\rho\beta)_f}{\rho_{nf}/\rho_f} Ri_a \sin\theta_a \left(\frac{dp_a}{dx_b} S + p_a(x_b) X_a\right) = \frac{2}{3} C(x_b) \frac{\partial}{\partial x_b}(F' K_a' - F'' K_a), \quad (27)$$

$$\begin{cases} \text{At } \eta_a = 0 : K_a(x_b, 0)\left(1 + 2\frac{dC}{dx_b}\right) = -2C(x_b)\frac{\partial K_a(x_b,0)}{\partial x_b}, \ K_a'(x_b, 0) = 0, \\ \text{As } \eta_a \to \infty : K_a'(x_b, \infty) \to 0, \end{cases} \quad (28)$$

$$\frac{\mu_{nf}/\mu_f}{\rho_{nf}/\rho_f} h_a'' + \frac{1}{3} F h_a' - \frac{2}{3}\frac{dC}{dx_b} F' h_a + \frac{1}{3} G' K_a\left(1 + 2\frac{dC}{dx_b}\right) +$$
$$\frac{(\rho\beta)_{nf}/(\rho\beta)_f}{\rho_{nf}/\rho_f} Ri_a \sin\theta_b \left(\frac{dq_a}{dx_b} S + q_a(x_b) X_a\right) = \frac{2}{3} C(x_b) \frac{\partial}{\partial x_b}(F' h_a - G' K_a), \quad (29)$$

$$\begin{cases} \text{At } \eta_a = 0 : h_a(x_b, 0) = 0, \\ \text{As } \eta_a \to \infty : h_a(x_b, \infty) \to 0, \end{cases} \quad (30)$$

$$\frac{k_{nf}}{k_f} X_a'' + \Pr_a \frac{(\rho c_p)_{nf}/(\rho c_p)_f}{3} F X_a' + \Pr_a \frac{(\rho c_p)_{nf}/(\rho c_p)_f}{3} K_a S'\left(1 + 2\frac{dC}{dx_b}\right) -$$
$$2\Pr_a \frac{(\rho c_p)_{nf}/(\rho c_p)_f}{3}\frac{dC}{dx_b} F' X_a + \frac{(k_{nf}/k_f)(\rho_{nf}/\rho_f)}{(\mu_{nf}/\mu_f)}\frac{dJ_a}{dx_b}(A_a e^{-\eta_a} + B_a) +$$
$$\frac{\mu_{nf}}{\mu_f}\Pr_a Ec_a \left(4\sin^2(\theta_a)\sin(2x_b)F'^2 + 8\sin^2(\theta_a)\sin^2(x_b)F'' K_a'' + 2\cos^2(\theta_a) G' h_a'\right) =$$
$$2\Pr_a \frac{(\rho c_p)_{nf}/(\rho c_p)_f}{3} C(x_b)\frac{\partial}{\partial x_b}(F' X_a - K_a S'), \quad (31)$$

$$\begin{cases} \text{At } \eta_a = 0 : \frac{k_{nf}}{k_f} X_a'(x_b, 0) = -Bi_a \frac{dL_a}{dx_b}(1 - S(x_b, 0)) + Bi_a L_a(x_b) X_a(x_b, 0), \\ \text{As } \eta_a \to \infty : X_a(x_b, \infty) \to 0, \end{cases} \quad (32)$$

when using explicit terms, the aforementioned equations RHS are overloaded with x_b derivatives. Equations (27)–(32) offer characteristics to the considered Equations (23)–(25) subject to BCs (26). For simultaneous treatment, the functions F, K_a, G and h_a are given in Equations (23), (24), (27) and (29). Similarly, functions S and X_a are presented in (25) and (31), requiring simultaneously the non-similar solution. To guarantee that all labels/classes in the leading equations and their BCs are retained, Equations (23)–(26) are not approximated. Hence, further proposed in the characteristic equations that the RHS of Equations (27), (29) and (30) are sufficiently small to allow for their neglection. As a result, the leading Equations (23), (24) and (26)–(30) for the problem of momentum fields in the CW and SW directions for the Equation-2 model (or LNSSM) can be combined as follows:

$$\frac{\mu_{nf}/\mu_f}{\rho_{nf}/\rho_f} F''' + \frac{1}{3} F F'' + \frac{\beta_a(x_b)}{3}(1 - F'^2) +$$
$$\frac{(\rho\beta)_{nf}/(\rho\beta)_f}{\rho_{nf}/\rho_f} Ri_a p_a(x_b) S \sin\theta_a = \frac{2C(x_b)}{3}(F' K_a' - F'' K_a), \quad (33)$$

$$\frac{\mu_{nf}/\mu_f}{\rho_{nf}/\rho_f} K_a''' + \frac{1}{3} F K_a'' + \frac{F'' K_a}{3}\left(1 + 2\frac{dC}{dx_b}\right) + \frac{1}{3}\frac{d\beta_a}{dx_b}(1 - F'^2) - \frac{2}{3} F' K_a'\left(\beta_a + \frac{dC}{dx_b}\right)$$
$$\frac{(\rho\beta)_{nf}/(\rho\beta)_f}{\rho_{nf}/\rho_f} Ri_a \sin\theta_a \left(\frac{dp_a}{dx_b} S + p_a(x_b) X_a\right) = 0, \quad (34)$$

$$\begin{cases} \text{At } \eta_a = 0.0 : F(x_b, 0) = F'(x_b, 0) = 0, \ K_a(x_b, 0) = 0, \ K_a'(x_b, 0) = 0, \\ \text{As } \eta_a \to \infty : F'(x_b, \infty) \to 1, \ K_a'(x_b, \infty) \to 0. \end{cases} \quad (35)$$

$$\frac{\mu_{nf}/\mu_f}{\rho_{nf}/\rho_f} G'' + \frac{1}{3} F G' + \frac{(\rho\beta)_{nf}/(\rho\beta)_f}{\rho_{nf}/\rho_f} Ri_a q_a(x_b) S \sin\theta_a = \frac{2C(x_b)}{3}(F' h_a - K_a G'), \quad (36)$$

$$\frac{\mu_{nf}/\mu_f}{\rho_{nf}/\rho_f} h_a'' + \frac{1}{3} F h_a' - \frac{2}{3}\frac{dC}{dx_b} F' h_a + \frac{1}{3} G' K_a\left(1 + 2\frac{dC}{dx_b}\right) +$$
$$\frac{(\rho\beta)_{nf}/(\rho\beta)_f}{\rho_{nf}/\rho_f} Ri_a \sin\theta_b \left(\frac{dq_a}{dx_b} S + q_a(x_b) X_a\right) = 0, \quad (37)$$

$$\begin{cases} \text{At} & \eta_a = 0 \quad : \quad G(x_b, 0) = 0, \ h_a(x_b, 0) = 0, \\ \text{As} & \eta_a \to \infty \quad : \quad G(x_b, \infty) \to 1, \ h_a(x_b, \infty) \to 0. \end{cases} \quad (38)$$

when x_b is taken into account as a uniform or fixed prescribable constraint at any fluid flow directions or locations, Equations (33)–(38), which are ODEs, can be handled as a requisite posted system by solving using the customary methods appropriate for similarity boundary layers. The leading Equations (25), (26), (31) and (32) for the thermal boundary layer similarly take place in the following form:

$$\frac{k_{nf}}{k_f} S'' + \Pr_a \frac{(\rho c_p)_{nf}/(\rho c_p)_f}{3} FS' + \frac{\mu_{nf}}{\mu_f} \Pr_a Ec_a \left(4\sin^2(x_b)\sin^2(\theta_a)F'^2 + \cos^2(\theta_a)G'^2\right) + \frac{(k_{nf}/k_f)(\rho_{nf}/\rho_{nf})}{(\mu_{nf}/\mu_f)} J_a(x_b)(A_a e^{-\eta_a} + B_a) = 2\Pr_a \frac{(\rho c_p)_{nf}/(\rho c_p)_f}{3} C(x_b)(F'X_a - K_a S'), \quad (39)$$

$$\frac{k_{nf}}{k_f} X_a'' + \Pr_a \frac{(\rho c_p)_{nf}/(\rho c_p)_f}{3} F X_a' + \Pr_a \frac{(\rho c_p)_{nf}/(\rho c_p)_f}{3} K_a S'\left(1 + 2\frac{dC}{dx_b}\right) - 2\Pr_a \frac{(\rho c_p)_{nf}/(\rho c_p)_f}{3} \frac{dC}{dx_b} F' X_a + \frac{(k_{nf}/k_f)(\rho_{nf}/\rho_{nf})}{(\mu_{nf}/\mu_f)} \frac{dJ_a}{dx_b}(A_a e^{-\eta_a} + B_a) + \frac{\mu_{nf}}{\mu_f} \Pr_a Ec_a \left(4\sin^2(\theta_a)\sin(2x_b)F'^2 + 8\sin^2(\theta_a)\sin^2(x_b)F''K_a'' + 2\cos^2(\theta_a)G'h_a'\right) = 0, \quad (40)$$

$$\begin{cases} \text{At} \ \eta_a = 0 & : \ \frac{k_{nf}}{k_f} S'(x_b, 0) = -Bi_a L_a(x_b)(1 - S(x_b, 0)), \\ & \quad \frac{k_{nf}}{k_f} X_a'(x_b, 0) = -Bi_a \frac{dL_a}{dx_b}(1 - S(x_b, 0)) + Bi_a L_a(x_b) X_a(x_b, 0), \\ \text{As} \ \eta_a \to \infty & : \ S(x_b, \infty) \to 0, \ X_a(x_b, \infty) \to 0. \end{cases} \quad (41)$$

Moreover, Equations (39)–(41) can be thought of as a group of ODEs for a fixed variable x_b. The system can be easily solved using the same method as for similarity boundary layers, much like Equation-1 model.

4.1. Local NS Technique (Equation-3 Model)

Equations (27), (29) and (31) are now differentiated with respect to x_b in the three-equation model before the additional functions are introduced.

$$N_a = \frac{\partial K_a}{\partial x_b} = \frac{\partial^2 F}{\partial x_b^2}, \ R_a = \frac{\partial h_a}{\partial x_b} = \frac{\partial^2 G}{\partial x_b^2} \ \text{and} \ H_a = \frac{\partial X_a}{\partial x_b} = \frac{\partial^2 S}{\partial x_b^2}. \quad (42)$$

The derivatives that directly involve x_b terms are clustered on the RHS and are seen to be

$$\frac{2}{3} C(x_b) \frac{\partial^2}{\partial x_b^2} (F'K_a' - F''K_a), \ \frac{2}{3} C(x_b) \frac{\partial^2}{\partial x_b^2} (F'h_a - G'K_a)$$

and

$$\frac{2}{3} \Pr_a \frac{(\rho c_p)_{nf}}{(\rho c_p)_f} C(x_b) \frac{\partial^2}{\partial x_b^2} (F'X_a - S'K_a).$$

Now it is expected that this amount is sufficiently small to be close to zero. However, no assumption is necessary to reach any of the terms in Equations (23)–(25), (27), (29) and (31). This result includes the following third truncation or the Equation-3 model for the momentum fields in the CW and SW directions, and the requisite energy equation with coupled BCs for convenience in subsequent computations are as follows:

$$\frac{\mu_{nf}/\mu_f}{\rho_{nf}/\rho_f} F''' + \frac{1}{3} FF'' + \frac{B_a(x_b)}{3}(1 - F'^2) + \frac{(\rho \beta)_{nf}/(\rho \beta)_f}{\rho_{nf}/\rho_f} Ri_a p_a(x_b) S \sin \theta_a = \frac{2C(x_b)}{3}(F'K_a' - F''K_a), \quad (43)$$

$$\frac{\mu_{nf}/\mu_f}{\rho_{nf}/\rho_f}K_a''' + \frac{1}{3}FK_a'' + \frac{F''K_a}{3}\left(1+2\frac{dC}{dx_b}\right) + \frac{1}{3}\frac{d\beta_a}{dx_b}(1-F'^2) - \frac{2}{3}F'K_a'\left(\beta_a + \frac{dC}{dx_b}\right)$$
$$\frac{(\rho\beta)_{nf}/(\rho\beta)_f}{\rho_{nf}/\rho_f}Ri_a\sin\theta_a\left(\frac{dp_a}{dx_b}S + p_a(x_b)X_a\right) = \frac{2}{3}C(x_b)(F'N_a' + K_a'^2 - F''N_a - K_aK_a''), \quad (44)$$

$$\frac{\mu_{nf}/\mu_f}{\rho_{nf}/\rho_f}N_a''' + \frac{2}{3}F''K_a\frac{d^2C}{dx_b^2} + \frac{1}{3}\left(1+4\frac{dC}{dx_b}\right)(F''N_a + K_aK_a'') + \frac{1}{3}(FN_a'' + K_aN_a'') +$$
$$\frac{1}{3}\frac{d^2\beta_a}{dx_b^2}(1-F'^2) - \frac{4}{3}\frac{d\beta_a}{dx_b}F'K_a' - \frac{2}{3}\frac{d^2C}{dx_b^2}F'K_a' - \frac{2}{3}(F'N_a' + K_a'^2)\left(\beta_a + 2\frac{dC}{dx_b}\right) +$$
$$\frac{(\rho\beta)_{nf}/(\rho\beta)_f}{\rho_{nf}/\rho_f}Ri_a\sin\theta_a\left(\frac{d^2p_a}{dx_b^2}S + 2\frac{dp_a}{dx_b}X_a + p_a(x_b)H_a\right) = 0, \quad (45)$$

$$\begin{cases} \text{At} \quad \eta_a = 0.0 \quad : \quad F'(x_b,0) = F(x_b,0) = 0, \quad K_a'(x_b,0) = K_a(x_b,0) = 0, \\ N'(x_b,0) = N_a(x_b,0) = 0, \\ \text{As} \quad \eta_a \to \infty \quad : \quad F'(x_b,\infty) \to 1, \quad K_a'(x_b,\infty) \to 0, \quad N_a'(x_b,\infty) \to 0 \end{cases} \quad (46)$$

$$\frac{\mu_{nf}/\mu_f}{\rho_{nf}/\rho_f}G'' + \frac{1}{3}FG' + \frac{(\rho\beta)_{nf}/(\rho\beta)_f}{\rho_{nf}/\rho_f}Ri_aq_a(x_b)S\sin\theta_a = \frac{2C(x_b)}{3}(F'h_a - K_aG'), \quad (47)$$

$$\frac{\mu_{nf}/\mu_f}{\rho_{nf}/\rho_f}h_a'' + \frac{1}{3}Fh_a' - \frac{2}{3}\frac{dC}{dx_b}F'h_a + \frac{1}{3}G'K_a\left(1+2\frac{dC}{dx_b}\right) +$$
$$\frac{(\rho\beta)_{nf}/(\rho\beta)_f}{\rho_{nf}/\rho_f}Ri_a\sin\theta_b\left(\frac{dq_a}{dx_b}S + q_a(x_b)X_a\right) = \frac{2}{3}C(x_b)(F'R_a + h_aK_a' - G'N_a - K_ah_a'), \quad (48)$$

$$\frac{\mu_{nf}/\mu_f}{\rho_{nf}/\rho_f}R_a'' + \frac{2}{3}\frac{d^2C}{dx_b^2}G'K_a + \frac{1}{3}FR_a' - \frac{4}{3}\frac{dC}{dx_b}(F'R_a + h_aK_a') + \frac{1}{3}h_a'K_a +$$
$$\frac{1}{3}\left(1+4\frac{dC}{dx_b}\right)(G'N_a + K_ah_a') - \frac{2}{3}\frac{d^2C}{dx_b^2}F'h_a +$$
$$\frac{(\rho\beta)_{nf}/(\rho\beta)_f}{\rho_{nf}/\rho_f}Ri_a\sin\theta_a\left(\frac{d^2q_a}{dx_b^2}S + 2\frac{dq_a}{dx_b}X_a + q_aH_a\right) = 0, \quad (49)$$

$$\begin{cases} \text{At} \quad \eta_a = 0 \quad : \quad G(x_b,0) = 0, \quad h_a(x_b,0) = 0, \quad R_a(x_b,0) = 0, \\ \text{As} \quad \eta_a \to \infty \quad : \quad G(x_b,\infty) \to 1, \quad h_a(x_b,\infty) \to 0, \quad R_a(x_b,\infty) \to 0 \end{cases} \quad (50)$$

$$\frac{k_{nf}}{k_f}S'' + \Pr_a\frac{(\rho c_p)_{nf}/(\rho c_p)_f}{3}FS' + \frac{\mu_{nf}}{\mu_f}\Pr_aEc_a\left(4\sin^2(x_b)\sin^2(\theta_a)F''^2 + \cos^2(\theta_a)G'^2\right) +$$
$$\frac{(k_{nf}/k_f)(\rho_{nf}/\rho_{nf})}{(\mu_{nf}/\mu_f)}J_a(x_b)(A_ae^{-\eta_a} + B_a) = 2\Pr_a\frac{(\rho c_p)_{nf}/(\rho c_p)_f}{3}C(x_b)(F'X_a - K_aS'), \quad (51)$$

$$\frac{k_{nf}}{k_f}X_a'' + \Pr_a\frac{(\rho c_p)_{nf}/(\rho c_p)_f}{3}FX_a' + \Pr_a\frac{(\rho c_p)_{nf}/(\rho c_p)_f}{3}K_aS'\left(1+2\frac{dC}{dx_b}\right) -$$
$$2\Pr_a\frac{(\rho c_p)_{nf}/(\rho c_p)_f}{3}\frac{dC}{dx_b}F'X_a + \frac{(k_{nf}/k_f)(\rho_{nf}/\rho_{nf})}{(\mu_{nf}/\mu_f)}\frac{dJ_a}{dx_b}(A_ae^{-\eta_a} + B_a) +$$
$$\frac{\mu_{nf}}{\mu_f}\Pr_aEc_a\left(4\sin^2(\theta_a)\sin(2x_b)F''^2 + 8\sin^2(\theta_a)\sin^2(x_b)F''K_a'' + 2\cos^2(\theta_a)G'h_a'\right) =$$
$$2\Pr_a\frac{(\rho c_p)_{nf}/(\rho c_p)_f}{3}C(x_b)(F'H_a + X_aK_a' - S'N_a - K_aX_a'), \quad (52)$$

$$\frac{k_{nf}}{k_f}H_a'' + \frac{2}{3}\Pr_a\frac{(\rho c_p)_{nf}}{(\rho c_p)_f}\frac{d^2C}{dx_b^2}K_aS' - \frac{4}{3}\Pr_a\frac{(\rho c_p)_{nf}}{(\rho c_p)_f}\frac{dC}{dx_b}(F'H_a + X_aK_a') +$$
$$\frac{1}{3}\Pr_a\frac{(\rho c_p)_{nf}}{(\rho c_p)_f}\left(1+4\frac{dC}{dx_b}\right)(K_aX_a' + S'N_a) + \frac{1}{3}\Pr_a\frac{(\rho c_p)_{nf}}{(\rho c_p)_f}(FH_a' + K_aX_a') -$$
$$\frac{2}{3}\Pr_a\frac{(\rho c_p)_{nf}}{(\rho c_p)_f}\frac{d^2C}{dx_b^2}F'X_a + \frac{(k_{nf}/k_f)(\rho_{nf}/\rho_{nf})}{(\mu_{nf}/\mu_f)}\frac{d^2J_a}{dx_b^2}(A_ae^{-\eta_a} + B_a) +$$
$$\frac{\mu_{nf}}{\mu_f}\Pr_aEc_a\begin{pmatrix} 8\sin^2(\theta_a)\cos(2x_b)F''^2 + 16\sin^2(\theta_a)\sin(2x_b)F''K_a'' + \\ 8\sin^2(\theta_a)\sin^2(x_b)F''N_a'' + 8\sin^2(\theta_a)\sin^2(x_b)K_a''^2 + \\ 2\cos^2(\theta_a)G'R_a' + 2\cos^2(\theta_a)h_a'^2 \end{pmatrix} = 0, \quad (53)$$

$$\begin{cases} \text{At} \quad \eta_a = 0 \quad : \quad \frac{k_{nf}}{k_f} S'(x_b, 0) = -Bi_a L_a(x_b)(1 - S(x_b, 0)), \\ \frac{k_{nf}}{k_f} X_a'(x_b, 0) = -Bi_a \frac{dL_a}{dx_b}(1 - S(x_b, 0)) + Bi_a L_a(x_b) X_a(x_b, 0), \\ \frac{k_{nf}}{k_f} H_a'(x_b, 0) = -Bi_a \frac{d^2 L_a}{dx_b^2}(1 - S(x_b, 0)) + 2Bi_a \frac{dL_a}{dx_b} X_a(x_b, 0) + Bi_a L_a(x_b) H_a(x_b, 0), \\ \text{As} \quad \eta_a \to \infty \quad : \quad S(x_b, \infty) \to 0, \quad X_a(x_b, \infty) \to 0, \quad H_a(x_b, \infty) \to 0. \end{cases} \quad (54)$$

In which

$$\beta_a(x_b) = -2\tan^2 x_b, \frac{d\beta_a}{dx_b} = -4\tan x_b \sec^2 x_b, \frac{d^2\beta_a}{dx_b^2} = -4(2\tan^2 x_b \sec^2 x_b + \sec^2 x_b),$$

$$q_a(x_b) = \frac{\sin x_b}{3\cos^2 x_b}, \frac{dq_a}{dx_b} = \frac{\cos^3 x_b + \sin 2x_b \sin x_b}{3\cos^4 x_b},$$

$$\frac{d^2 q}{dx_b^2} = \frac{\sin 4x_b + 2\cos^2 x_b \sin 2x_b + 2\cos^3 x_b \sin x_b + 8\sin^2 x_b \sin 2x_b}{6\cos^5 x_b}, \frac{d^2 C}{dx_b^2} = 2\sec^2 x_b \tan x_b,$$

$$p_a(x_b) = \frac{\sin x_b}{6\cos^3 x_b}, \frac{dp_a}{dx_b} = \frac{\cos^2 x_b + 3\sin^2 x_b}{6\cos^4 x_b}, \frac{d^2 p_a}{dx_b^2} = \frac{\cos x_b \sin 2x_b + 2(\sin x_b \cos^2 x_b + 3\sin^3 x_b)}{3\cos^5 x_b},$$

$$J_a(x_b) = \frac{2}{3}\tan x_b, \frac{dJ_a}{dx_b} = \frac{2}{3}\sec^2 x_b, \frac{d^2 J_a}{dx_b^2} = \frac{4}{3}\sec^2 x_b \tan x_b, C(x_b) = \tan x_b, \frac{dC}{dx_b} = \sec^2 x_b,$$

and

$$L_a(x_b) = \sqrt{\frac{2}{3}} \frac{\sqrt{\sin x_b}}{\cos x_b}, \frac{dL_a}{dx_b} = \frac{1 + \sin^2 x_b}{\sqrt{6}\cos^2(x_b)\sqrt{\sin x_b}},$$

$$\frac{d^2 L_a}{dx_b^2} = \frac{\cos^2 x_b \sin(2x_b)\sqrt{\sin x_b} - \left(\frac{\cos^3 x_b}{2\sqrt{\sin x_b}} - \sqrt{\sin x_b}\sin(2x_b)\right)(1 + \sin^2 x_b)}{\sqrt{6}\cos^4 x_b \sin x_b} \quad (55)$$

The terms in Equation (55) will once more be treated as known constant parameters at any fixed value of x_b, allowing Equations (43)–(54) to be viewed as a collection of connected ODEs of the similarity type. These equations were then solved numerically via a bvp4c. However, the thermal energy itself and the initial subsidiary equations are both still intact. The subsidiary secondary terms to the field of thermal and velocity are eliminated throughout the development of this system since the thermal and velocity equations are truncated twice to make the reductions of the errors. Therefore, as noted from the Equation-3 model outcomes it should be more precise from the upshots found from the Equation-1 and Equation-2 models.

4.2. Gradients

The gradients are demonstrated in the following form

$$C_F = \frac{2\mu_{nf}\left(\frac{\partial u_a}{\partial y_a}\right)\big|_{y_a=0}}{\rho_f u_\infty^2}, \quad C_G^* = \frac{2\mu_{nf}\left(\frac{\partial w_a}{\partial y_a}\right)\big|_{y=0}}{\rho_f u_\infty^2}, \quad Nu = \frac{-Ra k_{nf}\left(\frac{\partial T_a}{\partial y_a}\right)\big|_{y_a=0}}{k_f(T_f - T_\infty)}. \quad (56)$$

The aforementioned Equation (56) is further simplified to the following ND form by putting the requisite mentioned NST as follows:

$$Re^{1/2} C_F = 4\sqrt{3}\frac{\mu_{nf}}{\mu_f}\frac{\cos^2 x_b}{\sqrt{\sin x_b}} F_{\eta_a \eta_a}(x_b, 0), \quad Re^{1/2} C_G^* = 2\frac{\mu_{nf}}{\mu_f}\cot\theta_a \frac{\sqrt{3}\cos x_b}{\sqrt{\sin x_b}} G_{\eta_a}(x_b, 0),$$

$$Re^{-1/2} Nu = -\frac{k_{nf}}{k_f}\frac{\sqrt{3}\cos x_b}{\sqrt{\sin x_b}} S_{\eta_a}(x_b, 0), \quad (57)$$

where $Re^{1/2} C_F$ signifies the friction factor in the chord-wise direction and $Re^{1/2} C_G^*$ signifies in the span-wise direction, whilst the reduced heat transfer (RHT) represents $Re^{-1/2} Nu$ at the cylinder surface.

5. Analysis of the Results

The main goal of the current investigation was to produce precise numerical upshots for the mixed convective water-based graphene oxide nanoparticle flow via a YC experiencing the significant influence of the convective boundary conditions, viscous dissipation, and irregular heat source/sink. The impact of physical parameters such as solid nanoparticle volume fraction φ, yaw angle θ_a, mixed convection or buoyancy parameter Ri_a, Biot number Bi_a, heat source/sink parameter A_a, B_a, and the Eckert number Ec_a was profoundly measured when evaluating the given problem. Tables 1 and 2 represent the correlations and the experimental thermophysical data of the graphene oxide nanoparticles and the base (water) fluid, respectively. On the other hand, Table 3 shows the comparison of the given results obtained by the LNSSM (or the Equation-3 model) for the several values of the x_b and Ri_a with those of Kumari and Nath [31] in the absence of the yaw angle, viscous dissipation, convective boundary conditions, irregular heat source/sink parameter, and the nanoparticles volume fraction when $Pr_a = 0.7$. It was found that the results were extremely harmonious, which validates the current numerical scheme and the procedure's approach. Further, the utility of the new solution approach was strongly supported by a given exceptional comparison of the current and published results as shown in Table 3. Thus, it gives us confidence that the technique employed in the existing problem is accurate and also the new results found in the model are precisely useful and correct. However, the influence of these parameters on the velocity profiles in the chordwise and spanwise directions, temperature profiles, shear stresses in the chordwise and spanwise directions, and the heat transfer rate is graphically depicted in Figures 2–11, respectively. Meanwhile, the new quantitative data of the reduced shear stresses in both directions and heat transfer for the instance of buoyancy-assisting flow (BAF) and buoyancy-opposing flow (BOF) are given in the respective Tables 4–6. The value of the Prandtl number Pr_a is considered to be 6.2, which signifies water and is treated to be a fixed constant throughout the numerical scrutiny (see Refs. [32–35]). Additionally, the other influential parameters such as as $\varphi = 0.025$, $Bi_a = 0.5$, $\theta_a = 15^0$, $x_b = 0.5$, $Ec_a = 0.5$, $A_a = 0.1$ and $B_a = 0.1$ are executed as a default in the whole or complete numerical simulations. The black solid lines in all the graphs signify the outcomes for the case of BAF while the red solid lines denote the case of solutions for the BOF.

Table 3. Assessment of $Re^{1/2}C_F$ and $Re^{-1/2}Nu$ with the results of [31] when $\varphi = 0$, $\theta_a = 0^0$, $Ec_a = 0$, $Bi_a = 0$, $A_a = 0$, $B_a = 0$ and $Pr_a = 0.7$.

x_b	Ri_a	Kumari and Nath [31]		Current Results	
		$Re^{1/2}C_F$	$Re^{-1/2}Nu$	$Re^{1/2}C_F$	$Re^{-1/2}Nu$
0.0	0.0	1.3281	0.5854	1.3281	0.5854
	1.0	4.9663	0.8219	4.9663	0.8219
	2.0	7.7119	0.9302	7.7119	0.9302
1.0	0.0	1.9167	0.8666	1.9167	0.8666
	1.0	5.2578	1.0617	5.2578	1.0617
	2.0	7.8863	1.1685	7.8863	1.1685
2.0	0.0	2.3975	1.0963	2.3975	1.0963
	1.0	5.6993	1.2712	5.6993	1.2712
	2.0	8.3555	1.3741	8.3555	1.3741

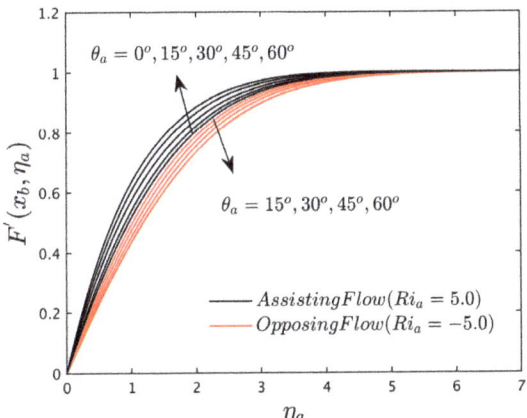

Figure 2. Influence of θ_a on $F'(x_b, \eta_a)$.

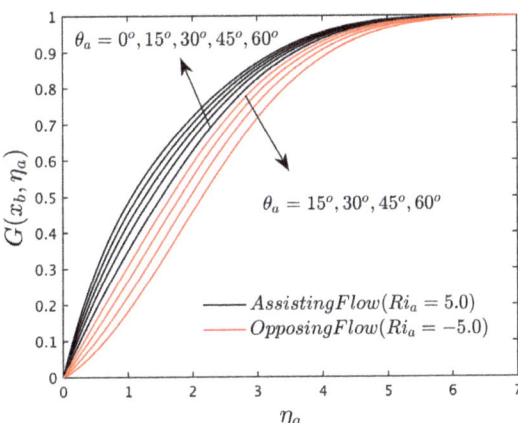

Figure 3. Influence of θ_a on $G(x_b, \eta_a)$.

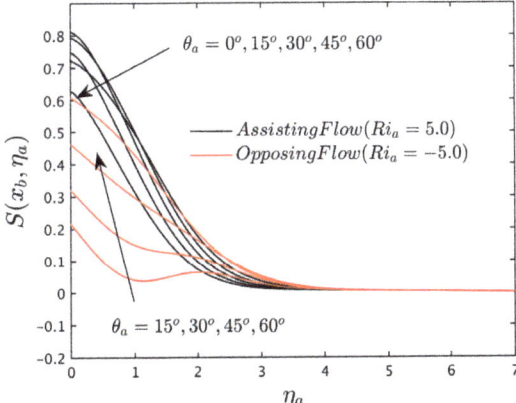

Figure 4. Influence of θ_a on $S(x_b, \eta_a)$.

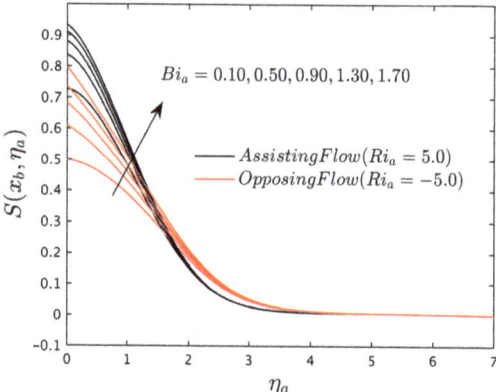

Figure 5. Influence of Biot number on $S(x_b, \eta_a)$.

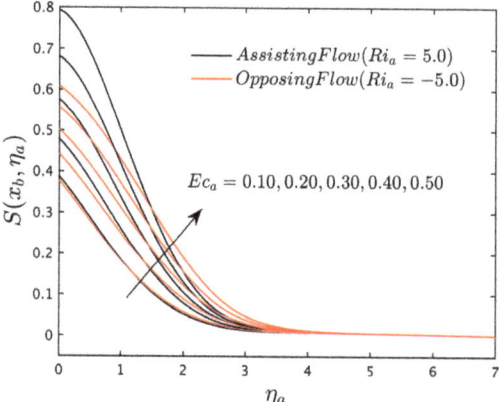

Figure 6. Influence of Eckert number on $S(x_b, \eta_a)$.

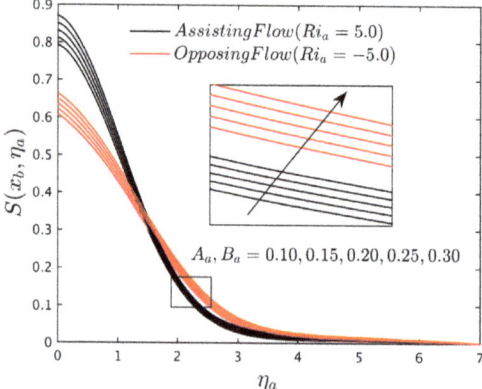

Figure 7. Influence of heat source factor on $S(x_b, \eta_a)$.

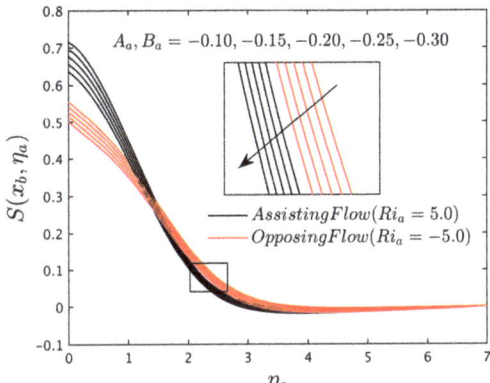

Figure 8. Influence of heat sink factor on $S(x_b, \eta_a)$.

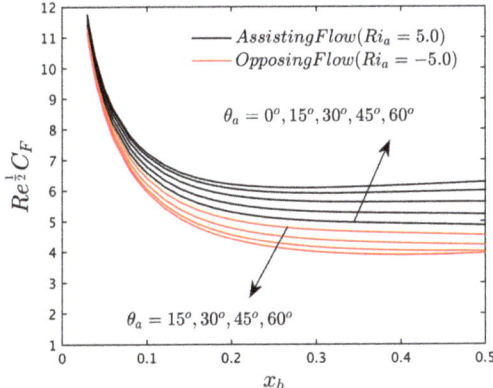

Figure 9. Influence of yaw angle on $\mathrm{Re}^{1/2}C_F$.

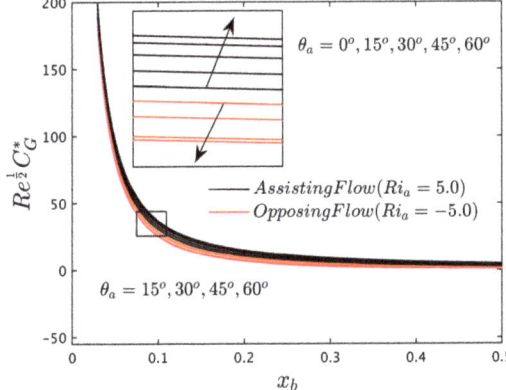

Figure 10. Influence of yaw angle on $\mathrm{Re}^{1/2}C_G^*$.

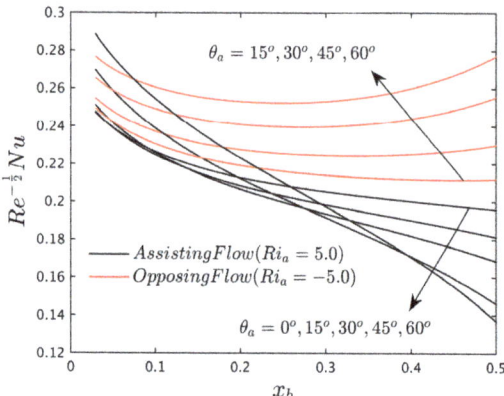

Figure 11. Influence of yaw angle on $Re^{-1/2}Nu$.

Table 4. Numerical values of $Re^{1/2}C_F$ and $Re^{1/2}C_G^*$ for the instance of BAF with variation in the values of φ and θ_a when $x_b = 0.5$, $A_a = 0.1$, $B_a = 0.1$, $Ec_a = 0.5$ and $Bi_a = 0.5$.

φ	θ_a	$Re^{1/2}C_F$	$Re^{1/2}C_G^*$
0.0	15°	5.6243618	3.5905573
0.030	-	5.6673802	3.6138515
0.035	-	5.7109067	3.6374371
0.025	15°	5.6243618	3.5905573
-	30°	6.2796404	4.6663076
-	45°	6.6225145	5.2286145

Table 5. Numerical values of $Re^{1/2}C_F$ and $Re^{1/2}C_G^*$ for the instance of BOF with variation in the values of φ and θ_a when $x_b = 0.5$, $A_a = 0.1$, $B_a = 0.1$, $Ec_a = 0.5$ and $Bi_a = 0.5$.

φ	θ_a	$Re^{1/2}C_F$	$Re^{1/2}C_G^*$
0.0	15°	4.2166338	1.2069270
0.030	-	4.2558146	1.2235525
0.035	-	4.2954308	1.2403568
0.025	15°	4.2166338	1.2069270
-	30°	4.0114303	0.8309678
-	45°	3.9431649	0.6997625

Table 6. The latest computational outputs of the RHT for the instances of BAF and BOF with variation in the values of several influential parameters when $x_b = 0.5$, $A_a = 0.1$, and $B_a = 0.1$.

φ	θ_a	Ec_a	Bi_a	Assisting Flow	Opposing Flow
0.025	15°	0.5	0.5	0.1448552	0.2758385
0.030	-	-	-	0.1455113	0.2765296
0.035	-	-	-	0.1461382	0.2771948
0.025	15°	0.5	0.5	0.1448552	0.2758385
-	30°	-	-	0.1368615	0.3102864
-	45°	-	-	0.1333446	0.3809967
0.025	15°	0.1	0.5	0.4327421	0.4409451
-	-	0.3	-	0.2996868	0.3525156
-	-	0.5	-	0.1461382	0.2758385
0.025	15°	0.5	0.1	0.0386987	0.0709445
-	-	-	0.3	0.1003166	0.1871035
-	-	-	0.5	0.1461382	0.2771948

Tables 4 and 5 illustrate the latest computational values of the reduced shear stresses (RSS) in both paths (chordwise and spanwise) for the case of BAF ($Ri_a = 5.0$) and OFs ($Ri_a = -5.0$) with diverse values of φ and θ_a when $x_b = 0.5$, $A_a = 0.1$, $B_a = 0.1$, $Pr_a = 6.2$, $Ec_a = 0.5$ and $Bi_a = 0.5$. Outcomes in the tabular form demonstrate that the RSS escalated for the case of BAF with varying values of φ and θ_a. On the other hand, for the case of BOF, the behavior of the reduced shear stresses looks similar to the case of BAF owing to the heightened values of φ but declined in both (CW and SW) directions for the higher values of yaw angle. Moreover, the shear stresses in both (chordwise and spanwise) directions upsurged by almost 0.76% and 0.65% for the case of BAF due to the deviations in the values of φ while it increased up to 11.65% and 29.96% with superior impacts of yaw angle. For the case of BOF, the RSS increased by almost 0.93% and 1.38% in the CW and SW directions due to the augmentation in the selected values of φ but decreased by almost 4.87% and 31.15% for the advanced values of θ_a.

Table 6 elucidates the numerical outputs of the RHT for the case of BAF and BOF with varying several parameters when $x_b = 0.5$, $A_a = 0.1$, and $B_a = 0.1$. Note that the results demonstrate that the heat transfer upsurging for both cases is due to the superior impact of the constraint φ and Bi_a while it is diminished for the case of BAF and BOF with upsurging values of Ec_a. Additionally, the larger values of the yaw angle decreased the importance of heat transfer for the phenomenon of BAF while it abruptly escalated for the case of BOF. percentage-wise, the heat transfer escalateD almost 0.45% and 59.18% for the case of BAF due to augmentation in the impacts of φ and Bi_a while for the case of BOF, it WAS enhanced up to 0.25% and 63.79%, respectively. Additionally, the rate of heat transfer decreased up to 5.52% and 30.76% for the situation of BAF with deviation in the values of yaw angle and Eckert number, respectively, while it improved up to the level of 12.47% for the case of BOF but decreased by almost 20.05%.

Figures 2–4 portray the impression of the yaw angle θ_a on the velocity field in both (chordwise and spanwise) directions and the temperature profiles of the water-based graphene oxide nanoparticles against η_a, respectively. Results were generated for the instance of BAF and BOF cases. Growing values of θ_a augmented the velocity curves in the CW direction as well as in the SW direction for the case of BAF while the flow behavior in both graphs (see Figures 3 and 4) behaved oppositely for the situation of OFs. In a general scenario, the cylinder inclines as the yaw angle rises, allowing the fluid to move quickly alongside the cylinder simultaneously. The cylinder's angular orientation, which affects the fluid's inner pressure, is the reason for this propensity. Furthermore, the CW direction and surface friction are both made worse by the cylinder's slope or gradient. When the yaw angle is zero, the nanofluid curves converge to a single curve which shows the ideal vertical cylinder. As a result, as the yaw angle increases, and it is possible to see more fluctuations/oscillations away from the wall than at the surface. Alternatively, the temperature curves decline slightly for the case of BAF due to the higher impacts of the yaw angle. Meanwhile, the behavior significantly progresses for the case of BOF, as shown graphically in Figure 4. Additionally, the gap in both solution curves is clear by increasing the impression of the yaw angle.

The impact of the Biot number Bi_a and the Eckert number Ec_a on the temperature distribution profiles of the water-based graphene oxide nanoparticles for the case of BAF and BOF are graphically depicted in Figures 5 and 6, respectively. It is shown that the ND temperature profile curves developed for both cases due to the higher implementation of Bi_a and Ec_a parameters. Generally, it can be noticed that the convective heating upsurged with increasing the values of the Bi_a, i.e., $Bi_a \to \infty$ stimulates the isothermal surface, shown in Figure 5, where, $S(x_b, 0) = 1$, as $Bi_a \to \infty$. The ND temperature difference between the surface and the nanofluid became stronger when the Biot number was higher because strong surface convection generated more heat that was transferred to the cylinder's surface. As a result, the TTBL and the temperature was augmented with superior impacts of the Biot number. In contrast, growing the Eckert number increased the fluid temperature because positive values of the Eckert number indicate an increment in kinetic energy, which contributes to the growing temperature distribution, as shown graphically in Figure 6.

Figures 7 and 8 indicate the temperature profiles of the water-based graphene oxide nanoparticles for the circumstance of BAF and BOF due to the variation in the internal heat source/sink factor, respectively. Outcomes reveal that the temperature solution curves escalated for both cases due to superior values of the internal heat source factor (IHSF) while decreased for both phenomena because of the higher internal heat sink factor. Moreover, the thickness of the temperature boundary layer was thicker in the case of BAF compared with the BOF. Physically, the rise of the IHSF generates more energy inside the boundary layer that causes the increment of the fluid temperature, see Figure 7. Meanwhile, the production of less energy inside the boundary layer is observed to determine the influence of the internal heat sink factor. As a result, the higher the internal heat sink factor influences the temperature deceleration, see Figure 8.

The influence of θ_a on the reduced shear stresses in both (chord-wise and span-wise) paths and the RHT versus x_b for the cases of BAF and BOF is illustrated in Figures 9–11, respectively. From Figures 9 and 10, it is practically seen that the shear stresses in the CW and SW directions increased for the situation of BAF while decreased for the case of BOF due to the higher employing values of the yaw angle. In addition, the shear stress is highly observed for the circumstance of BAF compared with the situation of the BOF. Additionally, for the case of BAF, the shear stresses in both directions increased by almost 11.65% and 29.96%, respectively, due to the higher values of yaw angle. However, the shear stresses decreased by up to 4.87% and 31.15% for the phenomenon of BOF. In contrast, the reduced heat transfer decelerated for the condition of BAF and increased for the BOF condition owing to advance values of the yaw angle θ_a (see Figure 11). Consequently, the reduced heat transfer upsurged by 12.47% for the case of BOF due to the higher impressions of the yaw angle but decreased by 5.52% for the case of BAF.

6. Conclusions

This study considered numerical solutions to the problem of a buoyancy or mixed convection nanofluid flow through a yawed cylinder with irregular heat sink/source effects. Additionally, the model comprised the impact of convective boundary conditions and viscous dissipation. The approach for finding LNSSs was exercised in the investigation. This technique's key characteristic is the retention of NS terms in the obtained equations without approximations. In that, the influence of varying strength of yaw angle, Biot number, the volume fraction of nanoparticle, Eckert number, internal heat sink/source factor, and the Richardson number on the velocity and shear stresses profile in both (CW and SW) directions, the non-dimensional temperature and heat transfer were analyzed employing distinct graphs and tables. After taking into consideration this thorough analysis, the results are described below:

- With variation in all the physical parameters, the solutions were obtained for the fixed constant values in the case of assisting and opposing flows.
- For enlarging magnitudes of the nanoparticles volume fraction, the shear stresses increased by about 0.76% and 0.93% for the instance of BAF, while for the case of BOF, they increased by almost 0.65% and 1.38%, respectively.
- Velocity and shear stress enhanced the case of assisting flow in both (chordwise and spanwise) directions due to the larger influences of the yaw angle, but both profile curves were diminished for the case of opposing flow.
- Shear stress decreased by almost 4.87% while it increased up to 11.65% for the case of opposing and assisting flows in the respective chordwise direction due to the higher impressions of yaw angle. However, it decreased and inclined spanwise by about 31.15% and 29.96%, respectively.
- The non-dimensional temperature upsurged for both assisting and opposing flow cases due to the higher values of Eckert and Biot numbers.

- For increasing values of Eckert number, the heat transfer rates decayed by about 30.76% and 20.05% for the phenomenon of AFs and OFs while increasing by around 0.45% and 0.25% with nanoparticles volume fraction.

Author Contributions: Conceptualization, S.E. and U.K.; methodology, S.E. and U.K.; software, S.E., A.M.A. and U.K.; validation, A.I., S.E., A.Z., S.A.B., S.K.S. and U.K.; formal analysis, A.Z., S.E. and A.M.A.; investigation, U.K. and A.M.A.; resources, A.I.; data curation, A.Z.; writing—original draft preparation, A.I., A.Z., U.K. and S.E.; writing—review and editing, A.I., A.Z., S.A.B., S.K.S. and A.M.A.; visualization, S.E.; supervision, A.I.; project administration, A.M.A. and A.I.; funding acquisition, A.I. and A.M.A. All authors have read and agreed to the published version of the manuscript.

Funding: This work was funded by Universiti Kebangsaan Malaysia (DIP-2020-001). Also, this work received support from Princess Nourah bint Abdulrahman University Researchers Supporting Project number (PNURSP2023R163), Princess Nourah bint Abdulrahman University, Riyadh, Saudi Arabia. In addition, this study is also funded by Prince Sattam bin Abdulaziz University project number (PSAU/2023/R/1444).

Data Availability Statement: Not applicable.

Acknowledgments: The authors are thankful for the support of Princess Nourah bint Abdulrahman University Researchers Supporting Project number (PNURSP2023R163), Princess Nourah bint Abdulrahman University, Riyadh, Saudi Arabia. Also, this work is supported via funding from Prince Sattam bin Abdulaziz University project number (PSAU/2023/R/1444).

Conflicts of Interest: The authors declare no conflict of interest.

Nomenclature

A_a	Exponentially decaying space coefficient
B_a	Temperature-dependent heat sink/source coefficient
Bi_a	Biot number
C_F	Friction factor in the chord-wise direction
C_G^*	Friction factor in the span-wise direction
Ec_a	Eckert number
G, F	Non-dimensionless velocities in span-wise and chord-wise directions, respectively
g	Gravity acceleration (m s^{-2})
h_f	Convective heat transfer
k_f	Thermal conductivity of base fluid (W m^{-1} K^{-1})
k_{nf}	Thermal conductivity of nanofluid (W m^{-1} K^{-1})
k_{snp}	Thermal conductivity of nanoparticles (W m^{-1} K^{-1})
Nu	Nusselt number
Pr_a	Prandtl number
Q^{bbb}	Irregular heat source/sink
R_a	Radius of the cylinder (m)
Ri_a	Mixed convective parameter
Re	Reynolds number
S	Non-dimesional temperature
T_a	Temperature of the nanofluid (K)
T_∞	Free-stream temperature (K)
T_f	Temperature of the hot fluid (K)
u_a	Velocity projection for the chordwise direction (m s^{-1})
v_a	Velocity projection for the normal direction (m s^{-1})
w_a	Velocity projection for the spanwise direction (m s^{-1})
w_E, u_E	Variable Free-stream velocities (m s^{-1})
w_∞, u_∞	Free-stream velocities in $z-$ and $x-$ directions (m s^{-1})
(x_a, y_a, z_a)	Curvilinear coordinates (m)

Greek symbols

β_{nf}	Nanofluid thermal expansion (K^{-1})
β_f	Base fluid thermal expansion (K^{-1})
β_s	Nanoparticle thermal expansion (K^{-1})
η_a, ζ_a	Transformed variables
μ_{nf}	Nanofluid dynamic viscosity (kg m s^{-1})
μ_f	Base fluid dynamic viscosity (kg m s^{-1})
ν_{nf}	Nanofluid kinematic viscosity (m^2 s^{-1})
ν_f	Kinematic viscosity of base fluid (m^2 s^{-1})
ρ_{nf}	Nanofluid density (kg m^{-3})
ρ_{snp}	Nanoparticles density (kg m^{-3})
ρ_f	Density of base fluid (kg m^{-3})
$(\rho c_p)_{nf}$	Nanofluid heat capacity (Jkg^{-1} K^{-1})
φ	Volume fraction of nanoparticles
ψ, F	Dimensional and non-dimensional stream functions

Acronyms

BAF	Bouyancy assisting flow
BCs	Boundary conditions
BOF	Bouyancy opposing flow
CBCs	Convective boundar conditions
CW	Chordwise
GO	Graphene oxide
LNS	Local non-similarity
MCN	Mixed convection
NDL	Non-dimensional
ODEs	Ordinary differential equations
PDEs	Partial differential equations
RHT	Rate of heat transfer
SP	Spanwise
TTBL	Thickness of the thermal boundary layer
YC	Yawed Cylinder

Subscripts

f	Condition at free stream
nf	Nanofluid
snp	Solid nanoparticles
w	Wall boundary condition
∞	Free-stream condition

Superscripts

$'$	Derivative w.r.t. η

References

1. Sears, W.R. The Boundary Layer of Yawed Cylinders. *J. Aeronaut. Sci.* **1948**, *15*, 49–52. [CrossRef]
2. Chiu, W.S.; Lienhard, J.H. On Real Fluid Flow Over Yawed Circular Cylinders. *J. Basic Eng.* **1967**, *89*, 851–857. [CrossRef]
3. Bücker, D.; Lueptow, R.M. The boundary layer on a slightly yawed cylinder. *Exp. Fluids* **1998**, *25*, 487–490. [CrossRef]
4. Roy, S. Non-uniform mass transfer or wall enthalpy into a compressible flow over yawed cylinder. *Int. J. Heat Mass Transf.* **2001**, *44*, 3017–3024. [CrossRef]
5. Subhashini, S.V.; Takhar, H.S.; Nath, G. Non-uniform multiple slot injection (suction) or wall enthalpy into a compressible flow over a yawed circular cylinder. *Int. J. Therm. Sci.* **2003**, *42*, 749–757. [CrossRef]
6. Roy, S.; Saikrishnan, P. Non-uniform slot injection (suction) into water boundary layer flow past yawed cylinder. *Int. J. Eng. Sci.* **2004**, *42*, 2147–2157. [CrossRef]

7. Saikrishnan, S. Boundary layer flow over a yawed cylinder with variable viscosity: Role of non-uniform double slot suction (injection). *Int. J. Numer. Methods Heat Fluid Flow* **2012**, *22*, 342–356.
8. Revathi, G.; Saikrishnan, P.; Chamkha, A.J. Non-similar solutions for unsteady flow over a yawed cylinder with non-uniform mass transfer through a slot. *Ain Shams Eng. J.* **2014**, *5*, 1199–1206. [CrossRef]
9. Patil, P.M.; Shashikant, A.; Roy, S.; Hiremath, P.S. Mixed convection flow past a yawed cylinder. *Int. Commun. Heat Mass Transf.* **2020**, *114*, 104582. [CrossRef]
10. Choi, S.U.S. Enhancing thermal conductivity of fluids with nanoparticle. In Proceedings of the 1995 International Mechanical Engineering Congress and Exhibition, San Francisco, CA, USA, 12–17 November 1995; pp. 99–105.
11. Avramenko, A.A.; Shevchuk, I.V. *Modelling of Convective Heat and Mass Transfer in Nanofluids with and without Boiling and Condensation*; Springer International Publishing: Cham, Switzerland, 2022. [CrossRef]
12. Ahmad, S.; Pop, I. Mixed convection boundary layer flow from a vertical flat plate embedded in a porous medium filled with nanofluids. *Int. Commun. Heat Mass Transf.* **2010**, *37*, 987–991. [CrossRef]
13. Nazar, R.; Tham, L.; Pop, I.; Ingham, D.B. Mixed Convection Boundary Layer Flow from a Horizontal Circular Cylinder Embedded in a Porous Medium Filled with a Nanofluid. *Transp. Porous Media* **2011**, *86*, 517–536. [CrossRef]
14. Rashad, A.M.; Chamkha, A.J.; Modather, M. Mixed convection boundary-layer flow past a horizontal circular cylinder embedded in a porous medium filled with a nanofluid under convective boundary condition. *Comput. Fluids* **2013**, *86*, 380–388. [CrossRef]
15. Turkyilmazoglu, M. Analytical solutions of single and multi-phase models for the condensation of nanofluid film flow and heat transfer. *Eur. J. Mech.-B/Fluids* **2015**, *53*, 272–277. [CrossRef]
16. Sulochana, C.; Naramgari, N. Stagnation point flow and heat transfer behavior of Cu-water nanofluid towards horizontal and exponentially stretching/shrinking cylinders. *Appl. Nanosci.* **2016**, *6*, 451–459. [CrossRef]
17. Rekha, M.B.; Sarris, I.E.; Madhukesh, J.K.; Raghunatha, K.R.; Prasannakumara, B.C. Impact of thermophoretic particle deposition on heat transfer and nanofluid flow through different geometries: An application to solar energy. *Chin. J. Phys.* **2022**, *80*, 190–205. [CrossRef]
18. Sarfraz, M.; Khan, M.; Yasir, M. Dynamics of water conveying iron oxide and graphene nanoparticles subject to stretching/spiraling surface: An asymptotic approach. *Ain Shams Eng. J.* **2022**, *18*, 102021. [CrossRef]
19. Mabood, F.; Yusuf, T.A.; Khan, W.A. Cu–Al_2O_3–H_2O hybrid nanofluid flow with melting heat transfer, irreversibility analysis and nonlinear thermal radiation. *J. Therm. Analy. Calor.* **2021**, *143*, 973–984. [CrossRef]
20. Wahid, N.S.; Arifin, N.M.; Khashi'Ie, N.S.; Pop, I. Mixed convection of a three-dimensional stagnation point flow on a vertical plate with surface slip in a hybrid nanofluid. *Chin. J. Phys.* **2021**, *74*, 129–143. [CrossRef]
21. Zangooee, M.R.; Hosseinzadeh, K.; Ganji, D.D. Hydrothermal analysis of hybrid nanofluid flow on a vertical plate by considering slip condition. *Theor. Appl. Mech. Lett.* **2022**, *12*, 100357. [CrossRef]
22. Malekshah, E.H.; Abed, A.M.; Aybar, H.Ş. Thermal analysis of multi-finned plate employing lattice Boltzmann method based on Taylor-series/least-squares. *Eng. Anal. Bound. Elem.* **2023**, *146*, 407–417. [CrossRef]
23. Bataller, R.C. Radiation effects for the Blasius and Sakiadis flows with a convective surface boundary condition. *Appl. Math. Comput.* **2008**, *206*, 832–840. [CrossRef]
24. Aziz, A. A similarity solution for laminar thermal boundary layer over a flat plate with a convective surface boundary condition. *Commun. Nonlinear Sci. Numer. Simul.* **2009**, *14*, 1064–1068. [CrossRef]
25. Makinde, O.D. Similarity solution of hydromagnetic heat and mass transfer over a vertical plate with a convective surface boundary condition. *Int. J. Phys. Sci.* **2010**, *5*, 700–710.
26. Wahid, N.S.; Arifin, N.M.; Khashi'Ie, N.S.; Pop, I.; Bachok, N.; Hafidzuddin, M.E.H. Unsteady MHD mixed convection flow of a hybrid nanofluid with thermal radiation and convective boundary condition. *Chin. J. Phys.* **2022**, *77*, 378–392. [CrossRef]
27. Malik, R.; Khan, M.; Munir, A.; Khan, W.A. Flow and Heat Transfer in Sisko Fluid with Convective Boundary Condition. *PLoS ONE* **2014**, *9*, e107989. [CrossRef]
28. Ramesh, G.K.; Gireesha, B.J.; Gorla, R.S.R. Boundary layer flow past a stretching sheet with fluid-particle suspension and convective boundary condition. *Heat Mass Transf.* **2015**, *51*, 1061–1066. [CrossRef]
29. Rashid, A.; Ayaz, M.; Islam, S.; Saeed, A.; Kumam, P.; Suttiarporn, P. Theoretical Analysis of the MHD Flow of a Tangent Hyperbolic Hybrid Nanofluid over a Stretching Sheet With Convective Conditions: A Nonlinear Thermal Radiation Case. *South Afr. J. Chem. Eng.* **2022**, *42*, 255–269. [CrossRef]
30. Prasad, K.V.; Vaidya, H.; Mebarek-Oudina, F.; Choudhari, R.; Nisar, K.S.; Jamshed, W. Impact of surface temperature and convective boundary conditions on a Nanofluid flow over a radially stretched Riga plate. *Proc. Inst. Mech. Eng. Part E* **2022**, *236*, 942–952. [CrossRef]
31. Khan, U.; Zaib, A.; Ishak, A.; Waini, I.; Pop, I.; Elattar, S.; Abed, A.M. Stagnation point flow of a water-based graphene-oxide over a stretching/shrinking sheet under an induced magnetic field with homogeneous-heterogeneous chemical reaction. *J. Magn. Magn. Mater.* **2023**, *565*, 170287. [CrossRef]
32. Khan, U.; Zaib, A.; Ishak, A.; Alotaibi, A.M.; Elattar, S.; Pop, I.; Abed, A.M. Impact of an Induced Magnetic Field on the Stagnation-Point Flow of a Water-Based Graphene Oxide Nanoparticle over a Movable Surface with Homogeneous–Heterogeneous and Chemical Reactions. *J. Magn. Magn. Mater.* **2022**, *8*, 155. [CrossRef]

33. Kumari, M.; Nath, G. Mixed convection boundary layer flow over a thin vertical cylinder with localized injection/suction and cooling/heating. *Int. J. Heat Mass Transf.* **2004**, *47*, 969–976. [CrossRef]
34. Rekha, M.B.; Sarris, I.E.; Madhukesh, J.K.; Raghunatha, K.R.; Prasannakumara, B.C. Activation energy impact on flow of AA7072-AA7075/Water-Based hybrid nanofluid through a cone, wedge and plate. *Micromachines* **2022**, *13*, 302. [CrossRef]
35. Ali, B.; Ahammad, N.A.; Awan, A.U.; Oke, A.S.; Tag-ElDin, E.M.; Shah, F.A.; Majeed, S. The Dynamics of Water-Based Nanofluid Subject to the Nanoparticle's Radius with a Significant Magnetic Field: The Case of Rotating Micropolar Fluid. *Sustainability* **2022**, *14*, 10474. [CrossRef]

Disclaimer/Publisher's Note: The statements, opinions and data contained in all publications are solely those of the individual author(s) and contributor(s) and not of MDPI and/or the editor(s). MDPI and/or the editor(s) disclaim responsibility for any injury to people or property resulting from any ideas, methods, instructions or products referred to in the content.

Article

Synergistic Lubrication and Antioxidation Efficacies of Graphene Oxide and Fullerenol as Biological Lubricant Additives for Artificial Joints

Qian Wu [1,†], Honglin Li [2,†], Liangbin Wu [1], Zihan Bo [1], Changge Wang [1], Lei Cheng [1], Chao Wang [1], Chengjun Peng [1,3], Chuanrun Li [1], Xianguo Hu [4], Chuan Li [2,*] and Bo Wu [1,*]

1 Pharmaceutical Engineering Technology Research Center, School of Pharmacy, Anhui University of Chinese Medicine, Hefei 230012, China
2 Key Laboratory of Resource Comprehensive Utilization, School of Chemistry and Material Engineering, Chaohu University, Hefei 238014, China
3 Anhui Province Key Laboratory of Pharmaceutical Preparation Technology and Application, Hefei 230012, China
4 School of Mechanical Engineering, Hefei University of Technology, Hefei 230009, China
* Correspondence: 053039@chu.edu.cn (C.L.); wubo@ahtcm.edu.cn (B.W.)
† These authors contributed equally to this work.

Citation: Wu, Q.; Li, H.; Wu, L.; Bo, Z.; Wang, C.; Cheng, L.; Wang, C.; Peng, C.; Li, C.; Hu, X.; et al. Synergistic Lubrication and Antioxidation Efficacies of Graphene Oxide and Fullerenol as Biological Lubricant Additives for Artificial Joints. *Lubricants* 2023, 11, 11. https://doi.org/10.3390/lubricants11010011

Received: 30 November 2022
Revised: 24 December 2022
Accepted: 27 December 2022
Published: 30 December 2022

Copyright: © 2022 by the authors. Licensee MDPI, Basel, Switzerland. This article is an open access article distributed under the terms and conditions of the Creative Commons Attribution (CC BY) license (https://creativecommons.org/licenses/by/4.0/).

Abstract: The service life of artificial joints has gradually failed to meet the needs of patients. Herein, the synergistic lubrication and antioxidant efficacies of graphene oxide (GO) and fullerenol (Fol) as biological lubricant additives for artificial joints were investigated. The lubrication mechanisms of biological lubricant containing GO and Fol at the friction interface of artificial joints were then revealed. Tribological tests showed that the average friction coefficients of Al_2O_3–Ti6Al4V pairs and Ti6Al4V–UHMWPE pairs for artificial joints could be reduced by 30% and 22%, respectively, when GO and Fol were used as biological lubricant additives simultaneously. The lubrication mechanism showed that some incommensurate sliding contact surfaces could be formed between the GO nanosheets and spherical Fol at the interface, which reduced the interaction forces of friction pairs. The maximum scavenging rates of •OH and DPPH free radicals by the biological lubricant containing GO and Fol were 35% and 45%, respectively, showing a good antioxidant efficacy of the biological lubricant. This can be attributed to the GO and Fol scavenging free radicals through electron transfer and hydrogen transfer. This study provides a theoretical basis for the development and application of carbon nanomaterials as biological lubricant additives for artificial joints in the future.

Keywords: biological lubrication additives; artificial joint; carbon nanomaterials; lubrication; antioxidation

1. Introduction

The incidence of the human joint system is rapidly increasing with age, genetics, obesity and other factors [1]. It is difficult for bone joints to heal after injury or pathological changes. Artificial joint replacement, as a safe and effective method for the treatment of advanced joint diseases, has been widely implemented all over the world. It is described as the process of a prosthetic joint component made of metal, ceramic or polymer replacing a damaged natural bone joint [2]. However, the prosthetic material is always in a state of direct contact friction due to the absence of synovial fluid secreted by the synovial membrane and cartilage tissue in the patient's natural joint system [3]. This leads to the wear of the prosthetic material inevitably and reduces the service life of the artificial joint greatly. In addition, metal and other prosthetic materials produce some wear debris in the process of friction [2]. These wear debris induce the production of reactive oxygen species (ROS) in the human tissue around the prosthesis and lead to a series of adverse biological reactions such as osteolysis, resulting in the failure of the artificial joint due to aseptic loosening [4–6]. Based on available clinical data, the safe service life of artificial joints in the human body is only ten to fifteen years [7]. With the reality of younger patients presenting with osteoarthritis, the current service life of artificial joints obviously cannot meet the actual needs of patients. Therefore, it is necessary to improve the lubrication of the artificial joint system and inhibit the wear of prosthetic materials to prolong the service life of artificial joints.

Extracorporeal injection of joint biological lubricant is the traditional way to improve the lubrication property of artificial joint surfaces. At present, the joint biological lubricant used in clinical environments is hyaluronic acid (HA) [8]. Studies have found that 30% to 40% of a gait cycle is in a state of fluid lubrication during normal activity of joints, and most of the rest time is in a state of boundary lubrication. HA can play the role of fluid lubrication at the natural cartilage interface by changing the viscosity of synovial fluid. However, HA molecules break and lose their good lubrication effect under the condition of boundary lubrication [9,10]. Consequently, HA cannot fully guarantee that the joint interface is in a good lubrication condition under various complex motion conditions. In addition, as an injection lubricant, exogenous HA is hydrolyzed in the artificial joint system and excreted from the body [11,12]; thus, it cannot play a long-term lubrication role in the interface of joint prostheses. Therefore, the development of a lubricant with a stable lubrication effect under a boundary lubrication condition is of great significance to improve the service life of artificial joints.

In recent years, graphene oxide (GO) with a lamellar structure and fullerenol (Fol) with a spherical structure as two typical representatives of carbon-based nanomaterials have attracted wide attention in the field of biological tribology due to their excellent physical and chemical properties and biocompatibility [13]. Researchers found that GO can reduce effectively friction and wear at the friction interface under a boundary lubrication condition by forming protective tribo-films and preventing the direct contact of the microconvex body [14–17]. Liu et al. reported that a water-soluble nanoparticle Fol as the lubrication additive could improve the friction performance and reduce the wear area significantly [18]. Chen et al. found that Fol could transform sliding friction into rolling friction during the friction process, reducing the friction coefficient and wear significantly [19]. More interestingly, researchers also found that GO and Fol have efficient free radical scavenging capabilities [20,21]. In particular, Fol has an antioxidant capacity hundreds of times higher than that of other antioxidants and acts as a free radical sponge in disease states associated with ROS overproduction [22–24].

These studies inspire us to use GO and Fol as biological lubrication additives for artificial joints at the same time. We hope that they can play a lubrication role at the interface of artificial joints, on the one hand, and play an antioxidant role in scavenging ROS in the body, on the other hand. In this way, the problems of wear and ROS increase in artificial joints can be solved simultaneously from two aspects, and the service life of artificial joints in the body can be prolonged as much as possible. In this study, the lubrication properties of GO and Fol at the interface of different artificial joint friction pairs were firstly investigated, and then the antioxidation properties of GO and Fol in scavenging reactive oxygen free radicals were evaluated. Finally, the lubrication and antioxidation mechanisms of GO and Fol were revealed. This study provides an important theoretical basis for the development and application of GO and Fol as biological lubrication additives for artificial joints in the future.

2. Materials and Methods

2.1. Materials and Reagents

GO was purchased from Nanjing XFNANO Materials Tech Co., Ltd. (Nanjing, China) Fol was purchased from Suzhou Dade Carbon Nanotechnology Co., Ltd. (Suzhou, China) 2, 2-diphenyl-1-picrylhydrazyl (DPPH), anhydrous ethanol, salicylic acid, ferrous sulfate ($FeSO_4$) and hydrogen peroxide (H_2O_2) were all purchased from Sinopharm Chemical Reagent Co., Ltd. (Shanghai, China) All reagents used were of analytical grade.

2.2. Preparation and Characterization of Biological Lubrication Additives

According to the mass ratio of 0.1 wt%, GO and Fol were added to ultrapure water in different proportions and ultrasonically dispersed for 60 min. Samples of the biological lubricant containing different proportions of GO and Fol additives were then obtained. Schematic diagram of experimental process was shown in Figure 1.

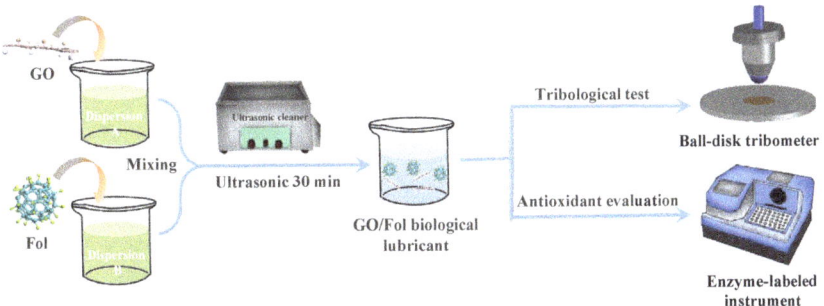

Figure 1. Schematic diagram of experimental process.

The morphology of the biological lubrication additives was studied by transmission electron microscopy (TEM, JEM–2100F, JEOL Ltd., Akishima-shi, Tokyo, Japan). The molecular structures of the biological lubrication additives were characterized by Fourier transform infrared spectroscopy (FTIR, Nicolet 6700, Thermo Fisher Scientific, Waltham, MA, USA). The accuracy of the FTIR was higher than 0.01 cm^{-1} and the resolution of the FTIR was higher than 0.09 cm^{-1}.

2.3. Dispersion Stability of Biological Lubrication Additives

The dispersion stability of different ratios of GO and Fol additives in biological lubricant was analyzed by an ultraviolet–visible spectrophotometer (UV-VIS, UV-8000, Shanghai Metash Instruments Co., Ltd., Shanghai, China). In general, the smaller the change in absorbance with time, the better the stability of GO and Fol dispersion in water [25]. The absorbance curves of biological lubrication additives GO, Fol, GO/Fol = 1/1, GO/Fol = 1/2 and GO/Fol = 2/1 in the wavelength range from 200 nm to 500 nm are shown in Figure 2a, and the changes in absorbance under the maximum absorption wavelength (λmax = 228 nm) with storage time are shown in Figure 2b–f. It can be clearly seen from Figure 2 that the absorbance of all biological lubrication additives remains almost constant for 8 days, and there is no obvious stratification in the photographs of lubricant samples. This indicates that all GO and Fol additives have excellent dispersion stability as biological lubricants.

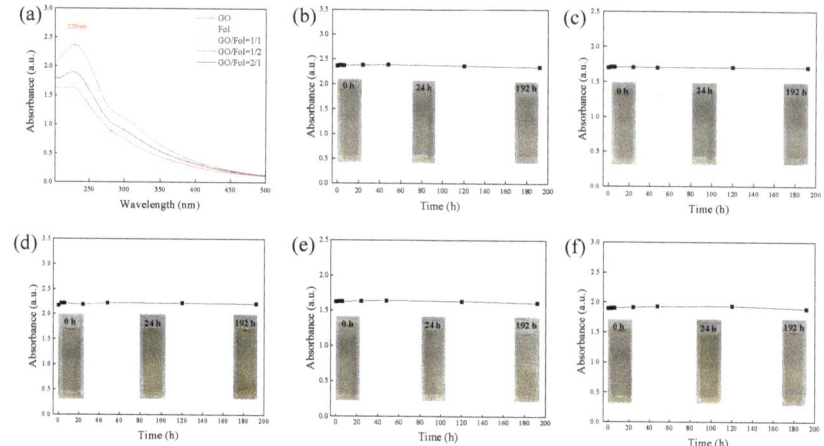

Figure 2. (a) The absorbance curves of biological lubrication additives GO, Fol, GO/Fol = 1/1, GO/Fol = 1/2 and GO/Fol = 2/1; changes in absorbance and photographs of biological lubrication additives (b) GO, (c) Fol, (d) GO/Fol = 1/1, (e) GO/Fol = 1/2 and (f) GO/Fol = 2/1 for different storage times.

2.4. Tribological Tests

Tribological tests were performed on a reciprocating ball–disk tribometer (CFT-I, Lanzhou Zhongkekaihua Technology Development Co., Ltd., Lanzhou, China). Before the tribological test, the ball and disk were cleaned ultrasonically for 30 min, and then the lubricant was dripped onto the ball–disk contact area. The stroke was set to 5 mm, the time was set to 30 min, the speed was set to 180 r/min, and the load was set to 1.5 N. Control experiments were performed on water without nanomaterials. Considering that metal, ceramic and polymer materials are widely used in existing artificial joints, we used two kinds of ball–disk friction pairs with different materials in this study. The materials used for the ball (Φ 6 mm) and disk (28 mm × 28 mm × 2.65 mm) are shown in Table 1. The friction coefficient (COF) was monitored in real time by the software provided with the tribometer, and the three-dimensional morphology of the wear tracks on the lower disk was photographed by a 3D laser scanning microscope (LSM, VK-X100, Keyence, Osaka, Japan). The chemical composition of the tribo-films on the wear surface was analyzed by Raman spectroscopy.

Table 1. Frictional pair materials.

Frictional Pair	Ball	Disk
Frictional pair 1	Al_2O_3 (R_a, 3 μm)	Ti6Al4V (R_a, 1 μm)
Frictional pair 2	Ti6Al4V (R_a, 3 μm)	Ultra-high-molecular-weight polyethylene (UHMWPE) (R_a, 1 μm)

2.5. Antioxidant Activity

2.5.1. DPPH Free Radical Scavenging Activity

The above-mentioned biological lubrication additives were diluted in different proportions to obtain sample solutions with concentrations of 0.05 mg/mL and 0.1 mg/mL. The DPPH was prepared with anhydrous ethanol to 0.1 mmol/L. Then, 100 μL of the sample solution and 100 μL of the DPPH solution were added to a 96-well plate and then left in the dark for 30 min. The absorbance at 517 nm was recorded with an enzyme-labeled instrument (Multiskan Spectrum, Thermo Fisher Scientific, Waltham, MA, USA). The inhibition rate (%) of DPPH free radical was calculated by the following Equation (1):

$$\text{DPPH free radical scavenging activity (\%)} = \frac{A_i - (A_x - A_{xo})}{A_i} \times 100\% \quad (1)$$

where A_i is the absorbance of ultrapure water instead of samples, A_x is the absorbance of added samples, and A_{xo} is the absorbance of anhydrous ethanol instead of DPPH solution.

2.5.2. Hydroxyl Free Radical Scavenging Activity

The scavenging ability of biological lubrication additives for hydroxyl radical was determined by the salicylic acid method. The above-mentioned biological lubrication additives were diluted in different proportions to obtain sample solutions with concentrations of 0.05 mg/mL and 0.1 mg/mL. A total of 50 µL of sample solution, 50 µL of 9.0 mM salicylic acid–ethanol solution, 50 µL of 9.0 mM $FeSO_4$ solution and 50 µL of 9 mM H_2O_2 solution were added to the 96-well plate. After incubation at 37 °C for 30 min, the absorbance at 510 nm was recorded with an enzyme-labeled instrument. The scavenging activity (%) of hydroxyl radical was calculated by the following Equation (2):

$$\text{Hydroxyl radical scavenging activity (\%)} = \frac{B_i - (B_x - B_{xo})}{B_i} \times 100\% \quad (2)$$

where B_i is the absorbance of ultrapure water instead of samples, B_x is the absorbance of added samples, and B_{xo} is the absorbance of ultrapure water instead of H_2O_2 solution.

3. Results and Discussion

3.1. Characterization

The microstructures of GO, Fol and GO/Fol were studied by transmission electron microscopy, as shown in Figure 3a–c. The morphology of GO showed an obvious ultra-thin nanosheet shape with some folds related to oxidation degree [26]. Spherical constructions with larger sizes can be easily found in Figure 3b due to the fact that Fol nanoparticles are easily agglomerated to form spherical clusters in water. However, smaller spherical Fol structures on the surface of GO were observed at the morphology of GO/Fol, shown in Figure 3c. This may be attributed to hydrogen bonding between GO and Fol, which inhibits the agglomeration of the Fol.

Figure 3d shows the functional groups of different biological lubrication additives. The peaks at about 1731 cm^{-1}, 1592 cm^{-1}, 1384 cm^{-1}, 1261 cm^{-1} and 1083 cm^{-1} belong to the characteristic absorption peaks of C=O stretching vibrations (νC=O), C=C stretching vibrations (νC=C), O–H in-plane deformation vibrations (δC–OH), C–O–C stretching vibrations (νC–O–C) and C–O stretching vibrations (νC–O), respectively. It can be seen from Figure 3d that Fol has the characteristic functional groups of νC=C, νC–O and δC–OH, while GO and GO/Fol have the characteristic functional groups of νC=C, νC=O, νC–O, νC–O–C and δC–OH. The above hydrophilic oxygenated functional groups give GO, Fol and GO/Fol excellent dispersion properties in water.

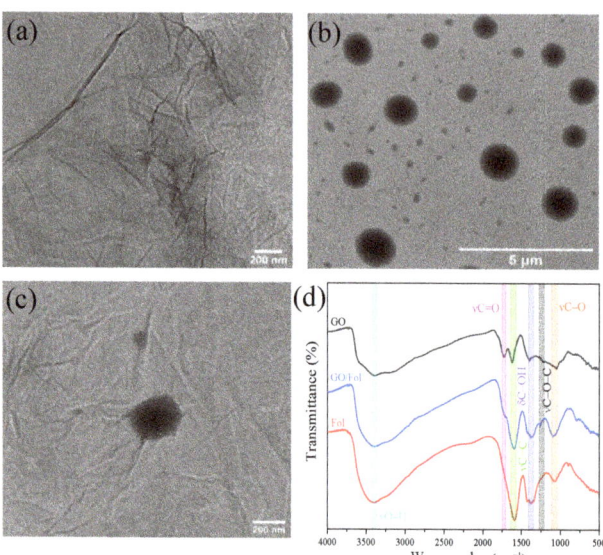

Figure 3. TEM images of (**a**) GO, (**b**) Fol and (**c**) GO/Fol; (**d**) FTIR spectra.

3.2. Evaluation of Lubrication Performance

3.2.1. Friction Reduction Properties of Biological Lubrication Additives

The friction reduction effect of GO/Fol biological lubrication additives with different proportions on two kinds of friction pairs was investigated on a reciprocating ball–disk tribometer. Figure 4a exhibits the relationship of the coefficient of friction (COF) of the pure water and the GO/Fol biological lubricant containing different proportions of additives on Al_2O_3–Ti6Al4V pairs with the change in test time. In the half-hour friction test, the COF values of biological lubricants GO, Fol, GO/Fol = 1/1, GO/Fol = 1/2 and GO/Fol = 2/1 were almost lower than water as a lubricant, which indicated that GO/Fol biological lubrication additives provided friction reduction performance. The average coefficient of friction (AVCOF) values of the pure water and the GO/Fol biological lubricant containing different proportions of additives are shown in Figure 4b. The AVCOF of the GO/Fol = 2/1 biological lubricant could be reduced by 30% compared to pure water, which is obviously lower than biological lubricant containing only GO or Fol additives. The result indicates that GO and Fol can play a synergistic role to significantly improve the friction reduction performance of biological lubricants.

The COF curves of the pure water and the GO/Fol biological lubricant containing different proportions of additives on Ti6Al4V–UHMWPE pairs are shown in Figure 4c. It can be seen that the COF of water was obviously higher than that of GO, Fol, GO/Fol = 1/1, GO/Fol = 1/2 and GO/Fol = 2/1 biological lubricants and presented an increasing trend with the extension in test time. However, the COF of GO, GO/Fol = 1/1, GO/Fol = 1/2 and GO/Fol = 2/1 biological lubricants presented a decreasing trend with the extension in test time oppositely, which confirmed that GO/Fol biological lubrication additives also had excellent friction reduction performance on Ti6Al4V–UHMWPE pairs. The AVCOF values of the pure water and the GO/Fol biological lubricant containing different proportions of additives are shown in Figure 4d. The AVCOF values of all GO/Fol biological lubricants were lower than pure water under the same conditions. Compared to pure water, the AVCOF values of GO/Fol = 1/1, GO/Fol = 1/2 and GO/Fol = 2/1 biological lubricants decreased by 24%, 24% and 22%, respectively. Moreover, the AVCOF values of biological lubricants containing GO/Fol were significantly lower than those of biological lubricants containing only GO or Fol. This demonstrated that GO and Fol can also play a synergistic role on the Ti6Al4V–UHMWPE pairs to significantly improve the friction reduction performance of biological lubricants.

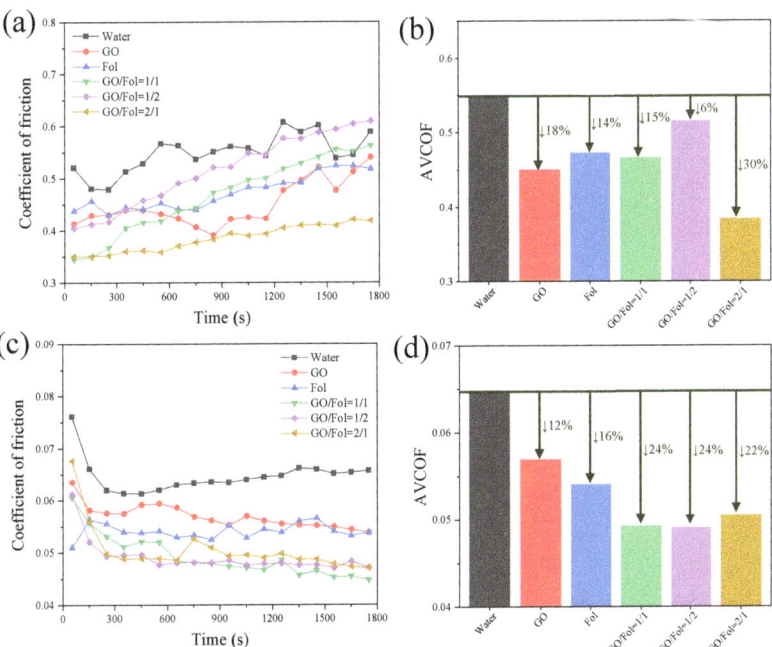

Figure 4. The coefficient of friction of the pure water and the GO/Fol biological lubricant containing different proportions of additives on (**a**) Al_2O_3–Ti6Al4V pairs and (**c**) Ti6Al4V–UHMWPE pairs; the AVCOF of the pure water and the GO/F biological lubricant containing different proportions of additives on (**b**) Al_2O_3–Ti6Al4V pairs and (**d**) Ti6Al4V–UHMWPE pairs.

3.2.2. Antiwear Properties of Biological Lubrication Additives

Microscopic images of wear scars on the Ti6Al4V disks lubricated by pure water and GO/Fol biological lubricants containing different proportions of additives are shown in Figure 5. When the pure water was used as lubricant, signs of deep grooves and serious scratches on the worn surface were observed. This may be caused by the long-term direct contact between the asperities on the surface of the upper and lower friction pairs, resulting in serious fatigue and abrasive wear. However, it can be clearly seen that the grooves became shallow and the wear track width decreased when the wear scar was lubricated by GO/Fol biological lubricants containing different proportions of additives as shown in Figure 5a. This indicated that the abrasive wear on the friction surface was inhibited by the GO/Fol biological lubrication additives to some extent. More specifically, the wear track width of biological lubricants containing both GO and Fol was obviously lower than that of biological lubricants containing only GO or Fol, which further confirmed the superior antiwear performance of GO/Fol biological lubricants owing to the synergistic effect of GO and Fol additives.

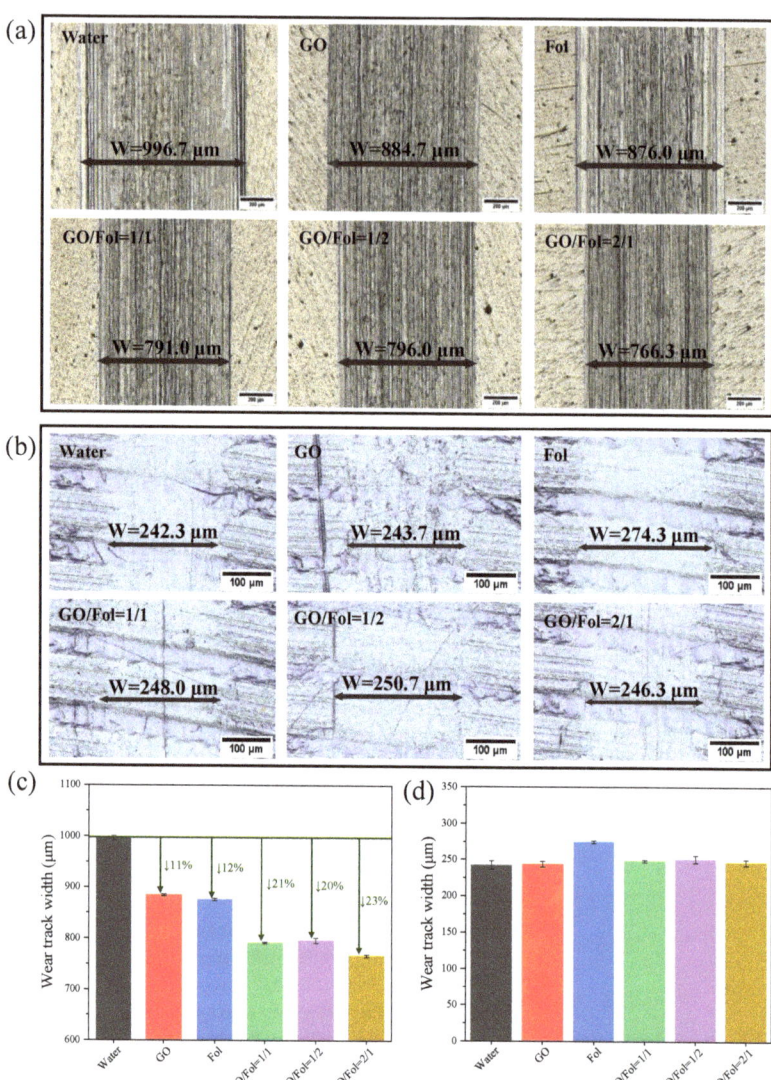

Figure 5. Microscopic images of wear scars on the (**a**) Ti6Al4V disks and (**b**) UHMWPE disks; the wear track width is indicated by the arrow line; the comparison of wear track width lubricated by pure water and GO/Fol biological lubricants containing different proportions of additives on the (**c**) Ti6Al4V disks and (**d**) UHMWPE disks.

Microscopic images of wear scars on the UHMWPE disks lubricated by pure water and GO/Fol biological lubricants containing different proportions of additives are shown in Figure 5b. The difference between the wear track width of biological lubricants and the pure water was small, indicating that the biological lubrication additives had little effect on the antiwear performance of pure water on the UHMWPE disks.

3.2.3. Composition Analysis of Wear Scars

The Raman spectra (Figure 6) and mappings (Figure 7) of the worn surfaces lubricated by pure water and different biological lubrication additives were measured to determine the carbon microstructure and composition of wear scars. The Raman spectrum of the worn surface on the Ti6Al4V disks is exhibited in Figure 6a. The carbon characteristic peaks of D peak (1335 cm^{-1}) and G peak (1590 cm^{-1}) on the Raman spectra of the worn surface lubricated by biological lubricants containing GO additives can be seen, and these were the typical signals of SP^3 and SP^2 hybrid carbon structures, respectively [27]. However, the corresponding D peak and G peak were not identified on the wear scar lubricated by pure water. This shed light on the fact that GO could easily adhere to the worn surface and form protective tribo-films during friction [28]. Furthermore, no evident D peak or G peak could be observed in the Raman spectra of the worn surface lubricated by biological lubricants containing only Fol additives. This means that Fol was not easily adsorbed on the friction interface to form tribo-films. Interestingly, the I_D/I_G intensity ratio of the worn surface lubricated by biological lubricant containing GO/Fol additives was much lower than that of the worn surface lubricated by biological lubricants containing only GO additives, which indicated that more graphitized carbon-structured tribo-films could be formed on the worn surface lubricated by biological lubricants containing both GO and Fol additives. This phenomenon might be due to the fact that the spherical Fol could be interspersed and loaded onto the GO lamellae by hydrogen bonding, which enabled both the GO and Fol to enter the friction interface to form carbon tribo-films on the worn surface. Raman mappings (Figure 7) further confirmed that the distribution area of the carbon tribo-films on the worn surface of the Ti6Al4V disks lubricated by biological lubricants containing GO/Fol additives was obviously wider than that of biological lubricants containing only GO additives. The Raman spectra of the worn surface on the UHMWPE disks are shown in Figure 6b. The carbon characteristic peaks of D peak and G peak were not observed on the worn surface, indicating that the biological lubrication additives could not form tribo-films on the worn surface of the UHMWPE disks. This is in accordance with the above-founded result that the biological lubrication additives had little effect on the antiwear performance of pure water on the UHMWPE disks.

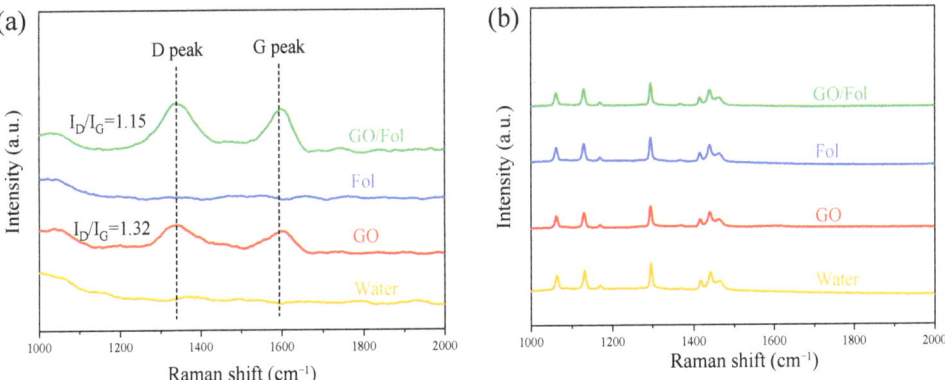

Figure 6. Raman spectra of the worn surfaces on the (**a**) Ti6Al4V disks and (**b**) UHMWPE disks lubricated by pure water and different biological lubrication additives.

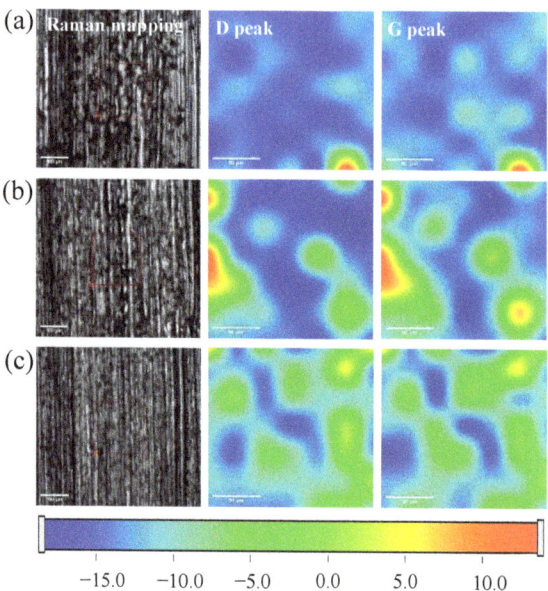

Figure 7. Raman mappings of the wear scars on the Ti6Al4V disks lubricated by (**a**) pure water, (**b**) biological lubricant containing only GO additives, (**c**) biological lubricant containing GO/Fol additives.

3.2.4. Lubrication Mechanism

According to the carbon structure analysis of the worn surface, the lubrication mechanism of the biological lubrication additives at the friction interface was speculated. The lubrication mechanism of the GO, Fol and GO/Fol additives on the Ti6Al4V disks is shown in Figure 8a. The uniformly dispersed lamellar GO can easily transfer into contact interfaces and form protective tribo-films during the friction process, which effectively inhibits the direct contact of the microconvex body to enhance the antiwear ability. At the same time, the lower van der Waals force between the GO nanosheets reduced the shear force of the sliding interface, which makes the interface much easier to slide and promotes friction reduction performance [27,29]. The spherical Fol can enter the contact interface of the friction pairs to separate the microconvex body and transform sliding friction into rolling friction. This means that the friction can be reduced due to the load distribution becomes uniform and the sliding shear force becomes small under the action of spherical Fol rolling at the interface [19]. Moreover, the spherical Fol could be interspersed and loaded onto the GO lamellae by hydrogen bonding when GO and Fol existed at the friction interface simultaneously, which enabled both the GO and Fol to enter the friction interface to form carbon protective tribo-films on the worn surface. Then, the rolling friction effect of Fol and the weak interlayer slip effect of GO could be exerted at the friction interface simultaneously to improve the tribological performance. In addition, some incommensurate sliding contact surfaces were formed between the GO nanosheets and spherical Fol at the interface, which further reduced the interaction forces of friction pairs [25].

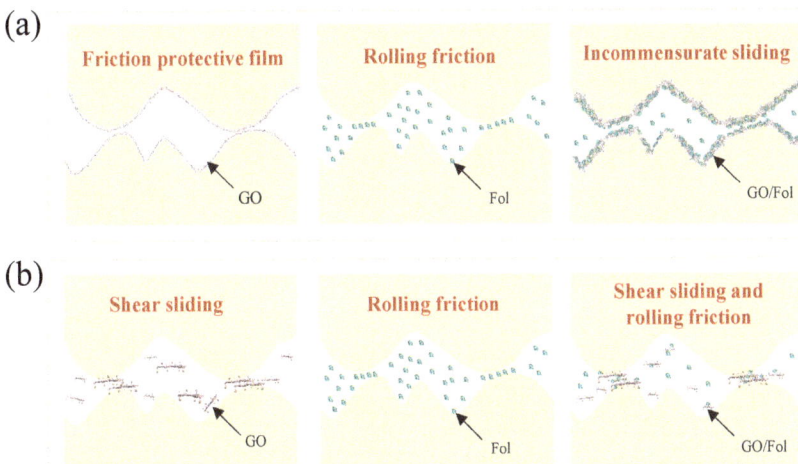

Figure 8. Lubrication mechanism of the GO, Fol and GO/Fol biological lubrication additives on (**a**) Ti6Al4V disks and (**b**) UHMWPE disks.

The lubrication mechanism of the GO, Fol and GO/Fol additives on the UHMWPE disks is shown in Figure 8b. Although GO and Fol did not easily form tribo-films on the worn surface of the UHMWPE disks, they could effectively enter the friction interface to reduce the interaction forces by the weak interlayer slip effect of GO and the rolling friction effect of Fol, respectively, to promote friction reduction performance. In addition, incommensurate sliding contact surfaces—formed between the GO nanosheets and spherical Fol at the interface—were also key to the good friction reduction performance of the GO/Fol additives on the UHMWPE disks.

3.3. Antioxidant Activity of Biological Lubrication Additives

As shown in Figure 9a,b, the free radical scavenging abilities of GO, Fol, GO/Fol = 1/1, GO/Fol = 1/2 and GO/Fol = 2/1 with different concentrations were compared. The results showed that both GO and Fol had scavenging ability on •OH and DPPH free radicals. The scavenging capacity was positively correlated with the concentration of the sample, which was consistent with results reported in the literature [30,31]. This is because Fol can effectively adsorb •OH and DPPH free radicals based on electron-deficient positions, and destroy these ROS by transferring electrons to fullerene cages (Figure 9c) [32]. At the same time, many OH groups on the surface of Fol can also remove •OH and DPPH free radicals by hydrogen transfer [33,34]. However, the ability of GO to scavenge free radicals is related to its SP^2 carbon structure, which plays a role through the formation of adducts or electron transfer. In addition, hydrogen atoms in oxygen-containing functional groups on the surface of GO can also participate in the neutralization of free radicals [20,35].

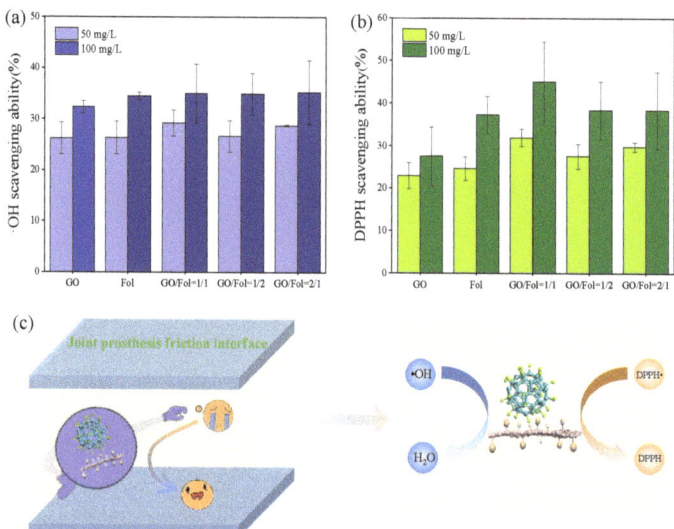

Figure 9. The (**a**) •OH and (**b**) DPPH free radical scavenging ability of GO, Fol and GO/Fol at different concentrations; (**c**) antioxidation mechanism of biological lubrication additives at joint prosthesis interface.

The scavenging ability of GO/Fol = 1/1 on •OH and DPPH free radicals has been improved compared with other groups, especially in the scavenging of DPPH free radical. When the concentration was 0.1 mg/mL, the scavenging effect of GO/Fol = 1/1 on •OH and DPPH free radicals reached 45% and 35%, respectively. The results showed that GO and Fol nanoparticles had a synergistic effect on free radical scavenging. This might be because both the spherical Fol and the lamellar GO tend to aggregate in aqueous solution to form large clusters, which would greatly reduce its ability to scavenge •OH and DPPH [35]. However, the spherical Fol could be interspersed and loaded onto the GO lamellae to inhibit agglomeration when GO and Fol existed in aqueous solution simultaneously, which provided more active sites for combining with free radicals and enhanced free radical scavenging ability.

4. Conclusions

In summary, GO and Fol were used as additives to prepare biological lubricants for artificial joints. The lubrication properties of biological lubricants containing different mass ratios of GO and Fol at two kinds of friction pairs were investigated comparatively. The lubrication mechanisms of GO and Fol as biological lubricant additives on the two friction pairs were revealed through the Raman characterization analysis of the friction interface. The antioxidation efficacies of the biological lubricants containing different mass ratios of GO and Fol were also evaluated. Tribological tests indicated that the synergistic effect of GO and Fol enabled the biological lubricant to exhibit superior friction reduction properties at the interface of both the Al_2O_3–Ti6Al4V and Ti6Al4V–UHMWPE friction pairs. The average friction coefficients of the Al_2O_3–Ti6Al4V pair and the Ti6Al4V–UHMWPE pair were reduced by 30% and 22%, respectively when GO and Fol were added to the biological lubricant with a mass ratio of 2 to 1. The superior friction reduction properties of GO and Fol can be attributed to the formation of some incommensurate sliding contact surfaces with low interfacial shear force between the GO nanosheets and spherical Fol at the friction interface. In addition, the biological lubricants containing GO and Fol also showed good antioxidant activity. The maximum scavenging rates of •OH and DPPH free radicals by biological lubricant containing GO and Fol were 35% and 45%, respectively. This can be attributed to the GO and Fol scavenging free radicals through electron transfer and hydrogen transfer. All these results show that GO and Fol have good application prospects as novel biological lubricant additives for artificial joints due to their synergistic lubrication and antioxidation efficacies.

Author Contributions: Conceptualization, Q.W. and B.W.; formal analysis, Q.W. and L.W.; funding acquisition, X.H., C.L. (Chuan Li), B.W. and H.L.; investigation, Q.W., L.W. and Z.B.; methodology, C.P., C.L. (Chuanrun Li) and B.W.; writing—original draft, Q.W., Z.B., C.W. (Chao Wang) and L.C.; writing—review and editing, H.L., L.W., C.W. (Changge Wang), C.W. (Chao Wang) and L.C. All authors have read and agreed to the published version of the manuscript.

Funding: This work is funded by the Key Project of Talent Support Plan of Anhui University of Chinese Medicine (No. 2022rczd001), the National Natural Science Foundation of China (No. 52075141), the Discipline Construction Quality Improvement Project of Chaohu University (No. kj20zsys02), the Scientific Research Planning Project of Anhui Provincial (No. 2022AH010092 and No. 2022AH051726) and the Support Program for Outstanding Young Talents in Anhui Province Colleges and Universities (No. gxyq2022079).

Acknowledgments: The authors thank Enzhu Hu and Kunhong Hu of Hefei University for their assistance in the experimental tests and data analyses.

Conflicts of Interest: The authors declare no conflict of interest.

References

1. Berenbaum, F.; Wallace, I.J.; Lieberman, D.E.; Felson, D.T. Modern-day environmental factors in the pathogenesis of osteoarthritis. *Nat. Rev. Rheumatol.* **2018**, *14*, 674–681. [CrossRef] [PubMed]
2. Merola, M.; Affatato, S. Materials for hip prostheses: A review of wear and loading considerations. *Materials* **2019**, *12*, 495. [CrossRef] [PubMed]
3. Rim, Y.A.; Ju, J.H. The role of fibrosis in osteoarthritis progression. *Life* **2020**, *11*, 3. [CrossRef] [PubMed]
4. Nine, M.J.; Choudhury, D.; Hee, A.C.; Mootanah, R.; Osman, N.A.A. Wear debris characterization and corresponding biological response: Artificial hip and knee joints. *Materials* **2014**, *7*, 980–1016. [CrossRef] [PubMed]
5. Yang, F.; Tang, J.; Dai, K.; Huang, Y. Metallic wear debris collected from patients induces apoptosis in rat primary osteoblasts via reactive oxygen species-mediated mitochondrial dysfunction and endoplasmic reticulum stress. *Mol. Med. Rep.* **2019**, *19*, 1629–1637. [CrossRef] [PubMed]
6. Hameister, R.; Kaur, C.; Dheen, S.T.; Lohmann, C.H.; Singh, G. Reactive oxygen/nitrogen species (ROS/RNS) and oxidative stress in arthroplasty. *J. Biomed. Mater. Res. Part B* **2020**, *108*, 2073–2087. [CrossRef] [PubMed]
7. Zhang, X.G.; Zhang, Y.L.; Jin, Z.M. A review of the bio-tribology of medical devices. *Friction* **2022**, *10*, 4–30. [CrossRef]
8. Méndez, P.A.; Ortiz, B.L.; Vásquez, G.M.; López, B.L. Mucoadhesive chitosan/OA nanoparticles charged with celecoxib inhibit prostaglandin E 2 LPS-induced in U 937 cell line. *J. Appl. Polym. Sci.* **2017**, *134*, 45288. [CrossRef]
9. Everhart, J.S.; DiBartola, A.C.; Swank, K.; Pettit, R.; Hughes, L.; Lewis, C.; Flanigan, D.C. Cartilage damage at the time of anterior cruciate ligament reconstruction is associated with weaker quadriceps function and lower risk of future ACL injury. *Knee Surg. Sport. Traumatol. Arthrosc.* **2020**, *28*, 576–583. [CrossRef]
10. Gleghorn, J.P.; Bonassar, L.J. Lubrication mode analysis of articular cartilage using Stribeck surfaces. *J. Biomech.* **2008**, *41*, 1910–1918. [CrossRef]
11. Huang, H.; Lou, Z.; Zheng, S.; Wu, J.; Yao, Q.; Chen, R.; Kou, L.; Chen, D. Intra-articular drug delivery systems for osteoarthritis therapy: Shifting from sustained release to enhancing penetration into cartilage. *Drug Deliv.* **2022**, *29*, 767–791. [CrossRef]
12. Šimek, M.; Nešporová, K.; Kocurková, A.; Foglová, T.; Ambrožová, G.; Velebný, V.; Kubala, L.; Hermannová, M. How the molecular weight affects the in vivo fate of exogenous hyaluronan delivered intravenously: A stable-isotope labelling strategy. *Carbohydr. Polym.* **2021**, *263*, 117927. [CrossRef] [PubMed]
13. Li, C.; Wu, B.; Chen, X.; Li, L.; Wang, X.; Gao, X.; Wang, X.; Hu, K.; Hu, X. Synergistic Lubricating Performance of Graphene Oxide and Modified Biodiesel Soot as Water Additives. *Lubricants* **2022**, *10*, 175. [CrossRef]
14. Song, H.-J.; Li, N. Frictional behavior of oxide graphene nanosheets as water-base lubricant additive. *Appl. Phys. A* **2011**, *105*, 827–832. [CrossRef]
15. Sarno, M.; Senatore, A.; Cirillo, C.; Petrone, V.; Ciambelli, P. Oil lubricant tribological behaviour improvement through dispersion of few layer graphene oxide. *J. Nanosci. Nanotechno.* **2014**, *14*, 4960–4968. [CrossRef]
16. Zhao, J.; Mao, J.; Li, Y.; He, Y.; Luo, J. Friction-induced nano-structural evolution of graphene as a lubrication additive. *Appl. Surf. Sci.* **2018**, *434*, 21–27. [CrossRef]
17. Chen, Z.; Liu, Y.; Luo, J. Tribological properties of few-layer graphene oxide sheets as oil-based lubricant additives. *Chin. J. Mech. Eng.* **2016**, *29*, 439–444. [CrossRef]

18. Liu, Y.; Wang, X.; Liu, P.; Zheng, J.; Shu, C.; Pan, G.; Luo, J. Modification on the tribological properties of ceramics lubricated by water using fullerenol as a lubricating additive. *Sci. China Technol. Sci.* **2012**, *55*, 2656–2661. [CrossRef]
19. Chen, S.; Ding, Q.; Gu, Y.; Quan, X.; Ma, Y.; Jia, Y.; Xie, H.; Tang, J. Study of tribological properties of fullerenol and nanodiamonds as additives in water-based lubricants for amorphous carbon (aC) coatings. *Nanomaterials* **2021**, *12*, 139. [CrossRef]
20. Zhou, Y.; Li, J.; Ma, H.; Zhen, M.; Guo, J.; Wang, L.; Jiang, L.; Shu, C.; Wang, C. Biocompatible [60]/[70] Fullerenols: Potent defense against oxidative injury induced by reduplicative chemotherapy. *ACS Appl. Mater. Interfaces* **2017**, *9*, 35539–35547. [CrossRef]
21. Voitko, K.V.; Goshovska, Y.V.; Demianenko, E.M.; Sementsov, Y.I.; Zhuravskyi, S.V.; Mys, L.A.; Korkach, Y.P.; Kolev, H.; Sagach, V.F. Graphene oxide nanoflakes prevent reperfusion injury of Langendorff isolated rat heart providing antioxidative activity in situ. *Free Radical Res.* **2022**, *56*, 328–341. [CrossRef] [PubMed]
22. Halenova, T.; Raksha, N.; Vovk, T.; Savchuk, O.; Ostapchenko, L.; Prylutskyy, Y.; Kyzyma, O.; Ritter, U.; Scharff, P. Effect of C60 fullerene nanoparticles on the diet-induced obesity in rats. *Int. J. Obes.* **2018**, *42*, 1987–1998. [CrossRef] [PubMed]
23. Forman, H.J.; Zhang, H. Targeting oxidative stress in disease: Promise and limitations of antioxidant therapy. *Nat. Rev. Drug Discov.* **2021**, *20*, 689–709. [CrossRef] [PubMed]
24. Goodarzi, S.; Da Ros, T.; Conde, J.; Sefat, F.; Mozafari, M. Fullerene: Biomedical engineers get to revisit an old friend. *Mater. Today* **2017**, *20*, 460–480. [CrossRef]
25. Wu, B.; Song, H.; Zhang, Q.; Hu, X. Controllable synthesis and friction reduction of ZnFe2O4@ C microspheres with diverse core-shell architectures. *Tribol. Int.* **2021**, *153*, 106614. [CrossRef]
26. Jaiswal, R.; Saha, U.; Goswami, T.H.; Srivastava, A.; Prasad, N.E. 'Pillar effect'of chemically bonded fullerene in enhancing supercapacitance performances of partially reduced fullerenol graphene oxide hybrid electrode material. *Electrochim. Acta* **2018**, *283*, 269–290. [CrossRef]
27. Xie, H.; Jiang, B.; Dai, J.; Peng, C.; Li, C.; Li, Q.; Pan, F. Tribological behaviors of graphene and graphene oxide as water-based lubricant additives for magnesium alloy/steel contacts. *Materials* **2018**, *11*, 206. [CrossRef]
28. Su, F.; Chen, G.; Huang, P. Lubricating performances of graphene oxide and onion-like carbon as water-based lubricant additives for smooth and sand-blasted steel discs. *Friction* **2020**, *8*, 47–57. [CrossRef]
29. Song, H.; Wang, Z.; Yang, J. Tribological properties of graphene oxide and carbon spheres as lubricating additives. *Appl. Phys. A* **2016**, *122*, 933. [CrossRef]
30. Hao, T.; Li, J.; Yao, F.; Dong, D.; Wang, Y.; Yang, B.; Wang, C. Injectable fullerenol/alginate hydrogel for suppression of oxidative stress damage in brown adipose-derived stem cells and cardiac repair. *ACS Nano* **2017**, *11*, 5474–5488. [CrossRef]
31. Abdelhalim, A.O.; Meshcheriakov, A.A.; Maistrenko, D.N.; Molchanov, O.E.; Ageev, S.V.; Ivanova, D.A.; Iamalova, N.R.; Luttsev, M.D.; Vasina, L.V.; Sharoyko, V.V. Graphene oxide enriched with oxygen-containing groups: On the way to an increase of antioxidant activity and biocompatibility. *Colloids Surf. B* **2022**, *210*, 112232. [CrossRef] [PubMed]
32. Wang, J.; Hu, Z.; Xu, J.; Zhao, Y. Therapeutic applications of low-toxicity spherical nanocarbon materials. *NPG Asia Mater.* **2014**, *6*, e84. [CrossRef]
33. Wang, Z.; Gao, X.; Zhao, Y. Mechanisms of antioxidant activities of fullerenols from first-principles calculation. *J. Phys. Chem. A* **2018**, *122*, 8183–8190. [CrossRef] [PubMed]
34. Podolsky, N.E.; Marcos, M.A.; Cabaleiro, D.; Semenov, K.N.; Lugo, L.; Petrov, A.V.; Charykov, N.A.; Sharoyko, V.V.; Vlasov, T.D.; Murin, I.V. Physico-chemical properties of C60 (OH) 22–24 water solutions: Density, viscosity, refraction index, isobaric heat capacity and antioxidant activity. *J. Mol. Liq.* **2019**, *278*, 342–355. [CrossRef]
35. Awan, F.; Bulger, E.; Berry, R.M.; Tam, K.C. Enhanced radical scavenging activity of polyhydroxylated C60 functionalized cellulose nanocrystals. *Cellulose* **2016**, *23*, 3589–3599. [CrossRef]

Disclaimer/Publisher's Note: The statements, opinions and data contained in all publications are solely those of the individual author(s) and contributor(s) and not of MDPI and/or the editor(s). MDPI and/or the editor(s) disclaim responsibility for any injury to people or property resulting from any ideas, methods, instructions or products referred to in the content.

Article

Comparative Study of Tribological Properties of Modified and Non-modified Graphene-Oil Nanofluids under Heated and Non-heated Conditions

Kean Pin Ng, Kia Wai Liew * and Elaine Lim

SIG: Machine Design and Tribology, Centre for Advanced Mechanical & Green Technology (CAMGT), Faculty of Engineering & Technology, Multimedia University, Melaka 75450, Malaysia
* Correspondence: kwliew@mmu.edu.my

Abstract: With the aim of achieving more effective friction and wear reduction in sliding bearing applications, surface-modified graphene, which exhibits better dispersion stability than non-modified graphene, was synthesized and applied in this study using various graphene allotropes, including graphene nanoplatelets (GNP), multiwalled carbon nanotubes (MWCNT) and nanostructured graphite (NSG). Friction and wear tests of each type of graphene allotrope under modified and non-modified conditions were studied using a pin-on-ring tribo tester. In addition, the dynamic viscosity of each synthesized nanofluid sample was measured using a falling-ball viscometer. A series of modified graphene-oil nanofluids and non-modified graphene-oil nanofluids were prepared and heated before their friction and wear performance was investigated at room temperature. Friction and wear behavior, as well as the dynamic viscosity of the heated nanofluids vary insignificantly when compared to those of the non-heated nanofluids. The results showed that the best friction and wear reduction was achieved by modified GNP with friction and wear reduction of 60.5% and 99.4%, respectively.

Keywords: graphene-oil nanofluids; heated and non-heated; tribological properties; graphene surface modification

Citation: Ng, K.P.; Liew, K.W.; Lim, E. Comparative Study of Tribological Properties of Modified and Non-modified Graphene-Oil Nanofluids under Heated and Non-heated Conditions. *Lubricants* 2022, *10*, 288. https://doi.org/10.3390/lubricants10110288

Received: 17 September 2022
Accepted: 25 October 2022
Published: 31 October 2022

Publisher's Note: MDPI stays neutral with regard to jurisdictional claims in published maps and institutional affiliations.

Copyright: © 2022 by the authors. Licensee MDPI, Basel, Switzerland. This article is an open access article distributed under the terms and conditions of the Creative Commons Attribution (CC BY) license (https://creativecommons.org/licenses/by/4.0/).

1. Introduction

Journal bearings are widely used in various types of mechanical systems, such as motors, wind turbine gearboxes and propeller shafts, especially in mechanisms with rotating shafts [1]. The presence of friction and wear within a journal bearing must be studied to ensure minimal friction and wear for better performance and longer bearing life [2,3]. For instance, minimum friction and wear of journal bearings in wind turbine gearboxes are essential to ensure maximum efficiency of the power to be delivered [4–6] and also to minimize the maintenance cost of repairing or replacing defective bearings of wind turbine gearboxes due to excessive friction and wear [7]. In order to minimize the friction and wear of journal bearings, numerous studies have been conducted in the past applying various types of advanced nano-lubrication concepts [8–10]. Among the studies conducted, graphene is found to be able to reduce friction and wear between two sliding surfaces effectively due to their unique self-lubricating tribo-film, whether under dry solid lubricating conditions [11] or in the form of liquid nano-lubricants [12,13]. When the lubricating surfaces slide over each other, the graphene layers can be sheared off easily and distributed over the lubricating surfaces forming the unique self-lubricating tribo-film owing to the weak Van der Waals force between the graphene layers [14,15]. Apart from that, graphene nanoparticles also perform a mending mechanism [16] by filling up the asperities gaps between lubricating surfaces. Furthermore, the graphene nanoparticles can also act as tiny ball bearings [17] which aids the sliding actions between two surfaces, lowering the friction coefficient trend.

In the application of journal bearing, liquid lubricating oil is used as a lubricating medium to achieve the condition of hydrodynamic lubrication [18]. Therefore, the dispersion stability of graphene in the base oil fluid is crucial to ensure the proper distribution of graphene for better friction and wear reduction [19,20]. However, not all graphene is suitable to be dispersed in base oil fluid to synthesize a stably dispersed graphene-oil nanofluid. Therefore, various approaches, including the addition of surfactant [21,22] and graphene surface modification processes [23], have been introduced in previous studies to improve the dispersion stability of graphene in base oil fluid.

The depletion of natural mineral resources is alarming and worries society because most commercial lubricating oils are made of petroleum-based mineral oil [24,25]. A new revolution of lubricating oil promoting the concept of eco-friendly and green tribology needs to be introduced. In this study, the method of lubrication with biodegradable vegetable oil is implemented. It is expected that using vegetable oil as the base oil for lubricating the journal bearings of the offshore wind turbine will help to minimize the problem of marine pollution because the oils are easily biodegradable and non-toxic.

One of the main functions of lubricating oil is to remove unwanted heat from mechanical systems [26] by absorbing the heat generated by the sliding movements between the shaft and the bearing housing, resulting in an increase in oil temperature. Studies on the thermophysical properties of nanofluids have also been conducted in the past to investigate the heat transfer performance of nanofluids compared to the normal base fluid [27–29]. It is necessary to ensure that the lubricating oil can operate efficiently at different operating temperatures. Hence, the effects of temperature on the thermophysical properties of graphene-oil nanofluids are investigated.

It is well known that multilayer graphene allotropes provide better friction and wear reduction [30,31]. In this research work, three types of most commonly found multilayer graphene allotropes, namely graphene nanoplatelets (GNP), multiwalled carbon nanotubes (MWCNT) and nanostructured graphite (NSG), are used in the formation of graphene-oil nanofluids to investigate the effects of different types of graphene nanoparticles on the tribological performance of the lubricating base fluid. In addition, the dynamic viscosity and tribological properties of the heated graphene-oil nanofluids are also investigated in this work. In short, the friction and wear performance of synthesized graphene-oil nanofluids using biodegradable vegetable oil with both unmodified and surface-modified graphene is investigated. In order to study the effects of frequent temperature cycling in a real application, the tribological performance of non-heated (fresh) and heated graphene-oil nanofluids is also investigated in this work.

2. Experimental Works

2.1. Tribo Specimens Preparation

Aluminum alloy 5083 (AL-5083) was selected as the pin specimen material to be used in this research work due to its excellent corrosive resistivity [32], which can closely simulate the journal bearings applications around marine environments such as wind turbines, propeller shafts of a ship and even a submarine. Moreover, aluminum alloy is widely applied as a bearing material due to its excellent friction performance when compared with other bearing materials such as white metal and bronze [33]. Table 1 shows the mechanical properties and chemical composition of AL-5083. Cylindrical pin specimens with a diameter of 5 mm and length of 20 mm were fabricated from AL-5083 rod. The arithmetic surface roughness, R_a of pin specimens, was controlled at 0.275 µm.

On the other hand, stainless steel 304 (SS-304) was selected as a counter ring material due to its strong mechanical properties [34,35], and it is commonly used as transmission shaft material in marine environments [36,37]. Counter rings with an inner diameter of 9 mm, an outer diameter of 22 mm and a thickness of 10 mm were machined from SS-304 rods using a CNC turning machine. Meanwhile, the R_a for all counter ring was controlled at 0.300 µm. The chemical composition and mechanical properties of SS-304 are shown in Table 2. Figure 1 shows the fabricated AL-5083 pin specimens and SS-304 counter ring.

Table 1. Chemical composition and mechanical properties of AL-5083.

Element	Compositions (wt%)
Si	0.10
Fe	0.16
Cu	0.02
Mn	0.57
Mg	4.15
Cr	0.11
Zn	0.02
Al	94.87
Mechanical Properties	
Tensile Strength	277 MPa
Yield Strength	130 MPa
Elongation Rate	13%

Table 2. Chemical composition and mechanical properties of SS-304.

Element	Compositions (wt%)
C	0.018
Mn	1.58
Si	0.28
S	0.025
P	0.035
Cr	18.14
Ni	8.07
Cu	0.61
Mo	0.31
Co	0.18
N	0.085
Fe	70.067
Mechanical Properties	
Tensile Strength	622 MPa
Yield Strength	584 MPa

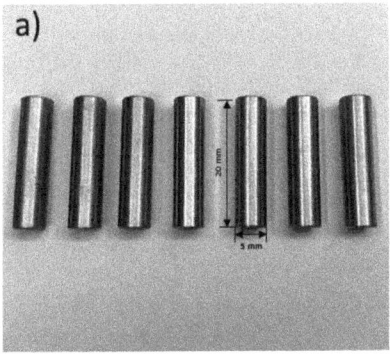
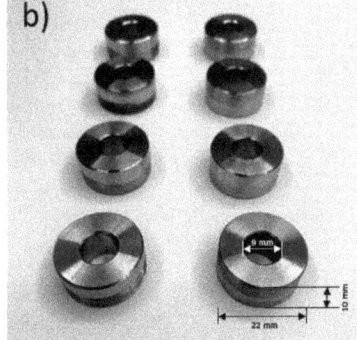

Figure 1. Photograph of (a) AL-5083 pin specimens and (b) SS-304 counter rings.

2.2. Modified and Non-modified Graphene-Oil Nanofluids Samples Preparation

Three types of multilayer graphene allotropes, namely, graphene nanoplatelets (GNP), multiwalled carbon nanotubes (MWCNT) and nanostructured graphite (NSG), were used to synthesize three types of graphene-oil nanofluids at 0.04 wt%. To improve the dispersion stability of graphene in oil, the graphene surface was modified using sodium dodecyl sulfate (SDS) and oleic acid (OA) in this work. The modification of graphene started by mixing graphene into an SDS solution. OA was then added to the sonicated mixture of SDS and graphene solution, and another round of sonication was performed. The final mixture of the graphene–SDS-OA solution was then heated in a furnace at 180 °C for 4 h until it was completely dried out and only modified graphene powders remained.

High oleic palm oil-based methyl ester (high oleic POME) was selected as the base oil for the synthesis of graphene-oil nanofluids throughout the research because the high oleic acid content improves the oxidative properties of the base oil when used at high operating temperatures [38]. Both non-modified graphene and modified graphene were separately added to high oleic POME and sonicated with a probe sonicator, Fisher Scientific, Model: FB 705, as shown in Figure 2, for two hours to synthesize non-modified graphene-oil nanofluids and modified graphene-oil nanofluids. The synthesized GNP-oil nanofluid is referred to as GNON, while the modified GNP-oil nanofluid is referred to as mGNON in the following. MWCNT-oil nanofluid is referred to as MWON, while modified MWCNT-oil nanofluid is referred to as mMWON hereafter. NSG-oil nanofluid is referred to as NSON, while modified NSG-oil nanofluid is referred to as mNSON in the following.

Figure 2. Sonication process of a graphene-oil nanofluid sample with a probe sonicator, Fisher Scientific, Model: FB 705.

2.3. Heating of Non-modified Graphene-Oil Nanofluids and Modified Graphene-Oil Nanofluids

Each synthesized nanofluid sample was placed in a beaker and heated with a hot plate, Faithful, model: SH-II-4C. The nanofluid in the beaker was heated from 30 °C to 85 °C in 5 °C steps, and each temperature interval was kept constant for five minutes. The heating process of the nanofluids was carried out up to 85 °C to simulate the real operating temperature of journal bearings in wind turbine gearboxes [33]. The nanofluids were then cooled naturally to room temperature. The non-heated nanofluid will be termed as "fresh" (F) in the following, while the heated and then cooled to room temperature will be referred to as "heated" (H).

2.4. Friction and Wear Test

A pin-on-ring (POR) tribo tester was used to conduct friction and wear tests in this research work via standard ASTM G-77 to simulate the real-life counter-formal contact [39] between the journal-bearing housing and transmission shaft. A schematic diagram of the POR tribo tester is shown in Figure 3. A stationary AL-5083 cylindrical pin was inserted into the specimen adaptor and tested against a rotating SS-304 counter ring, which was partially immersed into the oil container. As the counter ring started to revolve in a clockwise direction, it tended to pick up the graphene-oil nanofluids in the container and provide lubrication to the pin and ring contact surfaces. All sliding tests under boundary lubrication conditions were carried out at an ambient temperature and humidity environment with constant speed and a normal load of 1.36 m/s and 44.15 N, respectively. Modified and non-modified graphene-oil nanofluids subjected to heated and non-heated (fresh) conditions were used as lubricating oil in sliding tests. Prior to the sliding test, both pins and counter rings were cleaned using acetone.

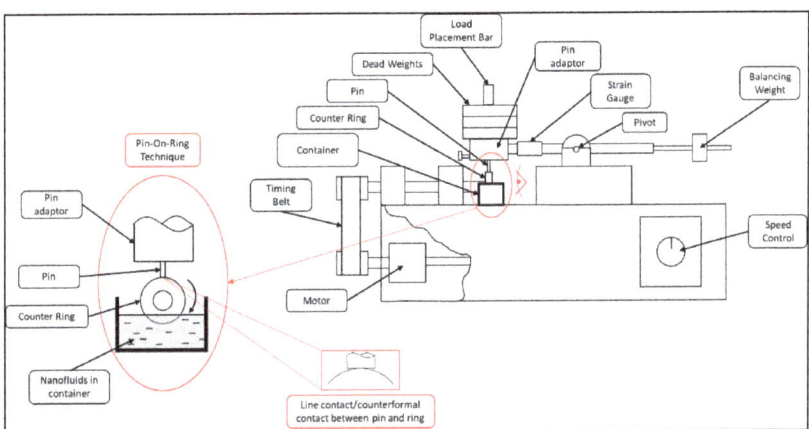

Figure 3. Schematic diagram of pin-on-ring tribo tester.

The magnitude of the tangential force F_T (N) on the sliding surface was measured with a pre-calibrated strain gauge mounted on the load level holding the pin specimen. The coefficient of friction (CoF) for the contact pair lubricated with different graphene-oil nanofluid samples was calculated using the ratio between the tangential force F_T (N) and the applied normal load F_N (N).

The weight loss of each pin specimen was determined by measuring the difference in weight after completion of the friction test using 0.1 mg electronic balance, SHIMADZU, Model: AW-220 (Shimadzu, Kyoto, Japan). The tribological performance of modified and non-modified graphene-oil nanofluids exposed to heated and non-heated conditions was compared. The worn surfaces of pin specimens were examined by a metallurgical microscope, Meiji, Model: MT7100 (Meiji Techno, San Jose, CA, USA).

2.5. Thermophysical Test—Dynamic Viscosity Measurement

The viscosity of the synthesized graphene-oil nanofluids was measured using a falling ball viscometer, HÖPPLER®, Model: KF 3.2, (Rheotest, Ottendorf-Okrilla, Germany), in the temperature range from 30 °C to 80 °C corresponding to the usual operating temperature of journal bearings in wind turbines [33]. The temperature range was measured and controlled using a water bath, Julabo®, Model: TW-12 (Julabo, PA, USA). Both the viscosity of non-heated (fresh) and heated graphene-oil nanofluids under non-modified and modified conditions were measured and compared.

3. Results and Discussions
3.1. Dispersion Stability

The dispersion stability of graphene-oil nanofluids is determined by observing how long it takes to observe the deposition of graphene in the sample bottles. In Figure 4, by comparing the fresh non-modified graphene-oil nanofluids, it was found that the worst dispersion stability was shown by F-MWON, where the deposition was already observed within an hour after the sonication process due to the tubular structural entanglements of the MWCNTs in the oil, which formed larger bulk clumps, as shown in the micrograph in Figure 5c. Due to the smaller nanoparticle size of NSG compared to GNP, as shown in the microscopic images in Figure 5a,b, respectively, F-NSON was observed to have better dispersion stability than F-GNON, with half of F-NSON still in black intensity, while F-GNON was already pale-gray in color on Day-7.

Figure 4. Dispersion stability test of fresh graphene-oil nanofluids after (**a**) completion of sonication, (**b**) one hour, (**c**) day-1, (**d**) day-3, (**e**) day-5 and (**f**) day-7 (arrangement from left: F-GNON, F-NSON and F-MWON) (Notes: F = fresh; H = heated) [40].

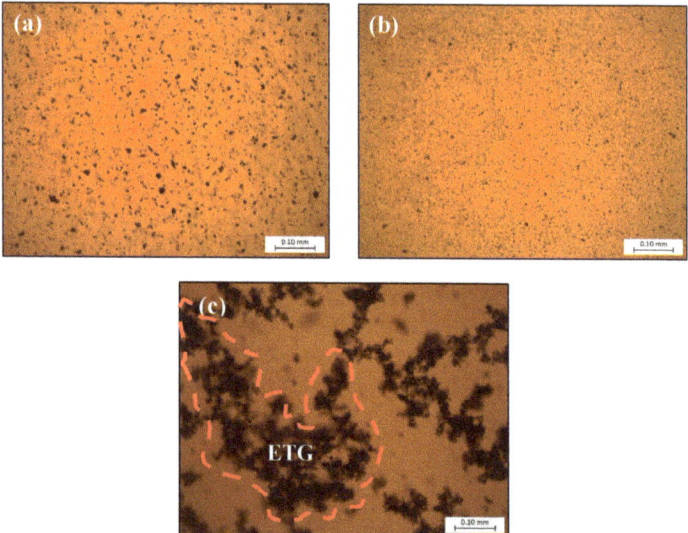

Figure 5. Micrographs of (**a**) F-GNON, (**b**) F-NSON and (**c**) F-MWON under 10× magnification (Notes: ETG = entanglement; F = fresh; H = heated).

For the fresh modified graphene-oil nanofluids (F-mGNON, F-mNSON and F-mMWON), it was observed that the dispersion stability improved significantly compared to the non-modified graphene-oil nanofluids, with the nanofluids still showing a completely dark-black intensity after 60 days, as shown in Figure 6. Figure 7 shows the microscopic images of fresh modified graphene-oil nanofluids under 10× magnification. The dispersion stability of modified graphene in high oleic POME has greatly improved, mainly due to the increase in hydrophobicity of graphene itself after it was surface modified with sodium dodecyl sulfate (SDS) and oleic acid (OA). Due to the mixing of SDS and graphene during the first modification step, clumps of SDS adhered to the graphene surface [23], leaving the sodium ion (Na$^+$) exposed on the outer graphene surface. Subsequent mixing of OA into the graphene–SDS solution leads to the association of the OH$^-$ group of OA with the Na$^+$ ion of the SDS compound, exposing the long hydrophobic carbon chain on the graphene surface, as illustrated in Figure 8.

Figure 6. Dispersion stability test of fresh modified graphene-oil nanofluids after (**a**) completion of sonication, (**b**) day-1, (**c**) day-10, (**d**) day-20, (**e**) day-40 and (**f**) day-60 (arrangement from left: F-mGNON, F-mNSON and F-mMWON) (Notes: F = fresh; H = heated).

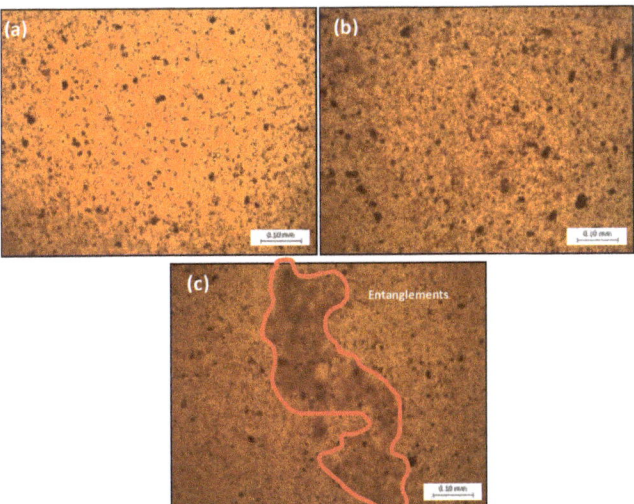

Figure 7. Micrographs of (**a**) F-mGNON, (**b**) F-mNSON and (**c**) F-mMWON under 10× magnification (Notes: F = fresh; H = heated).

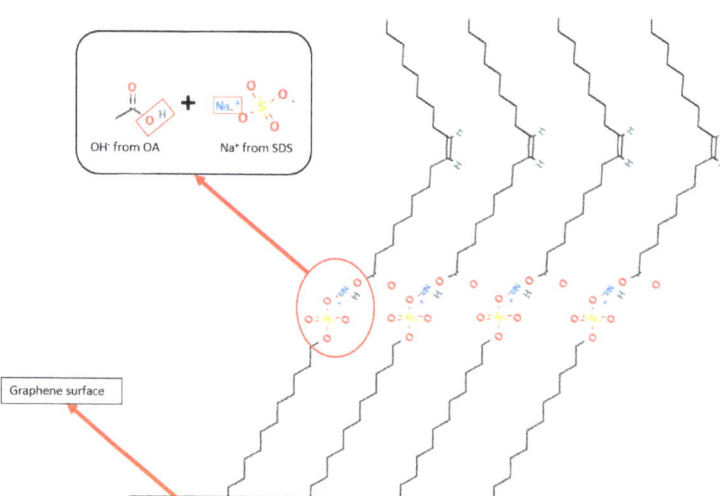

Figure 8. Reaction between sodium dodecyl sulfate (SDS) and oleic acid (OA), which adheres to the graphene surface and exposes the long hydrophobic carbon chain on the graphene surface to increase the hydrophobicity of graphene.

On the other hand, it was observed that the heated unmodified and heated modified graphene-oil nanofluids exhibited exactly the same dispersion stability as the fresh ones, as shown in Figure 9.

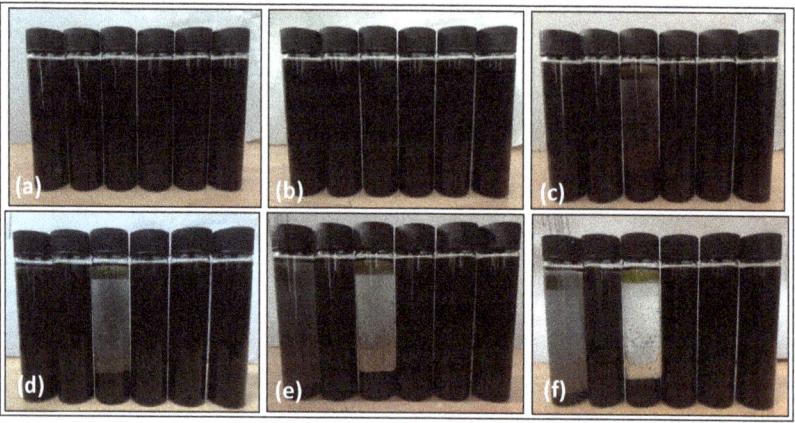

Figure 9. Dispersion stability test of heated graphene oil nanofluids and heated modified graphene-oil nanofluids after (**a**) completion of sonication, (**b**) one hour, (**c**) day-1, (**d**) day-3, (**e**) day-5 and (**f**) day-7 (arrangement from left: H-GNON, H-NSON, H-MWON, H-mGNON, H-mNSON and H-mMWON) (Notes: F = fresh; H = heated).

3.2. Dynamic Viscosity

All unmodified graphene-oil nanofluids have slightly higher dynamic viscosity than pure high oleic POME and show a decreasing tendency when the temperature gradually increases from 30 °C to 80 °C, as shown in Figure 10, mainly due to the thermal energy absorbed by the base oil molecules. This resulted in the base oil molecules having sufficient energy to overcome the intermolecular bonding between the molecules [41].

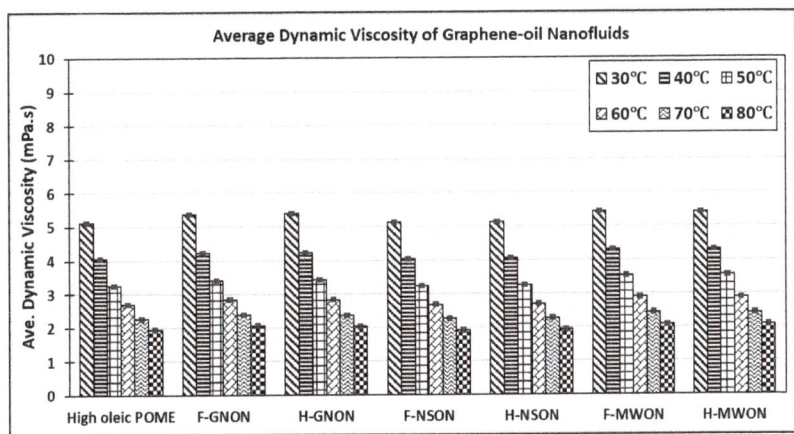

Figure 10. The average dynamic viscosity of non-heated (fresh) and heated non-modified graphene-oil nanofluids exposed to different temperatures ranges from 30 °C to 80 °C (Notes: F = fresh; H = heated).

The dynamic viscosity of non-modified graphene-oil nanofluids is closely related to the concept of solid fraction hindering the motions of liquid molecules [42], with MWON having the highest dynamic viscosity due to the fractured fragments of MWCNT formed [43] during sonication (largest solid fraction), followed by GNON and finally NSON, where GNP is slightly larger than NSG (solid fraction in GNON larger than NSON) as evidenced in Figure 5a,b. Figure 11 shows the presence of broken MWCNT fragments after the sonication process, leading to a significant increase in the solid fraction in MWON. When compared with heated non-modified graphene-oil nanofluids, no significant changes in dynamic viscosity were observed, as shown in Figure 10.

Figure 11. Equal amount of MWCNTs as solid fraction (**a**) before and (**b**) after sonication at complete sedimentation.

Meanwhile, all modified graphene-oil nanofluids show a significant increase in dynamic viscosity compared to the non-modified nanofluids, as shown in Figure 12. This is mainly due to the presence of SDS and OA compounds in the nanofluids, with both compounds also present as part of the solid fraction in the base oil. This higher viscosity trend indicates that the liquid film thickness of the modified graphene-oil nanofluids is thicker than that of the non-modified ones, which can effectively minimize the direct contact between two sliding surfaces and further reduce friction and wear [44]. Similarly, a trend of decreasing viscosity with increasing temperature was observed for all modified graphene allotropes, with the viscosity of mMWON being the highest, followed by mGNON and mNSON. When compared with heated modified graphene-oil nanofluids, no significant changes in viscosity were observed, as shown in Figure 12.

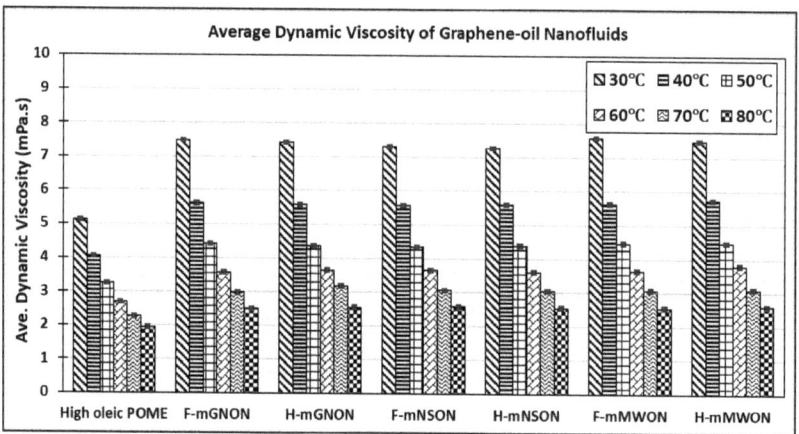

Figure 12. The average dynamic viscosity of fresh and heated modified graphene-oil nanofluids exposed to different temperature ranges from 30 °C to 80 °C (Notes: F = fresh; H = heated).

3.3. Friction and Wear Test Results

Comparing fresh and heated non-modified graphene-oil nanofluids as shown in Figure 13a,c,e, it was found that each graphene oil nanofluid, regardless of whether it is non-heated (fresh) or heated, shows no significant changes in the coefficient of friction (CoF) trend. NSON shows the largest CoF trend reduction, followed by GNON and finally MWON. Inconsistent fluctuations are also observed in the MWON trend, which may be due to the tubular structure of MWCNTs performing various lubrication mechanisms. In addition, MWCNTs may also break off and form sharp fragments, resulting in inconsistent roughness of the lubricating surfaces and contributing to a higher CoF trend. This was later confirmed by the microscopic images of the worn surfaces of the pins in Figure 17c. On the other hand, for GNON and NSON, due to the smaller size of the NSG nanoparticles, NSON tends to slip the NSG nanoparticles into the gap between the asperities more effectively than GNP, which is slightly larger, thus performing the mending mechanism more effectively and bringing significant reduction in friction and wear [16]. This is confirmed in Figure 17b,d by the micrographs of the worn surface of the pin with GNON and NSON lubrication, respectively, where the deep grooves of the pin with NSON lubrication are relatively smaller than those of the pin with GNON lubrication. The highest friction reduction of 52.03% was observed with NSON lubrication compared to plain high oleic POME (average CoF reduction from 0.12 to 0.06). Apart from this, NSON also shows the best reduction in pin weight loss of 59.27% compared to plain high oleic POME (reduction in pin weight loss from 32.9 mg to 13.4 mg). Figure 14 shows the results of pin weight loss reduction by different graphene-oil nanofluids in non-heated (fresh) and heated conditions.

Moreover, no significant changes are observed in the modified graphene-oil nanofluids in terms of CoF trend when comparing the non-heated (fresh) and heated conditions (see Figure 13b,d,f). This time, mGNON shows the best CoF trend reduction, followed by mNSON and, finally, mMWON. The lowest CoF trend is no longer achieved by NSG, as in the case of the non-modified category, which is mainly due to the nature of graphite, which is a compact stacked structure of graphene layers, giving it a relatively small surface area compared to graphene structures [45]. Thus, NSG provides less space for SDS to adhere to the surface of NSG. Similar observations were made in another study in which the authors investigated the adsorption of antibiotics on graphite and graphene structures. Later, it was found that graphene structures have better antibiotic adhesion than graphite structures because the surface area of the graphene layer is larger than that of graphite [46]. Thus, the increase of hydrophobicity in NSG compared to GNP during surface modification by SDS and OA is not significant. The structure of GNP is completely platelet-shaped, which provides a larger space for the adhesion of SDS on the GNP surface. Since part of SDS cannot adhere to the NSG surface, some parts remain in the solution of NSG–SDS–OA. When dried in the furnace, SDS tends to dry out and form crystallized solids, which lead to higher friction and wear due to abrasion of the third body [47,48] during sliding operations. Figure 15 shows the crystallized SDS observed in mNSON with a 400 mesh filter screen. Inconsistent fluctuations can still be observed in the mMWON trend because the modification of MWCNT does not change the tubular structure of MWCNT or prevent fragments from breaking.

All non-modified and modified graphene-oil nanofluids show a reduction CoF trend compared to pure high oleic POME, which is mainly due to the formation of a self-lubricating graphene tribo-film on the lubricating surfaces [23,49,50]. The best friction and wear reduction was found for mGNON, which exhibits an average CoF reduction of 60.5% compared to pure high oleic POME (average CoF reduction from 0.12 to 0.047). At the same time, as described in Section 3.2, mGNON, with its excellent fluid film thickness, also shows an excellent 99.4% reduction in pin weight loss compared to pure high oleic POME (pin weight loss reduction from 32.9 mg to 0.2 mg). Figure 16 shows the reduction in weight loss of pins by different modified graphene-oil nanofluids in non-heated (fresh) and heated conditions.

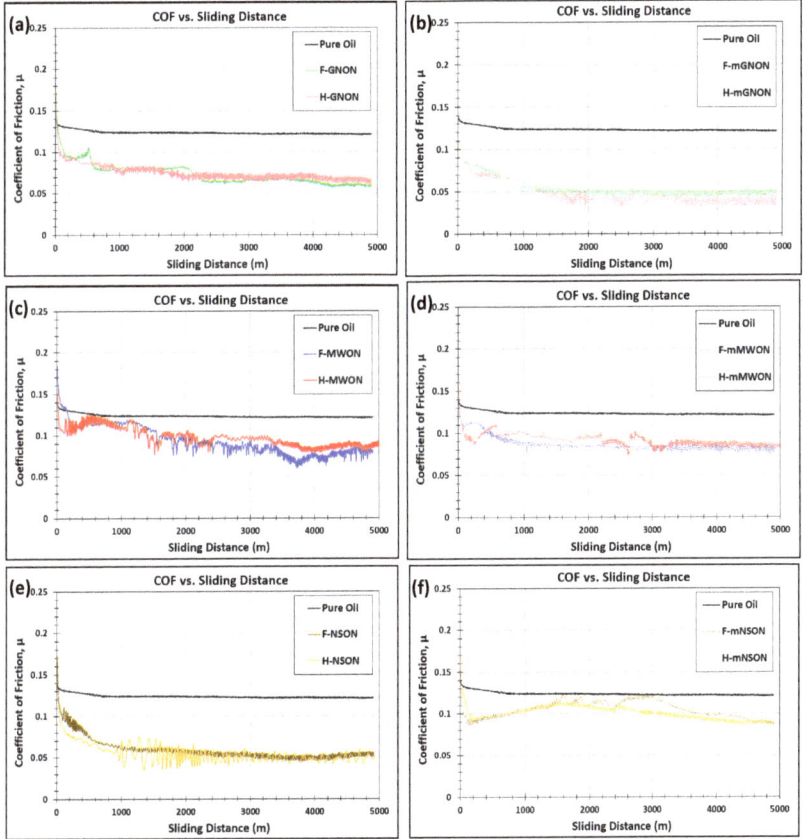

Figure 13. Friction coefficient for non-heated (fresh) and heated (a) GNON, (b) mGNON, (c) MWON, (d) mMWON, (e) NSON and (f) mNSON lubrication as compared to pure oil at constant normal load of 44.15 N and sliding speed of 1.36 m/s for 3600 s (Notes: F = fresh; H = heated).

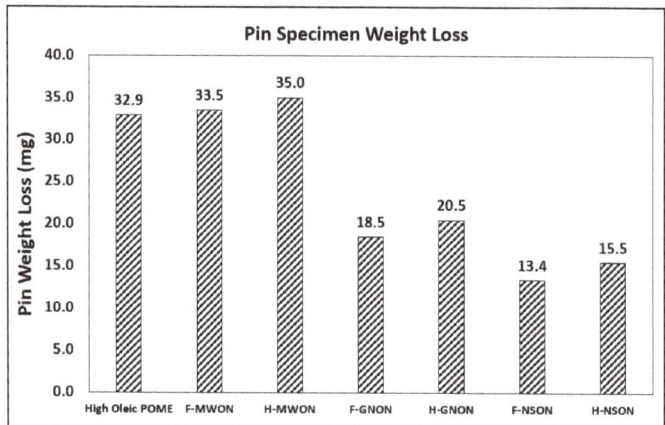

Figure 14. Pin weight loss for non-heated (fresh) and heated graphene-oil nanofluids as compared to high oleic POME at constant normal load of 44.15 N and sliding speed of 1.36 m/s for 3600 s (Notes: F = fresh; H = heated).

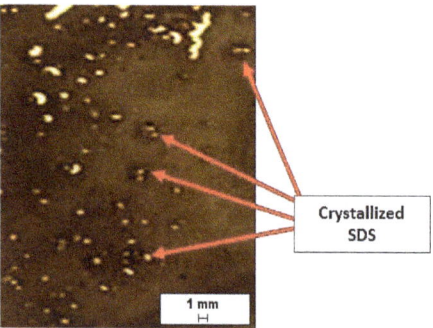

Figure 15. Crystallized SDS formed observed in mNSON after being filtered using 400 mesh filter screen.

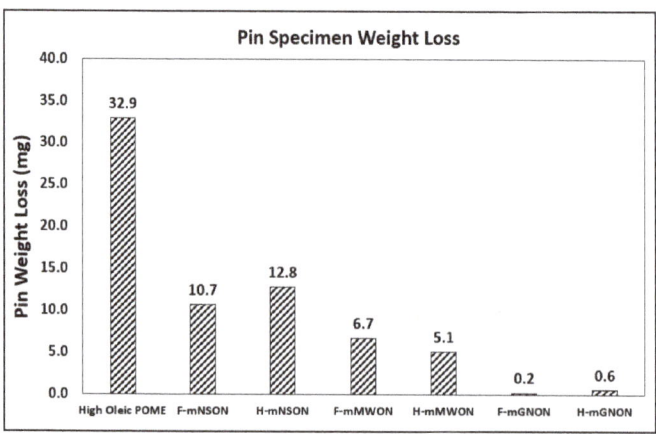

Figure 16. Pin weight loss for fresh and heated modified graphene-oil nanofluids as compared to high oleic POME at constant normal load of 44.15 N and constant sliding speed of 1.36 m/s for 3600 s (Notes: F = fresh; H = heated).

3.4. Pin Worn Surface Micrographs

Figure 17 shows the microscopic images of the worn surface of the pin for heated non-modified graphene-oil nanofluid lubrications, while Figure 18 shows microscopic images of the worn surface of the pin for heated modified graphene-oil nanofluid lubrications. Fewer scratches and grooves are observed in the modified graphene-oil nanofluid lubricated pin specimens, which can be attributed to the thicker fluid film thickness due to the higher dynamic viscosity of the modified graphene-oil nanofluid. This thicker fluid film thickness helps to minimize the probability of direct contact between the pin and the ring during the lubrication process [44].

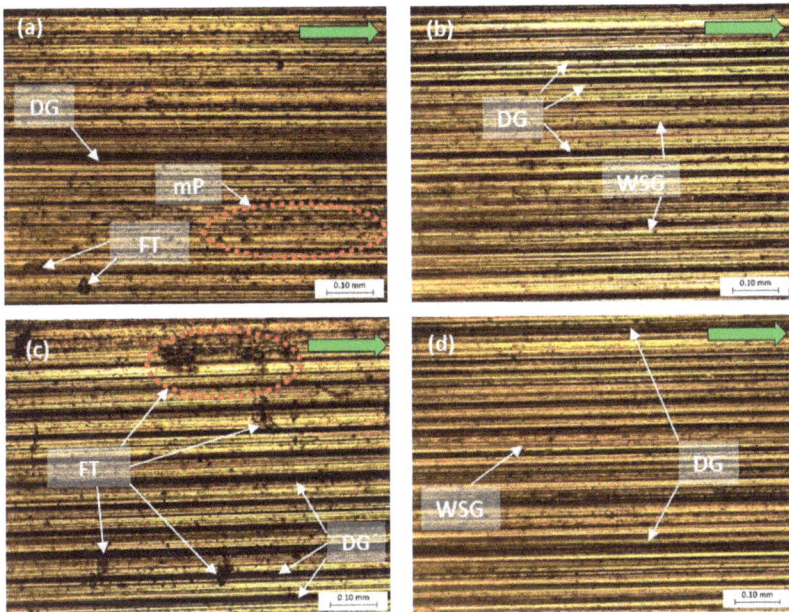

Figure 17. Micrographs of pin specimen worn surfaces tested with (**a**) high oleic POME, (**b**) GNON, (**c**) MWON and (**d**) NSON lubrication at constant normal load of 44.15 N and constant sliding speed of 1.36 m/s for 3600 s under 10× magnification (Notes: Green arrow = sliding direction; DG = deep groove; mP = micro-pits; WSG = wide-shallow groove; FT = fracture).

It can be observed that the worn surfaces of the pins in MWON lubrication in Figure 17c have several areas of pitting and severe scratches, which are mainly due to the scratching of MWCNT agglomerations formed after the sonication process [43]. Since modified MWCNTs in mMWON still tend to break off and form sharp fragments, some deep grooves can be observed on the worn surface of the pin (see Figure 18c), but they are not as pronounced as those on the non-modified one in Figure 17c due to the effective thickness of the fluid film in mMWON [44]. Some deep grooves can also be observed on the worn surface of the pin in mNSON lubrication (see Figure 18d), which is mainly due to the crystallized SDS in mNSON.

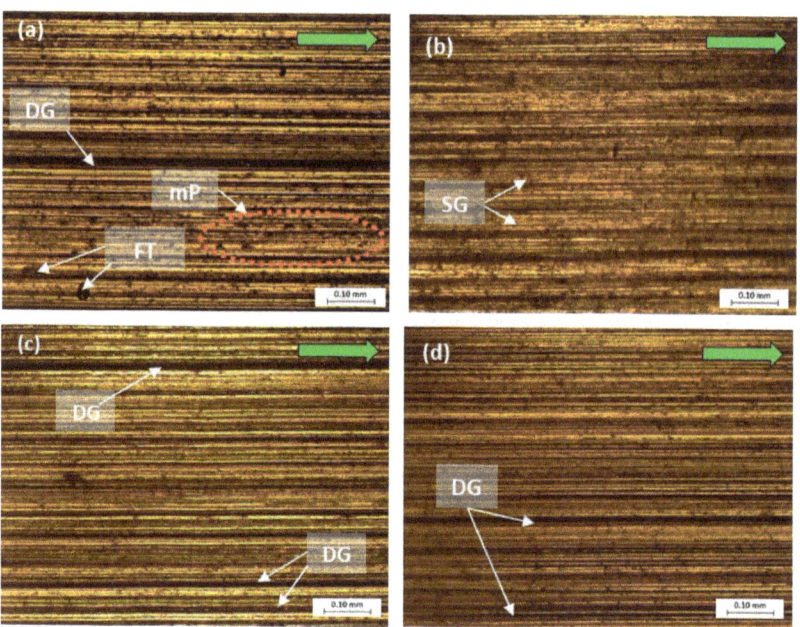

Figure 18. Micrographs of pin specimen worn surfaces tested with (**a**) high oleic POME, (**b**) mGNON, (**c**) mMWON and (**d**) mNSON lubrication at constant normal load of 44.15 N and constant sliding speed of 1.36 m/s for 3600 s under 101× magnification (Notes: Green arrow = sliding direction; DG = deep groove; mP = micro-pits; SG = shallow groove; FT = fracture).

4. Conclusions

This work proves that the synthesized graphene-oil nanofluids can significantly reduce friction and wear compared to pure base oil. Moreover, the modified graphene-oil nanofluids with excellent dispersion stability further reduce friction and wear in contact pairs. In addition, modified graphene-oil nanofluids with better dynamic viscosity have improved the wear resistance of sliding surfaces. mGNON shows an average friction reduction of 60.5% and a 99.4% reduction in the weight loss of pins.

Furthermore, the heated and non-heated synthesized (non-modified and modified) graphene-oil nanofluids show no drastic changes in dynamic viscosity, friction and wear performance. This proves that the thermophysical and tribological properties of graphene-oil nanofluids are not affected when they are heated and cooled back to room temperature. This suggests that the synthesized graphene-oil nanofluids have a longer lifetime and can be used in journal bearings applications without the need for frequent replacement.

Author Contributions: Conceptualization, K.W.L. and K.P.N.; methodology, K.W.L., K.P.N. and E.L.; formal analysis, K.W.L. and K.P.N.; investigation, K.P.N., K.W.L. and E.L.; resources, K.W.L. and E.L.; writing—original draft preparation, K.W.L. and K.P.N.; writing—review and editing, K.W.L., E.L. and K.P.N.; supervision, K.W.L. and E.L.; funding acquisition, E.L. and K.W.L. All authors have read and agreed to the published version of the manuscript.

Funding: This research was funded by Fundamental Research Grant Scheme (FRGS/1/2019/TK03/MMU/03/3).

Data Availability Statement: Available upon request.

Acknowledgments: This work was supported by Fundamental Research Grant Scheme (FRGS), Ministry of Higher Education, Malaysia. Special thanks to the Faculty of Engineering and Technology of Multimedia University, Malaysia for their support in allowing this research to be carried out.

Conflicts of Interest: The authors declare no conflict of interest.

References

1. Cameron, A.; Wood, W.L. The full journal bearing. *Proc. Inst. Mech. Eng.* **1949**, *161*, 59–72. [CrossRef]
2. Ünlü, B.S.; Atik, E. Determination of friction coefficient in journal bearings. *Mater. Des.* **2007**, *28*, 973–977. [CrossRef]
3. Litwin, W. Marine propeller shaft bearings under low-speed conditions: Water vs. oil lubrication. *Tribol. Trans.* **2019**, *62*, 839–849. [CrossRef]
4. de Azevedo, H.D.; Araújo, A.M.; Bouchonneau, N. A review of wind turbine bearing condition monitoring: State of the art and Challenges. *Renew. Sustain. Energy Rev.* **2016**, *56*, 368–379. [CrossRef]
5. Dutta, S. Positive Outlook for Wind-Turbine Lubricants. 2019. Available online: https://www.windpowerengineering.com/positive-outlook-for-wind-turbine-lubricants/ (accessed on 21 August 2022).
6. Feng, Y.; Qiu, Y.; Crabtree, C.J.; Long, H.; Tavner, P.J. Monitoring wind turbine gearboxes. *Wind Energy* **2012**, *16*, 728–740. [CrossRef]
7. Hennigan, G.; Page, C. Maximizing Maintenance Dollars In a Post-PTC World. 2018. Available online: https://www.mobil.com/en/industrial/~{}/media/7EDB938B727C4B59B7BE32EE8DB82053.ashx (accessed on 21 August 2022).
8. Moghadam, A.D.; Omrani, E.; Menezes, P.L.; Rohatgi, P.K. Mechanical and tribological properties of self-lubricating metal matrix nanocomposites reinforced by carbon nanotubes (CNTs) and graphene—A review. *Compos. Part B Eng.* **2015**, *77*, 402–420. [CrossRef]
9. Tevrüz, T. Tribological behaviours of carbon filled polytetrafluoroethylene (PTFE) dry journal bearings. *Wear* **1998**, *221*, 61–68. [CrossRef]
10. Kim, S.S.; Park, D.C.; Lee, D.G. Characteristics of carbon fiber phenolic composite for journal bearing materials. *Compos. Struct.* **2004**, *66*, 359–366. [CrossRef]
11. Srivyas, P.D.; Charoo, M.S. Friction and wear reduction by graphene nano platelets for hybrid nano aluminium matrix composite under dry sliding conditions. *Metall. Mater. Eng.* **2020**, *27*, 27–47. [CrossRef]
12. Zhang, G.; Xu, Y.; Xiang, X.; Zheng, G.; Zeng, X.; Li, Z.; Ren, T.; Zhang, Y. Tribological performances of highly dispersed graphene oxide derivatives in vegetable oil. *Tribol. Int.* **2018**, *126*, 39–48. [CrossRef]
13. Song, H.; Wang, Z.; Yang, J. Tribological properties of graphene oxide and carbon spheres as lubricating additives. *Appl. Phys. A* **2016**, *122*, 933. [CrossRef]
14. Senatore, A.; D'Agostino, V.; Petrone, V.; Ciambelli, P.; Sarno, M. Graphene oxide nanosheets as effective friction modifier for oil lubricant: Materials, methods, and tribological results. *ISRN Tribol.* **2013**, *2013*, 425809. [CrossRef]
15. Chengara, A.; Nikolov, A.D.; Wasan, D.T.; Trokhymchuk, A.; Henderson, D. Spreading of nanofluids driven by the structural disjoining pressure gradient. *J. Colloid Interface Sci.* **2004**, *280*, 192–201. [CrossRef] [PubMed]
16. Singh, R.K.; Dixit, A.R.; Sharma, A.K.; Tiwari, A.K.; Mandal, V.; Pramanik, A. Influence of graphene and multi-walled carbon nanotube additives on tribological behaviour of lubricants. *Int. J. Surf. Sci. Eng.* **2018**, *12*, 207–227. [CrossRef]
17. Fillon, M.; Bouyer, J. Thermohydrodynamic analysis of a worn plain journal bearing. *Tribol. Int.* **2004**, *37*, 129–136. [CrossRef]
18. Yunus, M.; Munshi, S.M. Performance Evaluation of Hydrodynamic Journal Bearing using Gearbox and Engine Oil (SAE90 and SAE20w50) by Experimental and Theoretical Methods. *Int. J. Mech. Eng. Inf. Technol.* **2015**, *3*, 1573–1583.
19. Bao, T.; Wang, Z.; Zhao, Y.; Wang, Y.; Yi, X. Long-term stably dispersed functionalized graphene oxide as an oil additive. *RSC Adv.* **2019**, *9*, 39230–39241. [CrossRef]
20. Lin, J.; Wang, L.; Chen, G. Modification of graphene platelets and their tribological properties as a lubricant additive. *Tribol. Lett.* **2010**, *41*, 209–215. [CrossRef]
21. Soudagar, M.E.; Nik-Ghazali, N.-N.; Kalam, M.A.; Badruddin, I.A.; Banapurmath, N.R.; Khan, T.M.Y.; Bashir, M.N.; Akram, N.; Farade, R.; Afzal, A. The effects of graphene oxide nanoparticle additive stably dispersed in dairy scum oil biodiesel-diesel fuel blend on CI engine: Performance, emission and combustion characteristics. *Fuel* **2019**, *257*, 116015–116031. [CrossRef]
22. Tummala, N.R.; Grady, B.P.; Striolo, A. Lateral confinement effects on the structural properties of surfactant aggregates: SDS on graphene. *Phys. Chem. Chem. Phys.* **2010**, *12*, 13137–13143. [CrossRef]
23. La, D.D.; Truong, T.N.; Pham, T.Q.; Vo, H.T.; Tran, N.T.; Nguyen, T.A.; Nadda, A.K.; Nguyen, T.T.; Chang, S.W.; Chung, W.J.; et al. Scalable fabrication of modified graphene nanoplatelets as an effective additive for engine lubricant oil. *Nanomaterials* **2020**, *10*, 877. [CrossRef]
24. Aleklett, K.; Höök, M.; Jakobsson, K.; Lardelli, M.; Snowden, S.; Söderbergh, B. The peak of the oil age—Analyzing the world oil production reference scenario in World Energy Outlook 2008. *Energy Policy* **2010**, *38*, 1398–1414. [CrossRef]
25. Mobarak, H.M.; Mohamad, E.N.; Masjuki, H.H.; Kalam, M.A.; al Mahmud, K.A.H.; Habibullah, M.; Ashraful, A.M. The prospects of biolubricants as alternatives in automotive applications. *Renew. Sustain. Energy Rev.* **2014**, *33*, 34–43. [CrossRef]
26. Udonne, J.D. A comparative study of recycling of used lubrication oils using distillation, acid and activated charcoal with clay methods. *J. Pet. Gas Eng.* **2011**, *2*, 12–19.
27. Naddaf, A.; Heris, S.Z.; Pouladi, B. An experimental study on heat transfer performance and pressure drop of nanofluids using graphene and multi-walled carbon nanotubes based on diesel oil. *Powder Technol.* **2019**, *352*, 369–380. [CrossRef]
28. McCash, L.B.; Akhtar, S.; Nadeem, S.; Saleem, S. Entropy analysis of the peristaltic flow of hybrid nanofluid inside an elliptic duct with sinusoidally advancing boundaries. *Entropy* **2021**, *23*, 732. [CrossRef]

29. Ali, A.; Saleem, S.; Mumraiz, S.; Saleem, A.; Awais, M.; Marwat, D.N.K. Investigation on tio2–cu/H2O hybrid nanofluid with slip conditions in MHD peristaltic flow of Jeffrey Material. *J. Therm. Anal. Calorim.* **2020**, *143*, 1985–1996. [CrossRef]
30. Berman, D.; Erdemir, A.; Sumant, A.V. Graphene: A new emerging lubricant. *Mater. Today* **2014**, *17*, 31–42. [CrossRef]
31. Li, Q.; Lee, C.; Carpick, R.W.; Hone, J. Substrate effect on thickness-dependent friction on graphene. *Phys. Status Solidi (B)* **2010**, *247*, 2909–2914. [CrossRef]
32. Davis, J.R. *Alloying: Understanding the Basics*; ASM International: Materials Park, OH, USA, 2011.
33. Meyer, T. Validation of Journal Bearings for Use in Wind Turbine Gearboxes. 2015. Available online: https://www.windsystemsmag.com/educational-opportunities-abound-at-windpower-2015/ (accessed on 21 August 2022).
34. Pantazopoulos, G.; Papaefthymiou, S. Failure and fracture analysis of austenitic stainless steel marine propeller shaft. *J. Fail. Anal. Prev.* **2015**, *15*, 762–767. [CrossRef]
35. Lorenzi, S.; Pastore, T.; Bellezze, T.; Fratesi, R. Cathodic protection modelling of a propeller shaft. *Corros. Sci.* **2016**, *108*, 36–46. [CrossRef]
36. Hantoro, R.H.; Utama, I.K.; Sulisetyono, A.S.; Erwandi, E. Validation of lumped mass lateral cantilever shaft vibration simulation on fixed-pitch vertical-axis ocean current turbine. *IPTEK J. Technol. Sci.* **2010**, *21*, 1–8. [CrossRef]
37. Kanwal, S.; Ali, N.Z.; Hussain, R.; Shah, F.U.; Akhter, Z. Poly-thiourea formaldehyde based anticorrosion marine coatings on type 304 stainless steel. *J. Mater. Res. Technol.* **2020**, *9*, 2146–2153. [CrossRef]
38. Jayadas, N.H.; Nair, K.P. Coconut oil as base oil for industrial lubricants—evaluation and modification of thermal, oxidative and low temperature properties. *Tribol. Int.* **2006**, *39*, 873–878. [CrossRef]
39. Bhushan, B. *Introduction to Tribology*; Wiley: New York, OH, USA, 2013.
40. Ng, K.P.; Liew, K.W.; Lim, E. Role of eco-friendly bio-based graphene-oil nanofluids on friction reduction for wind turbine application. In Proceedings of the IOP Conference Series: Earth and Environmental Science, Sapporo, Japan, 26–28 August 2021; p. 012012.
41. Krisnangkura, K.; Yimsuwan, T.; Pairintra, R. An empirical approach in predicting biodiesel viscosity at various temperatures. *Fuel* **2006**, *85*, 107–113. [CrossRef]
42. Konijn, B.J.; Sanderink, O.B.J.; Kruyt, N.P. Experimental study of the viscosity of suspensions: Effect of solid fraction, particle size and suspending liquid. *Powder Technol.* **2014**, *266*, 61–69. [CrossRef]
43. Zhang, L.; Pu, J.; Wang, L.; Xue, Q. Frictional dependence of graphene and carbon nanotube in diamond-like carbon/ionic liquids hybrid films in vacuum. *Carbon* **2014**, *80*, 734–745. [CrossRef]
44. Fang, Y.; Ma, L.; Luo, J. Modelling for water-based liquid lubrication with ultra-low friction coefficient in rough surface point contact. *Tribol. Int.* **2020**, *141*, 105901–105908. [CrossRef]
45. Ferrer, P.R.; Mace, A.; Thomas, S.N.; Jeon, J.-W. Nanostructured porous graphene and its composites for energy storage applications. *Nano Converg.* **2017**, *4*, 29. [CrossRef]
46. Carrales-Alvarado, D.H.; Rodríguez-Ramos, I.; Leyva-Ramos, R.; Mendoza-Mendoza, E.; Villela-Martínez, D.E. Effect of surface area and physical–chemical properties of graphite and graphene-based materials on their adsorption capacity towards metronidazole and trimethoprim antibiotics in aqueous solution. *Chem. Eng. J.* **2020**, *402*, 126155. [CrossRef]
47. Abdelbary, A. Sliding mechanics of polymers. In *Wear of Polymers and Composites*; Woodhead Publishing: Thorston, UK, 2014; pp. 37–66.
48. Halim, T.; Burgett, M.; Donaldson, T.K.; Savisaar, C.; Bowsher, J.; Clarke, I.C. Profiling the third-body wear damage produced in cocr surfaces by bone cement, COCR, and ti6al4v debris: A 10-cycle metal-on-metal simulator test. *Proc. Inst. Mech. Eng. Part H J. Eng. Med.* **2014**, *228*, 703–713. [CrossRef] [PubMed]
49. Rasheed, A.K.; Khalid, M.; Rashmi, W.; Gupta, T.C.S.M.; Chan, A. Graphene based nanofluids and nanolubricants—Review of recent developments. *Renew. Sustain. Energy Rev.* **2016**, *63*, 346–362. [CrossRef]
50. Tjong, S.C. Recent progress in the development and properties of novel metal matrix nanocomposites reinforced with carbon nanotubes and graphene nanosheets. *Mater. Sci. Eng. R Rep.* **2013**, *74*, 281–350. [CrossRef]

Review

Graphene-Family Lubricant Additives: Recent Developments and Future Perspectives

Yanfei Liu [1,2], Shengtao Yu [1], Qiuyu Shi [3], Xiangyu Ge [1,*] and Wenzhong Wang [1]

1. School of Mechanical Engineering, Beijing Institute of Technology, Beijing 100081, China
2. State Key Laboratory of Tribology in Advanced Equipment, Tsinghua University, Beijing 100084, China
3. State Grid Smart Grid Research Institute Co., Ltd., Beijing 102209, China
* Correspondence: gexy@bit.edu.cn

Abstract: Graphene-family materials have been investigated by researchers as promising additives for various lubrication systems due to their unique physical-chemical properties. It has been proven that graphene-family materials can lead to enhanced lubrication and wear-resistance performance, which have potential to reduce the energy losses and carbon emissions, and the wear of machines for industrial applications. Experimental, theoretical, and simulation studies have been performed to investigate the tribological behaviors of graphene-family materials as additives. The tribological properties of graphene-family materials, including graphene, reduced graphene oxide, functionalized graphene, and the combination of graphene-family materials and other materials as additives, and the fundamental mechanism are systematically reviewed and concluded. The authors also discuss the potential engineering applications of graphene-family materials as lubricating additives, and the unsolved issues and optimistic outlooks in the near future.

Keywords: graphene; friction; wear; additive

Citation: Liu, Y.; Yu, S.; Shi, Q.; Ge, X.; Wang, W. Graphene-Family Lubricant Additives: Recent Developments and Future Perspectives. *Lubricants* **2022**, *10*, 215. https://doi.org/10.3390/lubricants10090215

Received: 1 August 2022
Accepted: 31 August 2022
Published: 6 September 2022

Publisher's Note: MDPI stays neutral with regard to jurisdictional claims in published maps and institutional affiliations.

Copyright: © 2022 by the authors. Licensee MDPI, Basel, Switzerland. This article is an open access article distributed under the terms and conditions of the Creative Commons Attribution (CC BY) license (https://creativecommons.org/licenses/by/4.0/).

1. Introduction

The friction and wear of machine parts usually cause huge carbon emissions and economic losses, restricting the sustainable development of human society [1]. Under such a background, advanced lubrication techniques have been proposed by researchers to reduce energy dissipation and enhance durability [2–10]. Additives are usually added to lubricants to enhance the lubrication performance, viscosity under high temperature, antioxidation, wear-reduction properties, and so on. Among all the lubricant additives, carbon nanomaterials have attracted a lot of attention due to the low cost, eco-friendliness, and significant improvement in lubrication and anti-wear performances [11–13].

Graphene, as an atomically thin carbon material, exhibits extremely high mechanical strength, high thermal and electronic conductivities, and other unique properties, which make it highly attractive for numerous applications [14,15]. Graphene also exhibits excellent tribological performance. Even single-layer graphene can lead to a significant reduction in coefficient of friction (COF) and wear between friction pairs [16,17]. At the nanoscale or microscale, superlubricity (COF < 0.01) can be achieved between graphene layers with incommensurate contact [18–21]. Although graphene shows great potential for various applications, it should also be noticed that graphene has some inherent properties including zero bandgaps and chemical inertness, restricting the application of graphene. That is one of the reasons for researchers investigating the functionalization of graphene through the reaction with inorganic or organic molecules, chemical modification of the surface of graphene, and the noncovalent interaction with graphene [22–27]. Functionalized graphene (Figure 1) also exhibits different dispersion properties and tribological behaviors [28–30]. The excellent tribological behaviors and the feasible functional design of graphene-family materials make them promising additives for different lubrication systems. Within this work, the authors aim to review the recent achievements that have been realized with graphene-family materials as lubricating additives; and the functional mechanisms are also discussed in detail. Moreover, the unsolved issues and the optimistic outlooks for the graphene-family materials as additives are also discussed.

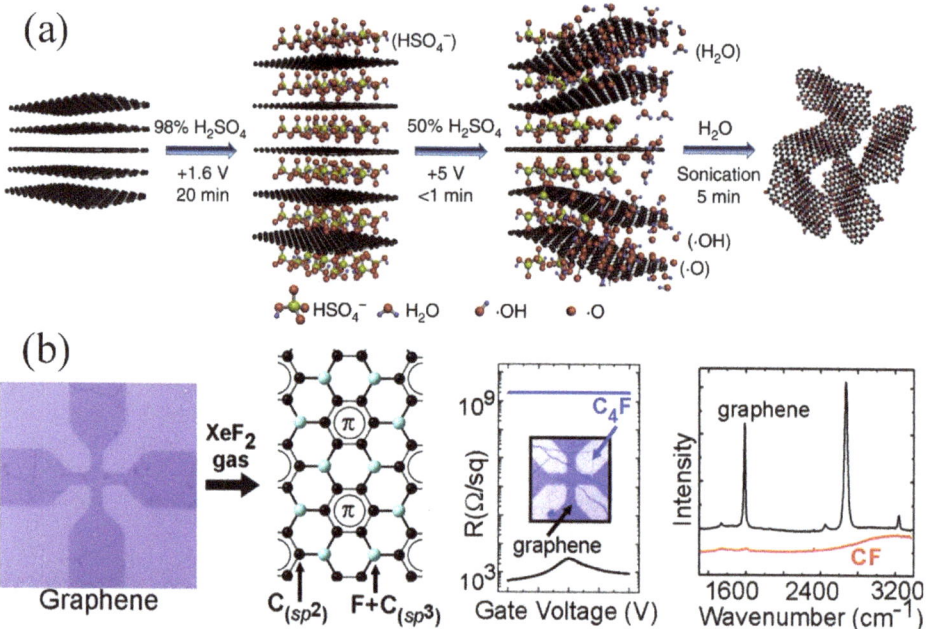

Figure 1. Functionalization of graphene-family materials including (a) oxidation [31] and (b) fluorination [32]. Reprinted with permission from Ref. [32]. American Chemical Society, 2010.

2. Tribological Behaviors of Graphene-Family Additives

Graphene-family materials have great potential as lubrication additives due to the atomic thin structure and high strength of graphene. Researchers have developed various lubricants containing graphene-family materials as additives, proving that graphene-family materials can effectively enhance lubrication and wear-resistance performances. However, different from the studies using graphene-family materials as solid lubricants, the detailed functional mechanisms of graphene-family materials as additives have not been elucidated due to the complexity of liquid-based lubricants. Hence, systematic studies are important for a better understanding of the tribological behaviors and the functional mechanisms of graphene-family materials as lubrication additives.

2.1. Pristine Graphene and Reduced Graphene Oxide (rGO)

Oil-based lubricants, including non-synthetic lubricating oils, synthetic lubricating oils, grease, and most recently, bio-based lubricating oil, are widely used as lubricants in both the market and the scientific research field. Typically, additives are needed for oil-based lubricants to improve the rheological, lubrication, anti-wear, and antioxidation behaviors. Different from traditional additives, nanomaterials have been used as additives due to the environmentally friendly, easy-processing properties and significantly improved lubrication and anti-wear performances [33]. Recently, graphene-family materials have attracted much attention from researchers as lubrication additives; and the performances of graphene-family materials as additives have been widely investigated. Among all the graphene-family materials, pristine graphene and reduced graphene oxide are ideal model materials for the understanding of the unique properties of graphene-family materials as lubrication additives. Eswaraiah et al. [34] prepared ultra-thin graphene through the exfoliation of graphite oxide. Different from the graphite oxide as starting material, the as-prepared ultra-thin graphene has a low concentration of oxide. Ultra-thin graphene nanoflakes were added to engine oil to prepare nanofluids with graphene concentration of 0.0125–0.0625 mg/mL (Figure 2a). It was found that the optimized graphene concentration of nanofluids is 0.0250 mg/mL, which can significantly reduce the COF by 80% (Figure 2b) and wear scar diameter of friction pairs. The performance of graphene as additive was attributed to the nano-bearing effect of graphene and the ultrahigh mechanical strength of graphene. It was found that the concentration of graphene as an additive has an important influence on tribological behaviors. Guo et al. [35] prepared PAO2 oil with graphene concentration from 0.05–0.5 wt%, finding that the lowest COF can be achieved with the lowest graphene concentration of 0.05 wt%. Cai et al. [36] found that even a low graphene concentration of 0.01 wt% can effectively reduce the COF by 78% and wear rate by 90% with textured bronze plates as friction pairs. The function mechanism of graphene nanoflakes was attributed to the nano-bearing effect and the formation of protective film on the friction pairs. Sanes et al. [37] investigated the influence of graphene nanoflakes and the ionic liquid (1-octyl-3-methylimidazolium tetrafluoroborate) as additives on the tribological behaviors of additive-free isoparaffinic base oil and formulated motor oil. It was found that the addition of 0.005 wt% of graphene nanoflakes can lead to a 70% reduction of COF of motor oil at the temperature of 150 °C, while for the additive-free base oil, ionic liquid dispersed with graphene nanoflakes can significantly reduce the friction and wear at room temperature. It has also been found the tribological behaviors of PAO6 [38–40], hydraulic oil [41], 4010 AL base oil [42], vegetable oil [43], cashew nut shells liquid [44], wax extracted from *Codonopsis pilosula* [45], or lithium complex and polypropylene-thickened greases [46] can be also improved with graphene as additives. However, there is still room for the improvement of graphene as a lubrication additive. The structure of the sliding interface has significant influence on the lubrication performance of graphene or other 2D materials as solid lubricants, which has been widely investigated at the nanoscale [47–49] or macroscale [5,50,51]. For graphene as an additive, Jin et al. [52] regulated the structure of graphene through ball-milling treatment, suppressing the commonly observed wrinkles and curved edge-site (Figure 3). With the ultra-flat graphene as additive, the COF and wear depth can be reduced by 49% and 93% at high contact pressure of 2.5 GPa, respectively.

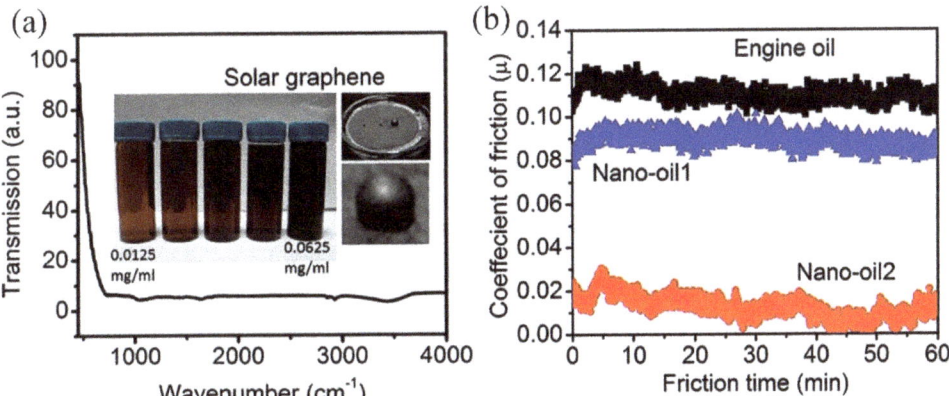

Figure 2. (**a**) Fourier transform infrared spectrum of as-prepared ultrathin graphene nanoflakes, where the insets are the photographs of the engine oil nanofluids with different concentrations of graphene, and the water droplet on graphene film; (**b**) COF versus time curves with engine oil and nanofluids. Reprinted with permission from Ref. [34]. American Chemical Society, 2011.

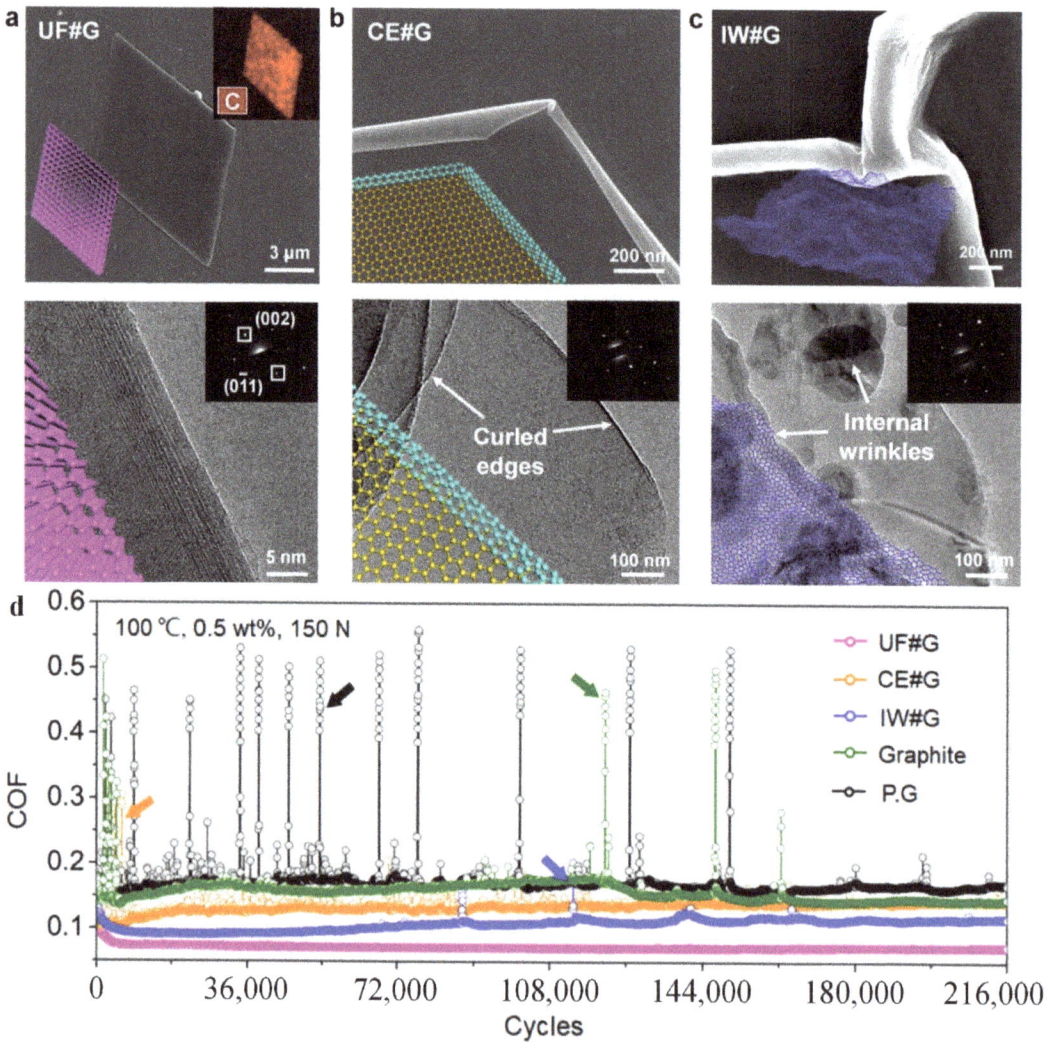

Figure 3. SEM and HRTEM images of (**a**) ultra-flat reduced graphene oxide, (**b**) curly edge reduced graphene oxide, and (**c**) internal wrinkle reduced graphene oxide; (**d**) COF versus time curves with different nanomaterials as additives. Reprinted with permission from Ref. [52]. Elsevier, 2022.

2.2. Functionalized Graphene

Pristine graphene or reduced graphene oxide exhibits excellent tribological behaviors as additives in the oil-based lubricant. However, those graphene-family materials are hard to be dispersed in water-based lubricants, restricting their application. In addition, the frictional and wear-resistant behaviors of graphene-family materials can be also regulated by functionalization. Hence, researchers have focused on the functionalization of graphene-family materials; and the tribological behaviors of functionalized graphene have also been systematically investigated. Wang et al. found that GO as an additive can effectively enhance the lubrication and wear-resistance performance of hexadecane-based oil [53]. Zhao et al. [38] compared the tribological behaviors of rGO and GO as additives in PAO6. It was found that PAO6 with GO exhibited much larger and unstable COF compared to that with rGO, which was attributed to the hindered interlayer sliding between lattice layers by oxygen-containing groups. For the achievement of better tribological behaviors, Fan et al. [54] prepared fluorinated graphene through direct fluorination with F_2. Fluorinated graphene with different concentrations of F was added into liquid paraffin oil as additives. It was found that fluorination with C/F ratio of 1.0 can reduce friction and wear by 50.4% and 90.9%, respectively. The formation of tough tribofilm was believed the dominant mechanism for the friction and wear reduction properties. Paul et al. [55] prepared dodecylamine functionalized graphene nanosheets as an additive, which can reduce the COF by up to 40% compared to the base engine oil. Li et al. [56] prepared graphene oxide grafted with titanate coupling agent (T-GO) as an additive to hydraulic oil. With 0.08 wt% of T-GO, the COF and wear can be reduced by up to 50% and 20%, and better extreme pressure performance compared to the original hydraulic oil. Fu et al. [57] prepared ionic liquid-modified graphene. The excellent tribological behaviors of the ionic liquid-modified graphene as an additive can be attributed to the adsorption of modified graphene and tribochemical reaction of contact surfaces. The alkylated graphene [58] and graphene oxide [59,60], phosphonium-organophosphate-modified graphene [61], CuAAC-modified graphene oxide [62], 3,5-ditert-butyl-4-hydroxybenzaldehyde-grafted graphene [63], octadecylamine- and dicyclohexylcarbodiimide-modified GO [64], octadecylamine-functionalized rGO [65], rGO doped with N and B species and polyisobutylene succinimide-grafted graphene [66] also exhibit excellent dispersibility and tribological behaviors as additives of oil-based lubricants.

Despite oil-based lubricants, functionalized graphene has been widely used as an additive to water-based lubricants due to its better dispersibility and stability [67–71]. Kinoshita et al. [72] found that with GO as an additive in pure water, the COF between tungsten carbide ball and stainless steel plate can be significantly reduced from around 0.4 to 0.05; and the wear of friction pairs can be also significantly suppressed. Adsorbed GO was found on the surface of friction pairs, which was believed to act as a protective coating for the lubrication and wear-resistance properties. He et al. [73] investigated the influence of pH value on the tribological behaviors of GO as an additive to water. It was found that the GO sheets can be broken down and chemically reduced under high pH value, leading to a higher COF. Wei et al. found that better lubrication behavior can be achieved with graphene oxide/polysaccharide copolymer nanohybrids [74]. With the nanohybrid as an additive in pure water, the COF can be reduced by 40% and 84% compared with pristine GO and individual copolymer as additives, respectively. Researchers have also designed various functionalized graphene to further enhance tribological behaviors. Fan et al. [75] prepared ionic liquids modified graphene oxide as an additive of multialkylated cyclopentanes, finding that the ionic liquids-modified GO can lead to a reduction of COF and wear by 27% and 74%, respectively. GO modified by imidazolyl dinitrile amine also exhibited improved tribological performance [76]. Liu et al. [77] prepared graphene grafted with polyethyleneimine and polyacrylic acid, which improves the multiple adsorption effects of graphene on the counterparts' surfaces, leading to the formation of tough and stable tribofilms. Researchers found that fluorinated graphene exhibits better tribological performance compared to pristine graphene at the macroscale [54,78], but the hydrophobic nature of fluorinated graphene makes it difficult to be used as an additive for water-based lubrication systems. Targeting this problem, Ye et al. [79] developed hydrophilic urea-modified fluorinated graphene. Using urea-modified fluorinated graphene as an additive to water with a concentration of 1 mg/mL, wear can be reduced by 64.4% compared to that lubricated by pure water. Min et al. [80] prepared fluorinated graphene oxide using hydrothermal reaction, obtaining excellent dispersibility in water and tremendous abrasion resistance performance. Fan et al. [81] prepared fluorinated graphene with relatively low F content under mild temperature conditions. The as-prepared fluorinated graphene oxide as an additive of water led to a 47% and 31% lower wear rate compared to that with the lubrication of pure water and water suspension of GO, respectively. Later, urea-modified fluorinated graphene oxide was also prepared as the additive for water lubrication [82].

With traditional lubricants, the COFs are usually in the range of 0.02–0.1 [81,83–86]. Under the background of energy saving and emission reduction, the concept of superlubricity (COF < 0.01) has attracted a lot of attention [1,87,88]. Ge et al. [89] found that macroscale superlubricity can be achieved by ethylene glycol water solution containing GO as an additive. The achievement of superlubricity was attributed to the synergy between hydrodynamic effect and boundary lubrication provided by GO tribofilm. Even better lubrication and wear-reduction performances were achieved with the combination of GO nanosheets and an ionic liquid as additives [90] (Figure 4). Superlubricity can be also achieved by GO and lithium salts as an additive for dihydric alcohol aqueous solutions [91]. The adsorbed GO and tribochemical reaction are both important for the achievement of superlubricity. The functional groups have a potential influence on the superlubricity behaviors with graphene-family as additives. Recently, GO-OH, GO–COOH, and GO–NH$_2$ were added to dihydric alcohols as additives [92]. It was found that robust superlubricity can be achieved using GO–NH$_2$ as an additive, which was believed attributed to the formation of the adsorption layer due to the high adhesion between GO–NH$_2$ and SiO$_2$ substrate. The achievement of superlubricity usually needs a running-in process with high COF and severe wear. Recently, Liu et al. [93] used GO quantum dots as nano-additive of ethylene glycol water solution. A superlubricity state with COF of 0.0068 can be achieved with an extremely short running-in process of 6 s. The tribofilms with adsorbed GOQDs were believed to be critical for the reduced running-in process, and the achievement of a superlubricity state at relatively high contact pressure, which was confirmed by the in-situ friction tests and surface characterization results. Graphene-family as additives can provide lubrication and wear-resistance performance. Superlubricity with high contact pressure and the extremely short running-in process can be achieved with water-based lubricants with graphene-family materials as additives. However, the choice of graphene-family materials for water-based superlubricity systems is largely restricted by dispersibility and stability. GO is commonly used for water-based lubrication systems, but the interlayer shearing strength of GO is higher than the pristine graphene and other functionalized graphene such as fluorinated graphene [94,95]; and the interlayer sliding can be further influenced by the adsorbed water molecules, leading to higher COF in the environments with the presence of water molecules [96,97]. Targeting the above-mentioned problems, Liu et al. [98] proposed a novel strategy to achieve superlubricity with water-based lubricant and hydrophobic graphene, where the coatings with different graphene-family coatings were firstly deposited on the SiO$_2$ substrate and the glycerol aqueous solution was then added on the coatings for lubrication. The performances of pristine graphene, GO, and fluorinated graphene was compared. All the deposited graphene coatings led to lower COF and wear of friction pairs. Among the graphene-family materials, pristine graphene coating exhibited the best lubrication performance at a sliding speed of 0.1 m/s, where macroscale superlubricity with COF of 0.004 can be achieved. The superlubricity behavior can be attributed to the hydrodynamic effect and the boundary lubrication provided by the tribofilm containing graphene nanoflakes (Figure 5). Combining graphene-family materials as additives and lubrication coatings is also a strategy to realize superlubricity. An extremely small COF of 0.002 was achieved on silicon-doped hydrogenated amorphous carbon film with the lubrication of ethylene glycol containing GO as an additive [99].

2.3. Synergy between Graphene-Family and Other Nanomaterials

Researchers also combined graphene-family materials and other nanomaterials including other kinks of 2D materials, carbon nanomaterials, metal or metal oxide nanoparticles, and silica nanoparticles as additives for further improved tribological behaviors. Xu et al. [100] investigated the synergetic effect between graphene and MoS$_2$ as additives for esterified bio-oil. The tribological behaviors of the bio-oil added with 0.5 wt% graphene, 0.5 wt% MoS$_2$, and 0.3 wt% graphene, and 0.2 wt% MoS$_2$ were compared, finding that the nanofluid with 0.3 wt% graphene and 0.2 wt% MoS$_2$ exhibited lower COF and wear rate comparing to other samples. The formation of tribofilm is critical for enhanced tribological behaviors. With the combination of graphene and MoS$_2$ nanoflakes, it was found that the oxidation and degradation of MoS$_2$ can be suppressed by graphene nanoflakes, and the structure of graphene and MoS$_2$ can be better maintained through the synergetic effect (Figure 6). Farsadi et al. [101] added functionalized GO and MoS$_2$ into petroleum-based oil as an additive, achieving improved tribological performances. MoS$_2$ can be also directly synthesized on graphene or GO through calcination [102] and hydrothermal reaction [103–105]. The obtained composited nanomaterials can effectively enhance the lubrication and wear resistance performance of polyalkylene glycol [102,104] and based oil [103,105]. The intrinsic incommensurate between graphene and MoS$_2$ was believed to be one of the reasons for the enhanced lubrication performance [105]. Graphene-family materials were also mixed with h-BN [106], or grafted with APTMS-modified h-BN to improve the tribological properties as additives [107].

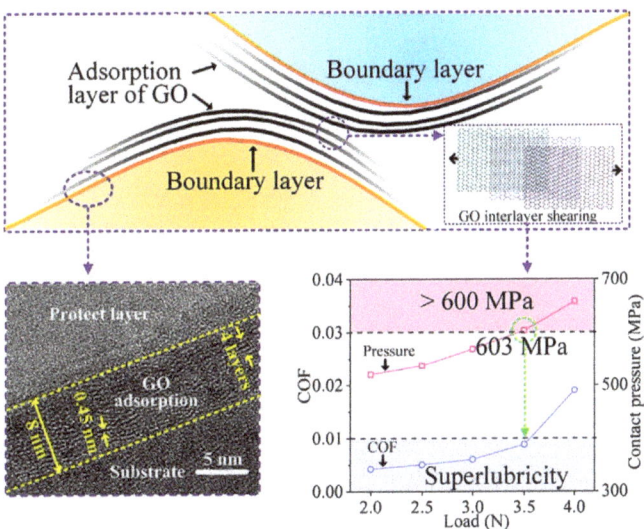

Figure 4. Superlubricity is achieved with the combination of GO nanosheets and an ionic liquid as additives in ethylene glycol water solution. Adsorbed GO can be observed onto the wear surface, which is critical for the achievement of superlubricity. Reprinted with permission from Ref. [90]. Elsevier, 2019.

The synergetic lubrication performance between graphene-family materials and nanoparticle or nanotubes was also investigated. A lubricant with GO and nanodiamonds as additives in pure water was developed by Wu et al. [108] to improve tribological performance. The low friction of approximately 0.03 can be obtained with 0.1 wt.% and 0.5 wt.% for GO and ND, respectively. Nanostructured tribofilm was found on the surface of friction pairs (Figure 7). The low shearing strength between graphene layers and the possible nano-bearing effect of ND was believed to be the mechanism for the remarkable lubrication performance. In addition ND, functionalized carbon spheres were also used combined with graphene, which led to an 18% reduction of COF compared to the CASTROL-20 W 40 engine oil [109]. Graphene-family materials have been combined with WS$_2$ [110], onion-like carbon [111], metal [112–120] or metal oxide [121–129] nanoparticles, silica nanoparticles [130–135], or carbon nanotubes [136–138] as additives for oil-based or water-based lubricants to provide better lubrication and wear-resistance performance.

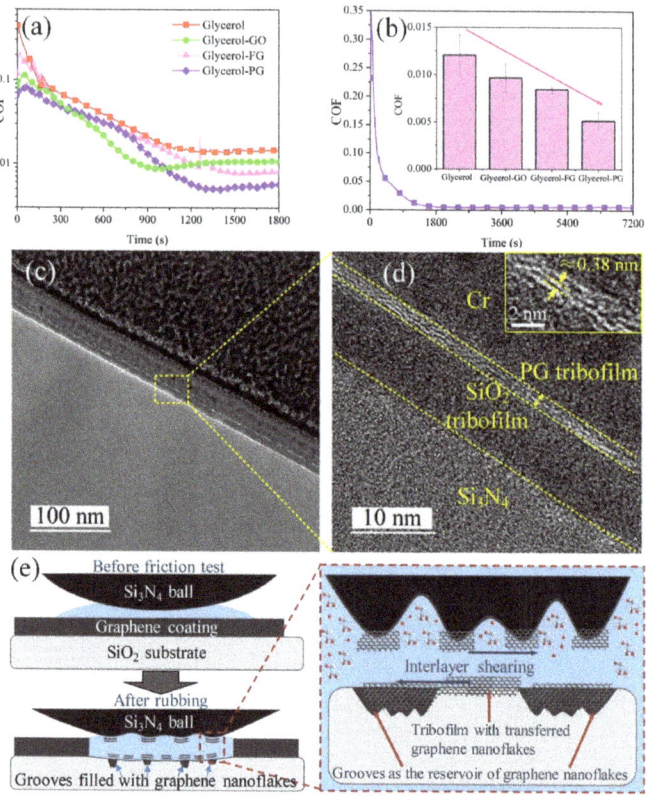

Figure 5. Superlubricity is achieved with the hydrophobic graphene coating and the glycerol aqueous solution. (**a**) COF versus time curves with different lubricants and the (**b**) long duration result with pristine graphene coating; (**c**) TEM and (**d**) HRTEM images of the tribofilm after the friction test with deposited pristine graphene coating; (**e**) schematic diagram of the superlubricity mechanism with hydrophobic graphene and water-based lubricant. Reprinted with permission from Ref. [98]. American Chemical Society, 2020.

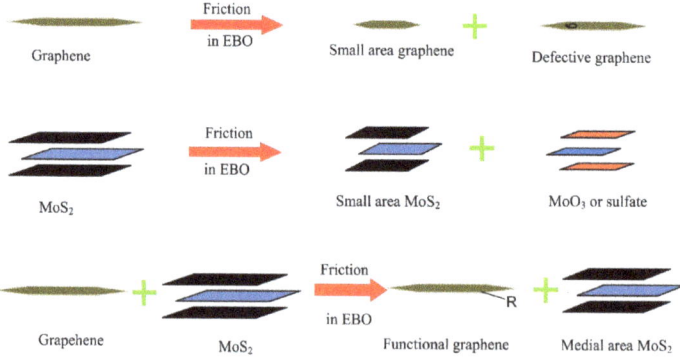

Figure 6. A proposed synergetic effect between graphene and MoS_2 nanoflakes as an additive in bio-oil. Schematic diagram of the structural evolution and chemical changes with graphene, MoS_2, and graphene combined with MoS_2 additives during friction process. Reprinted with permission from Ref. [100]. Elsevier, 2015.

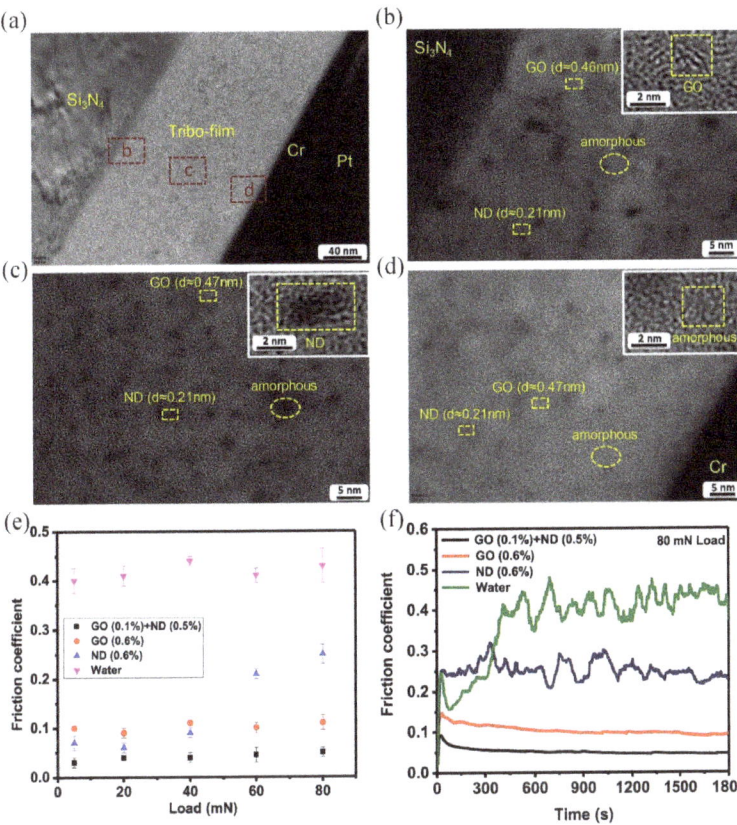

Figure 7. (**a–d**) Structure of the tribofilm containing GO and ND; (**e,f**) COFs with the lubrication of pure water, GO, ND, and the GO-ND suspensions. Reprinted with permission from Ref. [108]. Elsevier, 2019.

2.4. Lubrication Mechanisms of Graphene-Family Materials as Additives

The friction behavior with graphene-family materials as additives is a complex physical-chemical process, especially at the macroscale, leading to different tribological behaviors reported in the literature (Table 1). Various lubrication mechanisms have been proposed.

Table 1. Statistical data of studies related to graphene-family additives in liquid lubrication. Reproduced with permission. [15,139].

Additive (Fraction), in Base Liquid	Test Mode	Specimen Details	Test Parameters	Results		Lubrication Mechanisms
				COF Reduction	Wear Reduction	
GO (0.1 wt%), in 150 SN	Upper disk on the lower ball, Rotary	Ball: X45Cr13 steel, 52–54 HRC, Φ8 mm. Disk: X155CrVMo12-1 steel, 60 HRC, Ra 0.5 μm, Φ105 mm,	0.05–2.1 m/s; 30, 60, 90 N; 25, 50, 80 °C	30%	27%	Deposited GO protective layer [140]

Table 1. Cont.

Additive (Fraction), in Base Liquid	Test Mode	Specimen Details	Test Parameters	COF Reduction	Wear Reduction	Lubrication Mechanisms
Zinc borate/GO composite (2 wt%), in 500 SN	Four-ball	—	1200 rpm; 147 N; Test 2 h	48.2%	40%	Protective layer [141]
Cu nanoparticles decorated on polydopamine functionalized GO (0.1 wt%), in soybean oil	Ball on disk	Ball: GCr15 steel, Φ9.525 mm, 62 HRC. Disk: 45#steel	100–500 rpm; 1–12 N; Test 0.5 h	57%	27%	Tribolayer [113]
Modified graphene (0.075 wt%), in 350 SN	Four-ball	GCr15A steel, Φ12.7 mm, 61 HRC, Test standard: ASTM D4172-82	1200 rpm; 147 N; 75 ± 2 °C; Test 1 h	37%	—	Protective layer [142]
Single layer GO (0.06%), in water	Ball on three disks	Ball: Cr alloy steel, Ra 11.1 ± 0.4 nm, 64 HRC. Disk: AISI304 stainless steel, Ra 37.0 ± 6.2 nm, 92 HRB	50 mm/s; 20 N; sliding 7.5 m	44.4%	17.1%	Tribolayer [73]
Multilayer graphene (0.1 wt%), in benton grease	Ball on disk, reciprocation	Ball: AISI52100 steel, Φ10 mm, 710 HV. Disk: AISI52100, 664 HV	100–500 N; 10–50 Hz; Test 0.5 h	10.4%	25–50%	Tribolayer [143]
Graphene (0.01 wt%), in PAO4	Ball on disk	Ball: GCr15 steel, Φ9.525 mm. Disk: Bronze contained elliptical dimple (area ratios: 0, 5%, 10%, 20%)	5 mm/s; 5 N; Sliding 8 mm; 25, 60, 100, 150 °C Test 1.67 h	78%	90%	Tribolayer and texturing [36]
CeO_2-decorated graphene (0.06 wt%), in paraffin oil	Reciprocation, test method: ASTM D6425-05	Ball: GCr15 steel, Φ10 mm, Ra 20 nm, 62 HRC. Disk: GCr15 steel, 792 HV, Ra: 50 nm	6 cm/s; 15 Hz; 50 N; 17% RH; Test 0.5 h	52%	1.5%	Transfer layer and nanoparticle spacer between graphene sheets [129]
ZrO_2/rGO composite (0.06 wt%), in paraffin oil	Reciprocation, test method: ASTM D6425-05,	Ball: GCr15 steel, Φ10 mm, Ra 20 nm, 62 HRC. Disk: GCr15 steel, 790–820 HV, Ra 50 nm	6 cm/s; 50–450 N; 25 ± 5% RH; Test 0.5 h	56%	6.4%	Protective layer and ball bearing [123]
Graphene (23.8–110 μg/mL), in water	Ball on disk, rotary	Ball: GCr15 steel, Φ9.53 mm. Disk: GCr15 steel	62.8–251.2 mm/s; 2–15 N; 25 ± 5% RH; Test 0.5 h	81.3%	61.8%	Fluid adhesive layer and graphene protective layer [144]

Table 1. Cont.

Additive (Fraction), in Base Liquid	Test Mode	Specimen Details	Test Parameters	Results COF Reduction	Results Wear Reduction	Lubrication Mechanisms
Single layer graphene (1 mg/mL), in ethanol	Ball on disk	Ball: 100Cr6 steel, Φ4 mm. Disk: Iron (99.98% pure) and bronze (98% Cu and 2% Sn), Ra 30 nm	100 mm/s; 1 N; 50% RH	48%	—	Chemical passivation of iron by graphene [145]
Silica/GO composite (0.125 wt%), in EG	Pin on disk, rotary	Ball: AISI420 steel, Φ12.7 mm. Disk: AISI52100 steel, Φ40 mm, Ra < 0.1 µm	0.008 m/s; 68.9 N; 50% RH; Test 1 h	38%	31%	Tribolayer and ball bearing [135]
Modified GO, in oil/water emulsion	Ball on ring	Ball: bearing steel, Φ25.4 mm, 65 HRC, Ra 30 ± 10 nm. Ring: AISI52100 steel, 65 HRC, Ra 100 ± 50 nm	20–200 mm/s; 0.5 N; 25 ± 5% RH; sliding 100 m	18%	48%	Transfer layer, adsorption layer, and tribolayer; and the lubricity of emulsion droplets [146]
Graphene (0.01 wt%), in Span-80/PAO4	Ball on disk, reciprocation	Ball: GCr15 steel, Φ4 mm, 766 HV. Disk: RTCr2 alloy cast iron, 220 HV, Ra ≤ 0.03 µm	—	—	50% on normal surface, 90% on textured surface	Polishing, self-repairing, and tribolayer [147]
GO (2 mg/mL) in ethylene glycol aqueous solution	Ball on disk, rotary	Ball: Si_3N_4, Φ4 mm, Disk: SiO_2	0.1 m/s; 3 N; 10–25% RH	Superlubricity with COF ≈ 0.004	99.5% compared to ethylene glycol aqueous solution	Adsorbed GO; low shear stress between GO layers and the interface between GO and ethylene glycol aqueous solution, and the formation of hydrated networks [89]
GO quantum dots (1 mg/mL) in ethylene glycol aqueous solution	Ball on disk, rotary	Ball: Si_3N_4, Φ20 mm, Disk: sapphire	0.1 m/s; 15 N; 10–25% RH	Superlubricity with COF of 0.0068	90%	Formation of tribofilm containing GO quantum dots [93]

For the graphene-family additives, well-accepted lubrication and wear-resistance mechanism is the formation of graphene-containing tribofilm, which has been observed in both oil-based [38–43] and water-based lubrication systems [72,89]. However, the investigation of the formation and function of adsorbed graphene-family during friction is still at an early stage. There are still many opening questions to be solved. It was proven that the graphene-containing tribofilms can effectively reduce friction and wear, but the studies related to the formation of graphene-containing tribofilms are still insufficient. Ge et al. [92] found that the GO-NH_2 exhibited better lubrication performance compared to GO with -OH and -COOH groups, which was attributed to the better robustness of GO-NH_2 tribofilm arising from the larger adhesive force between functional groups of GO-NH_2 and contact surfaces. In addition to the adhesion force, tribochemical reactions between graphene-family materials also have potential influence on the tribofilm formation, thus the tribological behaviors. For fluorinated graphene, the formation of metal-F bonds during friction is believed to be an important mechanism for the formation of robust tribofilm and better macroscale lubrication performance [28]. In addition to the functional groups, it was found that the size of graphene-family materials also influences the formation of tribofilm and the lubrication behavior. Li et al. [148] investigated the influence of flake size and concentration of graphene as additive of linear alpha olefin on the lubrication behaviors using molecular dynamics simulation. It was found that when the size of graphene flakes is larger than 40 carbon atoms, the graphene flakes can be anchored on the surface of friction pairs to form a protective tribofilm, leading to a passivated and smoothened sliding interface and ultralow COF (Figure 8).

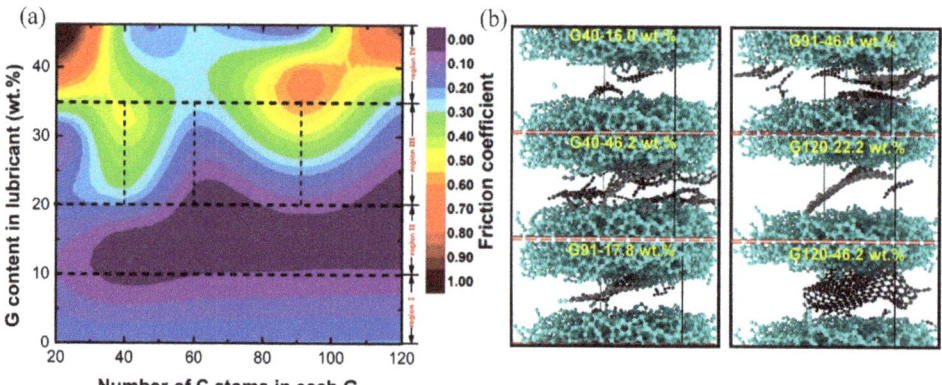

Figure 8. Effect of size and concentration of the graphene as additive on the COFs. (**a**) Change of COF with size and concentration of graphene; (**b**) interfacial structures containing graphene with different sizes and concentrations (the molecules of base oil are neglected for better observation of the structure of graphene). Reprinted with permission from Ref. [148]. American Chemical Society, 2020.

In addition to the formation of graphene-containing tribofilm, understanding how the graphene-containing tribofilms perform lubrication performance is also important. With the protection of graphene-containing tribofilm, the direct contact between asperities under boundary or mixed lubrication conditions can be prevented by the tribofilms, providing a low shear strength, and thus enhancing lubrication and wear-resistance performance. In addition, the wear interaction between graphene-containing tribofilm and lubricant molecules is also believed to be an important mechanism for the lubrication performance of graphene-family additives. A new strategy was proposed by Li et al. [149] to investigate the interaction between water molecules and graphene surfaces. A highly hydrophobic surface of self-assembled fluoroalkyl monolayers was prepared on the SiO_2 tip for the friction tests conducted using AFM. A superlubricity state with an extremely low COF of 0.0003 was obtained between the modified SiO_2 tip and graphene surface in a water environment, where the weak interaction between graphene surface and water molecules contributed to the achievement of superlubricity (Figure 9). The influence of weak interaction between graphene and liquid molecules on the lubrication performance was also verified by molecular dynamics simulations. It was found that the graphene tribofilm can promote the mobility of water molecules [150] or oil molecules [151,152], leading to a lower COF. In addition to those mechanisms, the nano-ball bearing [153], self-repairing [154], the micro-polishing effect [155], tribochemical reaction [57], and the incommensurate contact between graphene-family materials and other nanomaterials were also proposed as lubrication mechanisms for the graphene-family materials as additives.

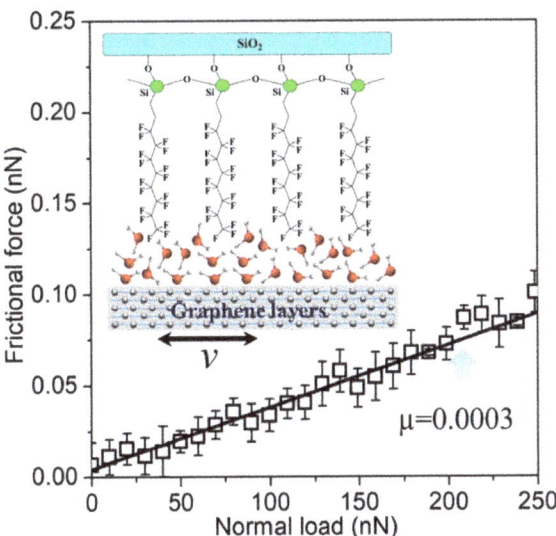

Figure 9. Superlubricity achieved by self-assembled fluoroalkyl monolayers and graphene surfaces in a water environment. Reprinted with permission from Ref. [149]. American Chemical Society, 2019.

2.5. Engineering Applications of Graphene-Family Materials as Additives

With decades of development, researchers have also focused on the commercialization and engineering applications of graphene-family materials [156,157]. The function of graphene-family materials as lubrication additives for engineering applications was also investigated. Rasheed et al. [158] investigated the heat transfer and tribological performance of engine oil with graphene as nano-additive using an internal combustion engine. It was found that with 0.01 wt.% of graphene, COF can be reduced by 21%, and the thermal conductivity of lubricant can be enhanced by 23%. In addition, heat transfer rate of the engine can be also enhanced by 70%. The wear of engine can be also suppressed with graphene nanoflakes as additives. The potential applications of graphene-family additives for machining process were also investigated. Baldin et al. [159] studied the effect of graphene addition in cutting fluids applied by minimum quantity lubrication (MQL) for end milling of AISI 1045 steel. It was found that the addition of graphene nanoflakes can enhance the lubrication performance of cutting fluids. For some cutting fluid, the addition of graphene nanoflakes leads to increased tool life, while for some cutting fluid, the addition of graphene nanoflakes can leads to reduced tool life. Li et al. [160] investigated the performance of vegetable oil-based cutting fluid dispersed with graphene nanoparticles for the MQL milling of Ti6Al4V. It was found that the graphene additive could enhance the cooling and lubrication performances of the oil film formed at the milling zone, leading to reduced milling temperature, milling force, tool wear, and enhanced tool life. The influence of graphene-family additive for drilling [161] and hard-turning [162] was also investigated, finding that graphene-family materials as additives could provide excellent lubrication, and improved surface roughness. Graphene-family additives have also been added into drilling fluid for the petroleum industry [157] to enhance the fluid loss control, rheology properties, and lubrication and wear-resistance performances. Those studies indicate that graphene-family materials are promising for various industrial applications.

3. Conclusions and Perspectives

Graphene-family materials have been investigated by researchers as promising additives for various lubrication systems due to the unique physical-chemical properties. It has been proven that graphene-family materials can lead to enhanced lubrication and wear-resistance performance. Experimental, theoretical, and simulation studies have been performed to investigate the tribological behaviors of graphene-family materials as additives, and various function mechanisms have been proposed. Over the years, many breakthroughs have been achieved, but there are still problems for both scientific research and engineering applications.

- For graphene-family additives, dispersion stability is an important issue related to the instability in oil or water-based lubrication systems. However, the high temperature introduced wear debris, and the material degradation all have potential influence on the long-term stability of graphene-family additives, which still need further investigation.
- The cost of graphene-family materials is still high for industrial applications. Developing a large-scale, low-cost preparation process is important for the practical application of graphene-family materials as additives.
- There is still a lack of widely accepted criteria for the designing of graphene-family materials as additives. For example, the optimized parameters such as particle size, layer numbers, types, and concentration of functional groups for specified application condition are still unclear. An in-depth investigation of the fundamental mechanisms and advanced techniques [163–165] for the guidance of designing and application of graphene-family materials as additives is needed in the future.

Author Contributions: Writing—original draft preparation, Y.L.; funding acquisition, Y.L.; writing—review and editing, Y.L., S.Y., Q.S., X.G. and W.W. All authors have read and agreed to the published version of the manuscript.

Funding: This work was financially supported by the National Natural Science Foundation of China (grant number 52005287), Beijing Institute of Technology Research Fund Program for Young Scholars, Tribology Science Fund of State Key Laboratory of Tribology in Advanced Equipment (SKLT) (No. SKLTKF21B14), and the Fund of the Key Laboratory of Advanced Materials of the Ministry of Education (No. ADV21-4).

Conflicts of Interest: The authors declare no conflict of interest.

References

1. Luo, J.; Liu, M.; Ma, L. Origin of friction and the new frictionless technology—Superlubricity: Advancements and future outlook. *Nano Energy* **2021**, *86*, 106092. [CrossRef]
2. Li, J.; Zhang, C.; Luo, J. Superlubricity Behavior with Phosphoric Acid–Water Network Induced by Rubbing. *Langmuir* **2011**, *27*, 9413–9417. [CrossRef] [PubMed]
3. Liu, Y.-F.; Liskiewicz, T.; Yerokhin, A.; Korenyi-Both, A.; Zabinski, J.; Lin, M.; Matthews, A.; Voevodin, A.A. Fretting wear behavior of duplex PEO/chameleon coating on Al alloy. *Surf. Coat. Technol.* **2018**, *352*, 238–246. [CrossRef]
4. Chen, X.; Zhang, C.; Kato, T.; Yang, X.-A.; Wu, S.; Wang, R.; Nosaka, M.; Luo, J. Evolution of tribo-induced interfacial nanostructures governing superlubricity in a-C:H and a-C:H:Si films. *Nat. Commun.* **2017**, *8*, 1675. [CrossRef]
5. Liu, Y.; Li, J.; Yi, S.; Ge, X.; Chen, X.; Luo, J. Enhancement of friction performance of nanocomposite fluorinated graphene and molybdenum disulfide coating by microdimple array. *Carbon* **2020**, *167*, 122–131. [CrossRef]
6. Berman, D.; Deshmukh, S.A.; Sankaranarayanan, S.K.R.S.; Erdemir, A.; Sumant, A.V. Macroscale superlubricity enabled by graphene nanoscroll formation. *Science* **2015**, *348*, 1118–1122. [CrossRef]
7. Liu, Y.; Yu, S.; Wang, W. Nanodiamond plates as macroscale solid lubricant: A "non-layered" two-dimension material. *Carbon* **2022**, *198*, 119–131. [CrossRef]
8. Liu, Y.; Li, J.; Li, J.; Yi, S.; Ge, X.; Zhang, X.; Luo, J. Shear-Induced Interfacial Structural Conversion Triggers Macroscale Superlubricity: From Black Phosphorus Nanoflakes to Phosphorus Oxide. *ACS Appl. Mater. Interfaces* **2021**, *13*, 31947–31956. [CrossRef]
9. Xie, Z.; Jiao, J.; Yang, K.; He, T.; Chen, R.; Zhu, W. Experimental and numerical exploration on the nonlinear dynamic behaviors of a novel bearing lubricated by low viscosity lubricant. *Mech. Syst. Signal Process.* **2023**, *182*, 109349. [CrossRef]
10. Xie, Z.; Zhu, W. Theoretical and experimental exploration on the micro asperity contact load ratios and lubrication regimes transition for water-lubricated stern tube bearing. *Tribol. Int.* **2021**, *164*, 107105. [CrossRef]
11. Xue, S.; Li, H.; Guo, Y.; Zhang, B.; Li, J.; Zeng, X. Water lubrication of graphene oxide-based materials. *Friction* **2022**, *10*, 977–1004. [CrossRef]
12. Kumar, V.B.; Sahu, A.K.; Rao, K.B. Development of Doped Carbon Quantum Dot-Based Nanomaterials for Lubricant Additive Applications. *Lubricants* **2022**, *10*, 144. [CrossRef]

13. Bai, Y.; Yu, Q.; Zhang, J.; Cai, M.; Liang, Y.; Zhou, F.; Liu, W. Soft-nanocomposite lubricants of supramolecular gel with carbon nanotubes. *J. Mater. Chem. A* **2019**, *7*, 7654–7663. [CrossRef]
14. Novoselov, K.S.; Fal'ko, V.I.; Colombo, L.; Gellert, P.R.; Schwab, M.G.; Kim, K. A roadmap for graphene. *Nature* **2012**, *490*, 192–200. [CrossRef] [PubMed]
15. Liu, Y.; Ge, X.; Li, J. Graphene lubrication. *Appl. Mater. Today* **2020**, *20*, 100662. [CrossRef]
16. Berman, D.; Deshmukh, S.A.; Sankaranarayanan, S.K.R.S.; Erdemir, A.; Sumant, A.V. Extraordinary Macroscale Wear Resistance of One Atom Thick Graphene Layer. *Adv. Funct. Mater.* **2014**, *24*, 6640–6646. [CrossRef]
17. Kim, K.-S.; Lee, H.-J.; Lee, C.; Lee, S.-K.; Jang, H.; Ahn, J.-H.; Kim, J.-H.; Lee, H.-J. Chemical Vapor Deposition-Grown Graphene: The Thinnest Solid Lubricant. *ACS Nano* **2011**, *5*, 5107–5114. [CrossRef]
18. Zheng, X.; Gao, L.; Yao, Q.; Li, Q.; Zhang, M.; Xie, X.; Qiao, S.; Wang, G.; Ma, T.; Di, Z.; et al. Robust ultra-low-friction state of graphene via moiré superlattice confinement. *Nat. Commun.* **2016**, *7*, 13204. [CrossRef]
19. Feng, X.; Kwon, S.; Park, J.Y.; Salmeron, M. Superlubric Sliding of Graphene Nanoflakes on Graphene. *ACS Nano* **2013**, *7*, 1718–1724. [CrossRef] [PubMed]
20. Li, J.; Gao, T.; Luo, J. Superlubricity of Graphite Induced by Multiple Transferred Graphene Nanoflakes. *Adv. Sci.* **2018**, *5*, 1700616. [CrossRef]
21. Li, J.; Li, J.; Luo, J. Superlubricity of Graphite Sliding against Graphene Nanoflake under Ultrahigh Contact Pressure. *Adv. Sci.* **2018**, *5*, 1800810. [CrossRef] [PubMed]
22. Ohta, T.; Bostwick, A.; Seyller, T.; Horn, K.; Rotenberg, E. Controlling the Electronic Structure of Bilayer Graphene. *Science* **2006**, *313*, 951–954. [CrossRef] [PubMed]
23. Elias, D.C.; Nair, R.R.; Mohiuddin, T.M.G.; Morozov, S.V.; Blake, P.; Halsall, M.P.; Ferrari, A.C.; Boukhvalov, D.W.; Katsnelson, M.I.; Geim, A.K.; et al. Control of Graphene's Properties by Reversible Hydrogenation: Evidence for Graphane. *Science* **2009**, *323*, 610–613. [CrossRef] [PubMed]
24. Wang, Q.H.; Hersam, M.C. Room-temperature molecular-resolution characterization of self-assembled organic monolayers on epitaxial graphene. *Nat. Chem.* **2009**, *1*, 206–211. [CrossRef]
25. Bostwick, A.; Ohta, T.; Seyller, T.; Horn, K.; Rotenberg, E. Quasiparticle dynamics in graphene. *Nat. Phys.* **2007**, *3*, 36–40. [CrossRef]
26. Bai, H.; Xu, Y.; Zhao, L.; Li, C.; Shi, G. Non-covalent functionalization of graphene sheets by sulfonated polyaniline. *Chem. Comm.* **2009**, *13*, 1667–1669. [CrossRef]
27. Georgakilas, V.; Otyepka, M.; Bourlinos, A.B.; Chandra, V.; Kim, N.; Kemp, K.C.; Hobza, P.; Zboril, R.; Kim, K.S. Functionalization of Graphene: Covalent and Non-Covalent Approaches, Derivatives and Applications. *Chem. Rev.* **2012**, *112*, 6156–6214. [CrossRef]
28. Liu, Y.; Li, J.; Chen, X.; Luo, J. Fluorinated Graphene: A Promising Macroscale Solid Lubricant Under Various Environments. *ACS Appl. Mater. Interfaces* **2019**, *11*, 40470–40480. [CrossRef]
29. Liu, Y.; Chen, X.; Li, J.; Luo, J. Enhancement of friction performance enabled by synergetic effect between graphene oxide and molybdenum disulfide. *Carbon* **2019**, *154*, 266–276. [CrossRef]
30. Wang, M.; Zhou, M.; Li, X.; Luo, C.; You, S.; Chen, X.; Mo, Y.; Zhu, H. Research progress of surface-modified graphene-based materials for tribological applications. *Mater. Res. Express* **2021**, *8*, 042002. [CrossRef]
31. Pei, S.; Wei, Q.; Huang, K.; Cheng, H.-M.; Ren, W. Green synthesis of graphene oxide by seconds timescale water electrolytic oxidation. *Nat. Commun.* **2018**, *9*, 145. [CrossRef] [PubMed]
32. Robinson, J.T.; Burgess, J.S.; Junkermeier, C.E.; Badescu, S.C.; Reinecke, T.L.; Perkins, F.K.; Zalalutdniov, M.K.; Baldwin, J.W.; Culbertson, J.C.; Sheehan, P.E.; et al. Properties of Fluorinated Graphene Films. *Nano Lett.* **2010**, *10*, 3001–3005. [CrossRef]
33. Esquivel-Gaon, M.; Nguyen, N.H.A.; Sgroi, M.F.; Pullini, D.; Gili, F.; Mangherini, D.; Pruna, A.I.; Rosicka, P.; Sevcu, A.; Castagnola, V. In vitro and environmental toxicity of reduced graphene oxide as an additive in automotive lubricants. *Nanoscale* **2018**, *10*, 6539–6548. [CrossRef]
34. Eswaraiah, V.; Sankaranarayanan, V.; Ramaprabhu, S. Graphene-Based Engine Oil Nanofluids for Tribological Applications. *ACS Appl. Mater. Interfaces* **2011**, *3*, 4221–4227. [CrossRef] [PubMed]
35. Guo, Y.-B.; Zhang, S.-W. The Tribological Properties of Multi-Layered Graphene as Additives of PAO2 Oil in Steel–Steel Contacts. *Lubricants* **2016**, *4*, 30. [CrossRef]
36. Cai, Z.B.; Zhao, L.; Zhang, X.; Yue, W.; Zhu, M.H. Combined Effect of Textured Patterns and Graphene Flake Additives on Tribological Behavior under Boundary Lubrication. *PLoS ONE* **2016**, *11*, e0152143. [CrossRef] [PubMed]
37. Sanes, J.; Avilés, M.-D.; Saurín, N.; Espinosa, T.; Carrión, F.-J.; Bermúdez, M.-D. Synergy between graphene and ionic liquid lubricant additives. *Tribol. Int.* **2017**, *116*, 371–382. [CrossRef]
38. Zhao, J.; Li, Y.; Wang, Y.; Mao, J.; He, Y.; Luo, J. Mild thermal reduction of graphene oxide as a lubrication additive for friction and wear reduction. *RSC Adv.* **2017**, *7*, 1766–1770. [CrossRef]
39. Ouyang, T.; Shen, Y.; Lei, W.; Xu, X.; Liang, L.; Waqar, H.S.; Lin, B.; Tian, Z.Q.; Shen, P.K. Reduced friction and wear enabled by arc-discharge method-prepared 3D graphene as oil additive under variable loads and speeds. *Wear* **2020**, *462–463*, 203495. [CrossRef]
40. Gu, W.; Chu, K.; Lu, Z.; Zhang, G.; Qi, S. Synergistic effects of 3D porous graphene and T161 as hybrid lubricant additives on 316 ASS surface. *Tribol. Int.* **2021**, *161*, 107072. [CrossRef]

41. Mao, J.; Zhao, J.; Wang, W.; He, Y.; Luo, J. Influence of the micromorphology of reduced graphene oxide sheets on lubrication properties as a lubrication additive. *Tribol. Int.* **2018**, *119*, 614–621. [CrossRef]
42. Wu, L.; Gu, L.; Jian, R. Lubrication mechanism of graphene nanoplates as oil additives for ceramics/steel sliding components. *Ceram. Int.* **2021**, *47*, 16935–16942. [CrossRef]
43. Kiu, S.S.K.; Yusup, S.; Soon, C.V.; Arpin, T.; Samion, S.; Kamil, R.N.M. Tribological investigation of graphene as lubricant additive in vegetable oil. *J. Phys. Sci.* **2017**, *28*, 257.
44. Bhaumik, S.; Paleu, V.; Pathak, R.; Maggirwar, R.; Katiyar, J.K.; Sharma, A.K. Tribological investigation of r-GO additived biodegradable cashew nut shells liquid as an alternative industry lubricant. *Tribol. Int.* **2019**, *135*, 500–509. [CrossRef]
45. Xie, M.; Cheng, J.; Huo, C.; Zhao, G. Improving the lubricity of a bio-lubricating grease with the multilayer graphene additive. *Tribol. Int.* **2020**, *150*, 106386. [CrossRef]
46. Larsson, E.; Westbroek, R.; Leckner, J.; Jacobson, S.; Rudolphi, Å.K. Grease-lubricated tribological contacts—Influence of graphite, graphene oxide and reduced graphene oxide as lubricating additives in lithium complex (LiX)- and polypropylene (PP)-thickened greases. *Wear* **2021**, *486–487*, 204107. [CrossRef]
47. Lee, C.; Li, Q.; Kalb, W.; Liu, X.-Z.; Berger, H.; Carpick, R.W.; Hone, J. Frictional Characteristics of Atomically Thin Sheets. *Science* **2010**, *328*, 76–80. [CrossRef]
48. Li, S.; Li, Q.; Carpick, R.W.; Gumbsch, P.; Liu, X.Z.; Ding, X.; Sun, J.; Li, J. The evolving quality of frictional contact with graphene. *Nature* **2016**, *539*, 541. [CrossRef]
49. Zhang, S.; Hou, Y.; Li, S.; Liu, L.; Zhang, Z.; Feng, X.-Q.; Li, Q. Tuning friction to a superlubric state via in-plane straining. *Proc. Natl. Acad. Sci. USA* **2019**, *116*, 24452–24456. [CrossRef] [PubMed]
50. Li, R.; Sun, C.; Yang, X.; Wang, Y.; Gao, K.; Zhang, J.; Li, J. Toward high load-bearing, ambient robust and macroscale structural superlubricity through contact stress dispersion. *Chem. Eng. J.* **2021**, *431*, 133548. [CrossRef]
51. Li, P.; Ju, P.; Ji, L.; Li, H.; Liu, X.; Chen, L.; Zhou, H.; Chen, J. Toward Robust Macroscale Superlubricity on Engineering Steel Substrate. *Adv. Mater.* **2020**, *32*, 2002039. [CrossRef] [PubMed]
52. Jin, B.; Zhao, J.; He, Y.; Chen, G.; Li, Y.; Zhang, C.; Luo, J. High-quality ultra-flat reduced graphene oxide nanosheets with super-robust lubrication performances. *Chem. Eng. J.* **2022**, *438*, 135620. [CrossRef]
53. Wang, W.; Zhang, G.; Xie, G. Ultralow concentration of graphene oxide nanosheets as oil-based lubricant additives. *Appl. Surf. Sci.* **2019**, *498*, 143683. [CrossRef]
54. Fan, K.; Chen, X.; Wang, X.; Liu, X.; Liu, Y.; Lai, W.; Liu, X. Toward Excellent Tribological Performance as Oil-Based Lubricant Additive: Particular Tribological Behavior of Fluorinated Graphene. *ACS Appl. Mater. Interfaces* **2018**, *10*, 28828–28838. [CrossRef] [PubMed]
55. Paul, G.; Shit, S.; Hirani, H.; Kuila, T.; Murmu, N.C. Tribological behavior of dodecylamine functionalized graphene nanosheets dispersed engine oil nanolubricants. *Tribol. Int.* **2019**, *131*, 605–619. [CrossRef]
56. Li, X.; Gan, C.; Han, Z.; Yan, H.; Chen, D.; Li, W.; Li, H.; Fan, X.; Li, D.; Zhu, M. High dispersivity and excellent tribological performance of titanate coupling agent modified graphene oxide in hydraulic oil. *Carbon* **2020**, *165*, 238–250. [CrossRef]
57. Fu, H.; Fan, X.; Li, W.; Zhu, M.; Peng, J.; Li, H. In situ modified multilayer graphene toward high-performance lubricating additive. *RSC Adv.* **2017**, *7*, 24399–24409. [CrossRef]
58. Zhang, L.; He, Y.; Zhu, L.; Yang, C.; Niu, Q.; An, C. In Situ Alkylated Graphene as Oil Dispersible Additive for Friction and Wear Reduction. *Ind. Eng. Chem. Res.* **2017**, *56*, 9029–9034. [CrossRef]
59. Mungse, H.P.; Gupta, K.; Singh, R.; Sharma, O.P.; Sugimura, H.; Khatri, O.P. Alkylated graphene oxide and reduced graphene oxide: Grafting density, dispersion stability to enhancement of lubrication properties. *J. Colloid Interface Sci.* **2019**, *541*, 150–162. [CrossRef]
60. Yin, S.; Ye, C.; Chen, Y.; Jin, C.; Wu, H.; Wang, H. Dependence of the lubrication enhancement of alkyl-functionalized graphene oxide and boric acid nanoparticles on the anti-oxidation property. *Colloids Surf. A Physicochem. Eng. Asp.* **2022**, *649*, 129521. [CrossRef]
61. Gan, C.; Liang, T.; Chen, D.; Li, W.; Fan, X.; Tang, G.; Lin, B.; Zhu, M. Phosphonium-organophosphate modified graphene gel towards lubrication applications. *Tribol. Int.* **2020**, *145*, 106180. [CrossRef]
62. Ismail, N.A.; Bagheri, S. Highly oil-dispersed functionalized reduced graphene oxide nanosheets as lube oil friction modifier. *Mater. Sci. Eng. B* **2017**, *222*, 34–42. [CrossRef]
63. Chouhan, A.; Mungse, H.P.; Sharma, O.P.; Singh, R.K.; Khatri, O.P. Chemically functionalized graphene for lubricant applications: Microscopic and spectroscopic studies of contact interfaces to probe the role of graphene for enhanced tribo-performance. *J. Colloid Interface Sci.* **2018**, *513*, 666–676. [CrossRef] [PubMed]
64. Wu, P.; Chen, X.; Zhang, C.; Zhang, J.; Luo, J.; Zhang, J. Modified graphene as novel lubricating additive with high dispersion stability in oil. *Friction* **2021**, *9*, 143–154. [CrossRef]
65. Mungse, H.P.; Khatri, O.P. Chemically Functionalized Reduced Graphene Oxide as a Novel Material for Reduction of Friction and Wear. *J. Phys. Chem. C* **2014**, *118*, 14394–14402. [CrossRef]
66. Bao, T.; Wang, Z.; Zhao, Y.; Wang, Y.; Yi, X. Long-term stably dispersed functionalized graphene oxide as an oil additive. *RSC Adv.* **2019**, *9*, 39230–39241. [CrossRef]
67. Chouhan, A.; Mungse, H.P.; Khatri, O.P. Surface chemistry of graphene and graphene oxide: A versatile route for their dispersion and tribological applications. *Adv. Colloid Interface Sci.* **2020**, *283*, 102215. [CrossRef]

68. Gan, C.; Liang, T.; Li, X.; Li, W.; Li, H.; Fan, X.; Zhu, M. Ultra-dispersive monolayer graphene oxide as water-based lubricant additive: Preparation, characterization and lubricating mechanisms. *Tribol. Int.* **2021**, *155*, 106768. [CrossRef]
69. Hu, Y.; Wang, Y.; Zeng, Z.; Zhao, H.; Ge, X.; Wang, K.; Wang, L.; Xue, Q. PEGlated graphene as nanoadditive for enhancing the tribological properties of water-based lubricants. *Carbon* **2018**, *137*, 41–48. [CrossRef]
70. Song, H.-J.; Li, N. Frictional behavior of oxide graphene nanosheets as water-base lubricant additive. *Appl. Phys. A* **2011**, *105*, 827–832. [CrossRef]
71. Kinoshita, H.; Kondo, M.; Nishina, Y.; Fujii, M. Anti-Wear Effect of Graphene Oxide in Lubrication by Fluorine-Containing Ionic Liquid for Steel. *Tribol. Online* **2015**, *10*, 91–95. [CrossRef]
72. Kinoshita, H.; Nishina, Y.; Alias, A.A.; Fujii, M. Tribological properties of monolayer graphene oxide sheets as water-based lubricant additives. *Carbon* **2014**, *66*, 720–723. [CrossRef]
73. He, A.; Huang, S.; Yun, J.-H.; Jiang, Z.; Stokes, J.; Jiao, S.; Wang, L.; Huang, H. The pH-dependent structural and tribological behaviour of aqueous graphene oxide suspensions. *Tribol. Int.* **2017**, *116*, 460–469. [CrossRef]
74. Wei, Q.; Fu, T.; Yue, Q.; Liu, H.; Ma, S.; Cai, M.; Zhou, F. Graphene oxide/brush-like polysaccharide copolymer nanohybrids as eco-friendly additives for water-based lubrication. *Tribol. Int.* **2021**, *157*, 106895. [CrossRef]
75. Fan, X.; Wang, L. High-performance lubricant additives based on modified graphene oxide by ionic liquids. *J. Colloid Interface Sci.* **2015**, *452*, 98–108. [CrossRef] [PubMed]
76. Han, Z.; Gan, C.; Li, X.; Feng, P.; Ma, X.; Fan, X.; Zhu, M. Electrochemical preparation of modified-graphene additive towards lubrication requirement. *Tribol. Int.* **2021**, *161*, 107057. [CrossRef]
77. Liu, J.; Wang, X.; Liu, Y.; Liu, X.; Fan, K. Bioinspired three-dimensional and multiple adsorption effects toward high lubricity of solvent-free graphene-based nanofluid. *Carbon* **2022**, *188*, 166–176. [CrossRef]
78. Matsumura, K.; Chiashi, S.; Maruyama, S.; Choi, J. Macroscale tribological properties of fluorinated graphene. *Appl. Surf. Sci.* **2018**, *432*, 190–195. [CrossRef]
79. Ye, X.; Ma, L.; Yang, Z.; Wang, J.; Wang, H.; Yang, S. Covalent Functionalization of Fluorinated Graphene and Subsequent Application as Water-based Lubricant Additive. *ACS Appl. Mater. Interfaces* **2016**, *8*, 7483–7488. [CrossRef]
80. Min, C.; He, Z.; Song, H.; Liang, H.; Liu, D.; Dong, C.; Jia, W. Fluorinated graphene oxide nanosheet: A highly efficient water-based lubricated additive. *Tribol. Int.* **2019**, *140*, 105867. [CrossRef]
81. Fan, K.; Liu, J.; Wang, X.; Liu, Y.; Lai, W.; Gao, S.; Qin, J.; Liu, X. Towards enhanced tribological performance as water-based lubricant additive: Selective fluorination of graphene oxide at mild temperature. *J. Colloid Interface Sci.* **2018**, *531*, 138–147. [CrossRef] [PubMed]
82. Zhu, W.; Wu, C.; Chang, Y.; Cheng, H.; Yu, C. Solvent-free preparation of hydrophilic fluorinated graphene oxide modified with amino-groups. *Mater. Lett.* **2019**, *237*, 1–4. [CrossRef]
83. Zhang, G.; Xu, J.; Xiang, X.; Zheng, G.; Zeng, X.; Li, Z.; Ren, T.; Zhang, Y. Tribological performances of highly dispersed graphene oxide derivatives in vegetable oil. *Tribol. Int.* **2018**, *126*, 39–48. [CrossRef]
84. Jaiswal, V.; Kalyani; Umrao, S.; Rastogi, R.B.; Kumar, R.; Srivastava, A. Synthesis, Characterization, and Tribological Evaluation of TiO$_2$-Reinforced Boron and Nitrogen co-Doped Reduced Graphene Oxide Based Hybrid Nanomaterials as Efficient Antiwear Lubricant Additives. *ACS Appl. Mater. Interfaces* **2016**, *8*, 11698–11710. [CrossRef]
85. Zhao, F.; Zhang, L.; Li, G.; Guo, Y.; Qi, H.; Zhang, G. Significantly enhancing tribological performance of epoxy by filling with ionic liquid functionalized graphene oxide. *Carbon* **2018**, *136*, 309–319. [CrossRef]
86. Wu, L.; Xie, Z.; Gu, L.; Song, B.; Wang, L. Investigation of the tribological behavior of graphene oxide nanoplates as lubricant additives for ceramic/steel contact. *Tribol. Int.* **2018**, *128*, 113–120. [CrossRef]
87. Luo, J.; Zhou, X. Superlubricitive engineering—Future industry nearly getting rid of wear and frictional energy consumption. *Friction* **2020**, *8*, 643–665. [CrossRef]
88. Chen, X.; Li, J. Superlubricity of carbon nanostructures. *Carbon* **2020**, *158*, 1–23. [CrossRef]
89. Ge, X.; Li, J.; Luo, R.; Zhang, C.; Luo, J. Macroscale Superlubricity Enabled by the Synergy Effect of Graphene-Oxide Nanoflakes and Ethanediol. *ACS Appl. Mater. Interfaces* **2018**, *10*, 40863–40870. [CrossRef]
90. Ge, X.; Li, J.; Wang, H.; Zhang, C.; Liu, Y.; Luo, J. Macroscale superlubricity under extreme pressure enabled by the combination of graphene-oxide nanosheets with ionic liquid. *Carbon* **2019**, *151*, 76–83. [CrossRef]
91. Ge, X.; Chai, Z.; Shi, Q.; Liu, Y.; Tang, J.; Wang, W. Liquid Superlubricity Enabled by the Synergy Effect of Graphene Oxide and Lithium Salts. *Materials* **2022**, *15*, 3546. [CrossRef]
92. Ge, X.; Chai, Z.; Shi, Q.; Li, J.; Tang, J.; Liu, Y.; Wang, W. Functionalized graphene-oxide nanosheets with amino groups facilitate macroscale superlubricity. *Friction* **2022**, 1–14. [CrossRef]
93. Liu, Y.; Yu, S.; Li, J.; Ge, X.; Zhao, Z.; Wang, W. Quantum dots of graphene oxide as nano-additive triggers macroscale superlubricity with an extreme short running-in period. *Mater. Today Nano* **2022**, *18*, 100219. [CrossRef]
94. Wang, L.-F.; Ma, T.-B.; Hu, Y.-Z.; Wang, H. Atomic-scale friction in graphene oxide: An interfacial interaction perspective from first-principles calculations. *Phys. Rev. B* **2012**, *86*, 125436. [CrossRef]
95. Wang, L.-F.; Ma, T.-B.; Hu, Y.-Z.; Wang, H.; Shao, T.-M. Ab Initio Study of the Friction Mechanism of Fluorographene and Graphane. *J. Phys. Chem. C* **2013**, *117*, 12520–12525. [CrossRef]
96. Arif, T.; Colas, G.; Filleter, T. Effect of Humidity and Water Intercalation on the Tribological Behavior of Graphene and Graphene Oxide. *ACS Appl. Mater. Interfaces* **2018**, *10*, 22537–22544. [CrossRef] [PubMed]

97. Saravanan, P.; Selyanchyn, R.; Tanaka, H.; Darekar, D.; Staykov, A.; Fujikawa, S.; Lyth, S.M.; Sugimura, J. Macroscale Superlubricity of Multilayer Polyethylenimine/Graphene Oxide Coatings in Different Gas Environments. *ACS Appl. Mater. Interfaces* **2016**, *8*, 27179–27187. [CrossRef]
98. Liu, Y.; Li, J.; Ge, X.; Yi, S.; Wang, H.; Liu, Y.; Luo, J. Macroscale Superlubricity Achieved on the Hydrophobic Graphene Coating with Glycerol. *ACS Appl Mater Interfaces* **2020**, *12*, 18859–18869. [CrossRef]
99. Yi, S.; Chen, X.; Li, J.; Liu, Y.; Ding, S.; Luo, J. Macroscale superlubricity of Si-doped diamond-like carbon film enabled by graphene oxide as additives. *Carbon* **2021**, *176*, 358–366. [CrossRef]
100. Xu, Y.; Peng, Y.; Dearn, K.D.; Zheng, X.; Yao, L.; Hu, X. Synergistic lubricating behaviors of graphene and MoS_2 dispersed in esterified bio-oil for steel/steel contact. *Wear* **2015**, *342–343*, 297–309. [CrossRef]
101. Farsadi, M.; Bagheri, S.; Ismail, N.A. Nanocomposite of functionalized graphene and molybdenum disulfide as friction modifier additive for lubricant. *J. Mol. Liq.* **2017**, *244*, 304–308. [CrossRef]
102. Gong, K.; Wu, X.; Zhao, G.; Wang, X. Nanosized MoS_2 deposited on graphene as lubricant additive in polyalkylene glycol for steel/steel contact at elevated temperature. *Tribol. Int.* **2017**, *110*, 1–7. [CrossRef]
103. Ismail, N.A.; Chowdhury, Z.Z.; Johan, M.R.; Zulkifli, N.W.M. MoS_2-Functionalized Graphene Composites—Potential Replacement for Lubricant Friction Modifier and Anti-Wear Additives. *Adv. Eng. Mater.* **2021**, *23*, 2100030. [CrossRef]
104. Gong, K.; Lou, W.; Zhao, G.; Wu, X.; Wang, X. MoS_2 nanoparticles grown on carbon nanomaterials for lubricating oil additives. *Friction* **2021**, *9*, 747–757. [CrossRef]
105. Hou, K.; Wang, J.; Yang, Z.; Ma, L.; Wang, Z.; Yang, S. One-pot synthesis of reduced graphene oxide/molybdenum disulfide heterostructures with intrinsic incommensurateness for enhanced lubricating properties. *Carbon* **2017**, *115*, 83–94. [CrossRef]
106. Qi, S.; Geng, Z.; Lu, Z.; Zhang, G.; Wu, Z. Synergistic Lubricating Behaviors of 3D Graphene and 2D Hexagonal Boron Nitride Dispersed in PAO4 for Steel/Steel Contact. *Adv. Mater. Interfaces* **2020**, *7*, 1901893. [CrossRef]
107. Samanta, S.; Sahoo, R.R. Covalently Linked Hexagonal Boron Nitride-Graphene Oxide Nanocomposites as High-Performance Oil-Dispersible Lubricant Additives. *ACS Appl. Nano Mater.* **2020**, *3*, 10941–10953. [CrossRef]
108. Wu, P.; Chen, X.; Zhang, C.; Luo, J. Synergistic tribological behaviors of graphene oxide and nanodiamond as lubricating additives in water. *Tribol. Int.* **2019**, *132*, 177–184. [CrossRef]
109. Radhika, P.; Sobhan, C.B.; Chakravorti, S. Improved tribological behavior of lubricating oil dispersed with hybrid nanoparticles of functionalized carbon spheres and graphene nano platelets. *Appl. Surf. Sci.* **2021**, *540*, 148402. [CrossRef]
110. Zheng, D.; Wu, Y.-p.; Li, Z.-y.; Cai, Z.-b. Tribological properties of WS_2/graphene nanocomposites as lubricating oil additives. *RSC Adv.* **2017**, *7*, 14060–14068. [CrossRef]
111. Su, F.; Chen, G.; Huang, P. Lubricating performances of graphene oxide and onion-like carbon as water-based lubricant additives for smooth and sand-blasted steel discs. *Friction* **2020**, *8*, 47–57. [CrossRef]
112. Meng, Y.; Su, F.; Chen, Y. Au/Graphene Oxide Nanocomposite Synthesized in Supercritical CO_2 Fluid as Energy Efficient Lubricant Additive. *ACS Appl. Mater. Interfaces* **2017**, *9*, 39549–39559. [CrossRef] [PubMed]
113. Song, H.J.; Wang, Z.Q.; Yang, J.; Jia, X.H.; Zhang, Z.Z. Facile synthesis of copper/polydopamine functionalized graphene oxide nanocomposites with enhanced tribological performance. *Chem. Eng. J.* **2017**, *324*, 51–62. [CrossRef]
114. Wang, L.; Gong, P.; Li, W.; Luo, T.; Cao, B. Mono-dispersed Ag/Graphene nanocomposite as lubricant additive to reduce friction and wear. *Tribol. Int.* **2020**, *146*, 106228. [CrossRef]
115. Gan, C.; Liang, T.; Li, W.; Fan, X.; Zhu, M. Amine-terminated ionic liquid modified graphene oxide/copper nanocomposite toward efficient lubrication. *Appl. Surf. Sci.* **2019**, *491*, 105–115. [CrossRef]
116. Zhang, Y.; Tang, H.; Ji, X.; Li, C.; Chen, L.; Zhang, D.; Yang, X.; Zhang, H. Synthesis of reduced graphene oxide/Cu nanoparticle composites and their tribological properties. *RSC Adv.* **2013**, *3*, 26086–26093. [CrossRef]
117. White, D.; Chen, M.; Xiao, C.; Huang, W.; Sundararajan, S. Microtribological behavior of Mo and W nanoparticle/graphene composites. *Wear* **2018**, *414–415*, 310–316. [CrossRef]
118. Meng, Y.; Su, F.; Chen, Y. Supercritical Fluid Synthesis and Tribological Applications of Silver Nanoparticle-decorated Graphene in Engine Oil Nanofluid. *Sci. Rep.* **2016**, *6*, 31246. [CrossRef]
119. Jia, Z.; Chen, T.; Wang, J.; Ni, J.; Li, H.; Shao, X. Synthesis, characterization and tribological properties of Cu/reduced graphene oxide composites. *Tribol. Int.* **2015**, *88*, 17–24. [CrossRef]
120. Meng, Y.; Su, F.; Chen, Y. Synthesis of nano-Cu/graphene oxide composites by supercritical CO_2-assisted deposition as a novel material for reducing friction and wear. *Chem. Eng. J.* **2015**, *281*, 11–19. [CrossRef]
121. Meng, Y.; Su, F.; Li, Z. Boundary and Elastohydrodynamic Lubrication Behaviors of Nano-CuO/Reduced Graphene Oxide Nanocomposite as an Efficient Oil-Based Additive. *Langmuir* **2019**, *35*, 10322–10333. [CrossRef]
122. Huang, S.; He, A.; Yun, J.-H.; Xu, X.; Jiang, Z.; Jiao, S.; Huang, H. Synergistic tribological performance of a water based lubricant using graphene oxide and alumina hybrid nanoparticles as additives. *Tribol. Int.* **2019**, *135*, 170–180. [CrossRef]
123. Zhou, Q.; Huang, J.; Wang, J.; Yang, Z.; Liu, S.; Wang, Z.; Yang, S.J.R.A. Preparation of a reduced graphene oxide/zirconia nanocomposite and its application as a novel lubricant oil additive. *RSC Adv.* **2015**, *5*, 91802–91812. [CrossRef]
124. Zhao, J.; Li, Y.; He, Y.; Luo, J. In Situ Green Synthesis of the New Sandwichlike Nanostructure of Mn_3O_4/Graphene as Lubricant Additives. *ACS Appl. Mater. Interfaces* **2019**, *11*, 36931–36938. [CrossRef] [PubMed]
125. Wang, J.; Zhuang, W.; Yan, T.; Liang, W.; Li, T.; Zhang, L.; Wei, X. Tribological performances of copper perrhenate/graphene nanocomposite as lubricating additive under various temperatures. *J. Ind. Eng. Chem.* **2021**, *100*, 296–309.

126. Ren, B.; Gao, L.; Xie, B.; Li, M.; Zhang, S.; Zu, G.; Ran, X. Tribological properties and anti-wear mechanism of ZnO@graphene core-shell nanoparticles as lubricant additives. *Tribol. Int.* **2020**, *144*, 106114. [CrossRef]
127. Alghani, W.; Karim, M.S.A.; Bagheri, S.; Amran, N.A.M.; Gulzar, M. Enhancing the Tribological Behavior of Lubricating Oil by Adding TiO_2, Graphene, and TiO_2/Graphene Nanoparticles. *Tribol. Trans.* **2019**, *62*, 452–463. [CrossRef]
128. Sun, J.; Ge, C.; Wang, C.; Li, S. Tribological behavior of graphene oxide-Fe_3O_4 nanocomposites for additives in water-based lubricants. *Fuller. Nanotub. Carbon Nanostruct.* **2022**, *30*, 863–872. [CrossRef]
129. Bai, G.; Wang, J.; Yang, Z.; Wang, H.; Wang, Z.; Yang, S. Preparation of a highly effective lubricating oil additive–ceria/graphene composite. *RSC Adv.* **2014**, *4*, 47096–47105. [CrossRef]
130. Guo, Y.; Guo, L.; Li, G.; Zhang, L.; Zhao, F.; Wang, C.; Zhang, G. Solvent-free ionic nanofluids based on graphene oxide-silica hybrid as high-performance lubricating additive. *Appl. Surf. Sci.* **2019**, *471*, 482–493. [CrossRef]
131. Guo, P.; Chen, L.; Wang, J.; Geng, Z.; Lu, Z.; Zhang, G. Enhanced Tribological Performance of Aminated Nano-Silica Modified Graphene Oxide as Water-Based Lubricant Additive. *ACS Appl. Nano Mater.* **2018**, *1*, 6444–6453. [CrossRef]
132. Shen, Y.; Lei, W.; Tang, W.; Ouyang, T.; Liang, L.; Tian, Z.Q.; Shen, P.K. Synergistic friction-reduction and wear-resistance mechanism of 3D graphene and SiO_2 nanoblend at harsh friction interface. *Wear* **2022**, *488–489*, 204175. [CrossRef]
133. Song, W.; Yan, J.; Ji, H. Tribological Performance of an Imidazolium Ionic Liquid-Functionalized SiO_2@Graphene Oxide as an Additive. *ACS Appl. Mater. Interfaces* **2021**, *13*, 50573–50583. [CrossRef] [PubMed]
134. Xiong, S.; Zhang, B.; Luo, S.; Wu, H.; Zhang, Z. Preparation, characterization, and tribological properties of silica-nanoparticle-reinforced B-N-co-doped reduced graphene oxide as a multifunctional additive for enhanced lubrication. *Friction* **2021**, *9*, 239–249. [CrossRef]
135. Singh, V.K.; Elomaa, O.; Johansson, L.-S.; Hannula, S.-P.; Koskinen, J. Lubricating properties of silica/graphene oxide composite powders. *Carbon* **2014**, *79*, 227–235. [CrossRef]
136. Mohamed, A.; Tirth, V.; Kamel, B.M. Tribological characterization and rheology of hybrid calcium grease with graphene nanosheets and multi-walled carbon nanotubes as additives. *J. Mater. Res. Technol.* **2020**, *9*, 6178–6185. [CrossRef]
137. Kamel, B.M.; El-Kashif, E.; Hoziefa, W.; Shiba, M.S.; Elshalakany, A.B. The effect of MWCNTs/GNs hybrid addition on the tribological and rheological properties of lubricating engine oil. *J. Dispers. Sci. Technol.* **2021**, *42*, 1811–1819. [CrossRef]
138. Ouyang, T.; Tang, W.; Pan, M.; Tang, J.; Huang, H. Friction-reducing and anti-wear properties of 3D hierarchical porous graphene/multi-walled carbon nanotube in castor oil under severe condition: Experimental investigation and mechanism study. *Wear* **2022**, *498–499*, 204302. [CrossRef]
139. Paul, G.; Hirani, H.; Kuila, T.; Murmu, N.C.J.N. Nanolubricants dispersed with graphene and its derivatives: An assessment and review of the tribological performance. *Nanoscale* **2019**, *11*, 3458–3483. [CrossRef]
140. Senatore, A.; D'Agostino, V.; Petrone, V.; Ciambelli, P.; Sarno, M. Graphene oxide nanosheets as effective friction modifier for oil lubricant: Materials, methods, and tribological results. *J. ISRN Tribol.* **2013**, *2013*, 425809. [CrossRef]
141. Cheng, Z.-L.; Li, W.; Liu, Z. Preparation, characterization, and tribological properties of oleic diethanolamide-capped zinc borate-coated graphene oxide composites. *J. Alloy. Compd.* **2017**, *705*, 384–391. [CrossRef]
142. Lin, J.; Wang, L.; Chen, G. Modification of Graphene Platelets and their Tribological Properties as a Lubricant Additive. *Tribol. Lett.* **2011**, *41*, 209–215. [CrossRef]
143. Fan, X.; Xia, Y.; Wang, L.; Li, W. Multilayer Graphene as a Lubricating Additive in Bentone Grease. *Tribol. Lett.* **2014**, *55*, 455–464. [CrossRef]
144. Liang, S.; Shen, Z.; Yi, M.; Liu, L.; Zhang, X.; Ma, S. In-situ exfoliated graphene for high-performance water-based lubricants. *Carbon* **2016**, *96*, 1181–1190. [CrossRef]
145. Marchetto, D.; Restuccia, P.; Ballestrazzi, A.; Righi, M.C.; Rota, A.; Valeri, S. Surface passivation by graphene in the lubrication of iron: A comparison with bronze. *Carbon* **2017**, *116*, 375–380. [CrossRef]
146. Wu, Y.; Zeng, X.; Ren, T.; de Vries, E.; van der Heide, E. The emulsifying and tribological properties of modified graphene oxide in oil-in-water emulsion. *Tribol. Int.* **2017**, *105*, 304–316. [CrossRef]
147. Zheng, D.; Cai, Z.-b.; Shen, M.-x.; Li, Z.-y.; Zhu, M.-h. Investigation of the tribology behaviour of the graphene nanosheets as oil additives on textured alloy cast iron surface. *Appl. Surf. Sci.* **2016**, *387*, 66–75. [CrossRef]
148. Li, X.; Zhang, D.; Xu, X.; Lee, K.-R. Tailoring the Nanostructure of Graphene as an Oil-Based Additive: Toward Synergistic Lubrication with an Amorphous Carbon Film. *ACS Appl. Mater. Interfaces* **2020**, *12*, 43320–43330. [CrossRef]
149. Li, J.; Cao, W.; Li, J.; Ma, M.; Luo, J. Molecular Origin of Superlubricity between Graphene and a Highly Hydrophobic Surface in Water. *J. Phys. Chem. Lett.* **2019**, *10*, 2978–2984. [CrossRef]
150. Li, C.; Tang, W.; Tang, X.-Z.; Yang, L.; Bai, L. A molecular dynamics study on the synergistic lubrication mechanisms of graphene/water-based lubricant systems. *Tribol. Int.* **2022**, *167*, 107356. [CrossRef]
151. Li, X.; Xu, X.; Zhou, Y.; Lee, K.-R.; Wang, A. Insights into friction dependence of carbon nanoparticles as oil-based lubricant additive at amorphous carbon interface. *Carbon* **2019**, *150*, 465–474. [CrossRef]
152. Li, X.; Xu, X.; Qi, J.; Zhang, D.; Wang, A.; Lee, K.-R. Insights into Superlow Friction and Instability of Hydrogenated Amorphous Carbon/Fluid Nanocomposite Interface. *ACS Appl. Mater. Interfaces* **2021**, *13*, 35173–35186. [CrossRef] [PubMed]
153. Dou, X.; Koltonow, A.R.; He, X.; Jang, H.D.; Wang, Q.; Chung, Y.-W.; Huang, J. Self-dispersed crumpled graphene balls in oil for friction and wear reduction. *Proc. Natl. Acad. Sci. USA* **2016**, *113*, 1528. [CrossRef]

154. Kong, L.; Sun, J.; Bao, Y. Preparation, characterization and tribological mechanism of nanofluids. *RSC Adv.* **2017**, *7*, 12599–12609. [CrossRef]
155. Zin, V.; Agresti, F.; Barison, S.; Colla, L.; Fabrizio, M. Influence of Cu, TiO$_2$ Nanoparticles and Carbon Nano-Horns on Tribological Properties of Engine Oil. *J. Nanosci. Nanotechnol.* **2015**, *15*, 3590–3598. [CrossRef]
156. Zurutuza, A.; Marinelli, C. Challenges and opportunities in graphene commercialization. *Nat. Nanotechnol.* **2014**, *9*, 730–734. [CrossRef]
157. Neuberger, N.; Adidharma, H.; Fan, M. Graphene: A review of applications in the petroleum industry. *J. Pet. Sci. Eng.* **2018**, *167*, 152–159. [CrossRef]
158. Rasheed, A.K.; Khalid, M.; Javeed, A.; Rashmi, W.; Gupta, T.C.S.M.; Chan, A. Heat transfer and tribological performance of graphene nanolubricant in an internal combustion engine. *Tribol. Int.* **2016**, *103*, 504–515. [CrossRef]
159. Baldin, V.; da Silva, L.R.R.; Houck, C.F.; Gelamo, R.V.; Machado, Á.R. Effect of Graphene Addition in Cutting Fluids Applied by MQL in End Milling of AISI 1045 Steel. *Lubricants* **2021**, *9*, 70. [CrossRef]
160. Li, M.; Yu, T.; Zhang, R.; Yang, L.; Li, H.; Wang, W. MQL milling of TC4 alloy by dispersing graphene into vegetable oil-based cutting fluid. *Int. J. Adv. Manuf. Technol.* **2018**, *99*, 1735–1753. [CrossRef]
161. Yi, S.; Li, G.; Ding, S.; Mo, J. Performance and mechanisms of graphene oxide suspended cutting fluid in the drilling of titanium alloy Ti-6Al-4V. *J. Manuf. Processes* **2017**, *29*, 182–193. [CrossRef]
162. Singh, R.K.; Sharma, A.K.; Dixit, A.R.; Tiwari, A.K.; Pramanik, A.; Mandal, A. Performance evaluation of alumina-graphene hybrid nano-cutting fluid in hard turning. *J. Clean. Prod.* **2017**, *162*, 830–845. [CrossRef]
163. Kałużny, J.; Świetlicka, A.; Wojciechowski, Ł.; Boncel, S.; Kinal, G.; Runka, T.; Nowicki, M.; Stepanenko, O.; Gapiński, B.; Leśniewicz, J.; et al. Machine Learning Approach for Application-Tailored Nanolubricants' Design. *Nanomaterials* **2022**, *12*, 1765. [CrossRef] [PubMed]
164. Ryu, B.; Wang, L.; Pu, H.; Chan, M.K.Y.; Chen, J. Understanding, discovery, and synthesis of 2D materials enabled by machine learning. *Chem. Soc. Rev.* **2022**, *51*, 1899–1925. [CrossRef]
165. Hasan, M.S.; Kordijazi, A.; Rohatgi, P.K.; Nosonovsky, M. Machine learning models of the transition from solid to liquid lubricated friction and wear in aluminum-graphite composites. *Tribol. Int.* **2022**, *165*, 107326. [CrossRef]

MDPI AG
Grosspeteranlage 5
4052 Basel
Switzerland
Tel.: +41 61 683 77 34

Lubricants Editorial Office
E-mail: lubricants@mdpi.com
www.mdpi.com/journal/lubricants

Disclaimer/Publisher's Note: The title and front matter of this reprint are at the discretion of the . The publisher is not responsible for their content or any associated concerns. The statements, opinions and data contained in all individual articles are solely those of the individual Editors and contributors and not of MDPI. MDPI disclaims responsibility for any injury to people or property resulting from any ideas, methods, instructions or products referred to in the content.

www.ingramcontent.com/pod-product-compliance
Lightning Source LLC
LaVergne TN
LVHW070718100526
838202LV00013B/1123